162
Structure and Bonding

Aims and Scope

The series *Structure and Bonding* publishes critical reviews on topics of research concerned with chemical structure and bonding. The scope of the series spans the entire Periodic Table and addresses structure and bonding issues associated with all of the elements. It also focuses attention on new and developing areas of modern structural and theoretical chemistry such as nanostructures, molecular electronics, designed molecular solids, surfaces, metal clusters and supramolecular structures. Physical and spectroscopic techniques used to determine, examine and model structures fall within the purview of *Structure and Bonding* to the extent that the focus is on the scientific results obtained and not on specialist information concerning the techniques themselves. Issues associated with the development of bonding models and generalizations that illuminate the reactivity pathways and rates of chemical processes are also relevant

The individual volumes in the series are thematic. The goal of each volume is to give the reader, whether at a university or in industry, a comprehensive overview of an area where new insights are emerging that are of interest to a larger scientific audience. Thus each review within the volume critically surveys one aspect of that topic and places it within the context of the volume as a whole. The most significant developments of the last 5 to 10 years should be presented using selected examples to illustrate the principles discussed. A description of the physical basis of the experimental techniques that have been used to provide the primary data may also be appropriate, if it has not been covered in detail elsewhere. The coverage need not be exhaustive in data, but should rather be conceptual, concentrating on the new principles being developed that will allow the reader, who is not a specialist in the area covered, to understand the data presented. Discussion of possible future research directions in the area is welcomed.

Review articles for the individual volumes are invited by the volume editors.

In references *Structure and Bonding* is abbreviated *Struct Bond* and is cited as a journal.

More information about this series at
http://www.springer.com/series/430

D. Michael P. Mingos
Editor

Gold Clusters, Colloids and Nanoparticles II

With contributions by

I. Atanasov · R.L. Chantry · H.-J. Freund · S.L. Horswell ·
R.L. Johnston · G.N. Khairallah · Z.Y. Li · D.M.P. Mingos ·
N. Nilius · R.A.J. O'Hair · T. Risse · S. Shaikhutdinov ·
M. Sterrer · A. Zavras

 Springer

Editor
D. Michael P. Mingos
Inorganic Chemistry Laboratory
University of Oxford
Oxford
United Kingdom

ISSN 0081-5993   ISSN 1616-8550 (electronic)
ISBN 978-3-319-07844-1   ISBN 978-3-319-07845-8 (eBook)
DOI 10.1007/978-3-319-07845-8
Springer Cham Heidelberg New York Dordrecht London

Library of Congress Control Number: 2014949194

Printed on acid-free paper

Springer is part of Springer Science+Business Media (www.springer.com)

Preface

Gold has a special place in the history of mankind – its chemical inertness and enduring physical qualities make it an ideal metal for the fabrication of high value coins and jewellery. They retain their bright appearance even after exposure to extreme conditions for tens or hundreds of years. Indeed as I write some intrepid gold hunters are seeking to bring up a billion dollars worth of gold bullion from the bottom of the Atlantic. It was originally mined during the Californian gold rush of the 1850s and lost at sea 150 miles East of the Carolinas in 1857 whilst being transported to New York. Colloidal gold has been known since ancient times and was used for making highly coloured glasses. Red and mauve glasses were particularly highly valued by the Romans. The realisation that colloids of the more noble metals could be studied and understood using scientific methods can be traced back to Michael Faraday's research at the Royal Institution in the 1850s. The term colloid was proposed in 1861 by Graham, by which time Faraday had established that gold, silver, copper, platinum, tin, iron, lead, zinc, palladium, aluminium, rhodium, iridium, mercury and arsenic shared the ability to form dilute clear solutions and thin films, which could be detected and studied by their ability to interact with light. This suggested that their dimensions were of the same order of magnitude as the wavelength of light.

In the mid-twentieth century the first examples of structurally characterised molecular cluster compounds of gold were reported by Malatesta and Mason and as a result of research in England and Holland the field was expanded and higher nuclearity examples of these metal–metal bonded compounds were established by crystallographic techniques. These highly coloured compounds were stabilised by tertiaryphosphine ligands, and their relevance to nano-chemistry and colloids was recognised at an early stage. In the twenty-first century the development of a very flexible synthetic route to cluster compounds of gold based on thiolato-ligands provided an important impetus to the field. This coincided with the recognition that gold which has previously been thought to be inferior to the platinum metals as a catalyst for the transformation of organic feedstocks was shown to be active as a homogenous and heterogeneous catalyst. These exciting developments have made it timely to publish a pair of volumes of *Structure and Bonding* devoted to gold clusters, colloids and nano-particles.

The first volume opens with a historical overview of the area and provides not only a broad introduction to the area and defines more clearly the characteristics of gold clusters, colloids and nano-particle. It also addresses their characterisation and their chemical and physical properties. The potential applications of these species are also discussed. Professor Konishi provides a review of recent developments in "Phosphine-coordinated Pure-Gold Clusters: Diverse Geometrical Structures and their Unique Optical Properties". This area originated in the 1960s, and Professor Konishi has made notable contributions to this area in recent years. This is followed by three chapters by Professors Jin, Chen and Dass, and their co-authors which describe the synthesis of gold clusters based on organothiolato-ligands. An area which blossomed following the report of a widely applicable synthetic method by Brust, Schiffrin and their coworkers in 2006. Although the synthesis of these compounds is relatively straightforward the methods result in mixtures of cluster compounds and these chapters discuss the techniques which have been developed in order to produce mixtures which either contain fewer components or are amenable to modern separation techniques. The development of more sophisticated purification techniques has led to single crystals of many of these clusters which have been structurally characterised to atomic resolutions. These structures have revealed some novel architectures and also led to the important realisation that in these clusters the gold atoms have a dual role. They not only contribute to the central core, but also contribute novel gold-thiolato-ligands, which protect the surface of the cluster. Professor Schmid, who was an important contributor to the original development of the syntheses of high nuclearity gold clusters, has written with Professor Simon and Dr. Broda a chapter on the "Size and Ligand Specific Bioresponse of Gold Clusters and Nanoparticles". They emphasise new research which suggests that the high nuclearity gold phosphine clusters have important applications in the areas of biochemistry and medicine. These chapters generally deal with the synthesis of clusters prepared in water or organic solvents and the balance is restored by an important chapter by Professor Fielicke and Dr. Woodham which discusses the synthesis and properties of gold clusters in the gas phase.

The second volume starts with historically based account of the bonding in gold clusters. In the late twentieth century empirical molecular orbital calculations provided some important insights into the bonding in these clusters and led to the development of bonding models which were sufficiently robust to predict the structures of specific clusters and provide a broad framework for understanding the structures and reactions of molecular cluster compounds. The great increase in computer power since that time has made it possible to apply ab initio molecular orbital calculations to large gold clusters and the results have provided important insights into the bonding interactions which are responsible for their stabilities and structures. The relevance of DFT calculations is emphasised in the chapter by Professors Johnston and Li on gold nano-alloys and clusters. The importance of gold clusters in catalysis is stressed in the chapter by Professor Freund and his co-authors titled "Model Catalysts Based on Au Clusters and Nano Particles". The final chapter by Professor O'Hair and his co-authors underlines the importance of mass-spectrometry in the characterisation of gold clusters in their chapter titled "Gas Phase Formation, Structure and Reactivity of Gold Cluster Ions".

This is an exciting and rapidly developing area of nano-science and is attracting chemists, physicists and material scientists. The resulting interdisciplinary research continues to throw up many interesting structures and applications. I thank all the authors and the editorial staff at Springer for contributing to a volume which helps to define the field for those who are outside it and the stimulation to those in the field to make it one of the important areas of science in this century.

Oxford, UK   Michael Mingos
May 2014

Contents

Structural and Bonding Issues in Clusters and Nano-clusters 1
D. Michael P. Mingos

Interfacial Structures and Bonding in Metal-Coated Gold Nanorods 67
Ruth L. Chantry, Ivailo Atanasov, Sarah L. Horswell,
Z.Y. Li, and Roy L. Johnston

Model Catalysts Based on Au Clusters and Nanoparticles 91
Niklas Nilius, Thomas Risse, Shamil Shaikhutdinov, Martin Sterrer,
and Hans-Joachim Freund

Gas Phase Formation, Structure and Reactivity of Gold Cluster Ions ... 139
Athanasios Zavras, George N. Khairallah, and Richard A.J. O'Hair

Index 231

Struct Bond (2014) 162: 1–66
DOI: 10.1007/430_2014_141
© Springer International Publishing Switzerland 2014
Published online: 26 March 2014

Structural and Bonding Issues in Clusters and Nano-clusters

D. Michael P. Mingos

Abstract The study of gold cluster compounds origin/ated from Malatesta's syn-theses of tertiary phosphine derivatives in the 1960s and was greatly extended between 1970 and 2000. Single crystal X-ray studies defined the major structural classes and led to the development of a theoretical model which accounted for their closed shell requirements in terms of their topological features. This model proved to be sufficiently flexible to be extended to related heteronuclear cluster com-pounds. Since the turn of the century the range of gold cluster compounds has been greatly extended by the study of organothiolato-gold cluster compounds and unligated gold clusters. The structures of the organothiolato-compounds have revealed that some of the gold atoms combine with the ligands to generate a novel class of metallo-organothiolato-ligands which protect and stabilise the inner core of gold atoms. The smaller cluster cone angles of the organothiolato-based ligands have accessed clusters with higher nuclearities (>100 gold atoms) some of which have been structurally characterised. These developments originally suggested that the phosphine and organothiolato-clusters defined quite distinct classes of gold clusters, but recent structural and theoretical developments have reconciled many of these differences. This review summarises the structures of the clusters and suggests a unified theoretical model which accounts for the broad structural properties of the two classes of compound.

Keywords Capping principle · Clusters · Condensed clusters · Gold · Jellium model · Nanoclusters · Oblate · Organothiolato · Phosphine · Polyhedral skeletal electron pair theory · Prolate · Sphericity · Tensor surface harmonic · Toroidal

D.M.P. Mingos (✉)
Inorganic Chemistry Laboratory, Oxford University, South Parks Road, Oxford OX1 3QR, UK
e-mail: michael.mingos@seh.ox.ac.uk

Contents

1   Introduction 3
2   Single Crystal X-Ray Structural Determinations 4
3   Theoretical Studies 11
    3.1   Brief Historical Introduction to Theoretical Models 11
    3.2   Bonding in Simple Gold Clusters 12
    3.3   Introduction of Interstitial Atoms 18
    3.4   Bonding in Condensed and Fused Clusters 23
    3.5   Introduction of Interstitial Atoms 34
    3.6   Bonding Interrelationships Between Organothiolato- and Gold Phosphine Clusters . 35
    3.7   Construction of Organothiolato-Clusters from Phosphine Cluster Building Blocks . . 38
    3.8   Inherent Structure Rule and the Superatom Model 49
    3.9   Spherical and Close-Packed Arrangements 52
    3.10 Summary 56
References 60

Abbreviations

ANP	Gold nano-particle
Ar	Aryl
ccp	Cubic close packed
ccs	Collision cross section
DFT	Density functional theory
dmf	Dimethylformamide
dppe	Bis(diphenylphosphino)ethane
dppm	Bis(diphenylphosphino)methane
dppp	Bis(diphenylphosphino)propane
Et	Ethyl
hcp	Hexagonal close packed
HOMO	Highest occupied molecular orbital
HRTEM	High resolution transmission electron microscopy
IM-MS	Combined ion mobility mass spectrometry
IMS	Ion mobility mass spectrometry
i-Pr	Isopropyl
LUMO	Lowest unoccupied molecular orbital
Me	Methyl
Mes	Mesityl, 2,4,6-trimethylphenyl (not methanesulfonyl)
MS	Mass spectrometry
octyl	n-Octyl
Pc	Phthalocyanine
pec	Polyhedral skeletal electron pair count
Ph	Phenyl
Pr	Propyl
sec	Skeletal electron count

SEC   Size exclusion chromatography
SEM   Scanning electron microscopy
SERS   Surface enhanced Raman spectroscopy
SR   Organothiolato
t-Bu   *tert*-butyl
TEM   Transmission electron microscopy
TGA   Thermogravimetric analysis
Tio   Tiopronin
TOAB   Tetra(*n*-octyl)ammonium bromide
Tol   4-Methylphenyl
XRD   X-ray diffraction

1   Introduction

The introduction to the preceding volume defined more clearly the differences between clusters, colloids and nanoparticles and highlighted the important synthetic routes to these interesting materials and their characterisation. It also suggested that there are grey areas in-between these classes, where the differences remain less clearly defined. The original synthesis of gold cluster compounds, based on the reduction of linear gold(I) phosphine complexes, originated in the late 1960s from Malatesta's group in Milan [1–5] and was prompted by the contemporary characterisation of related hydrido-complexes of platinum. The area was greatly extended by Mingos [6–15] and Steggerda [16–19] in the 1970s and 1980s. The molecular cluster compounds which they characterised unambiguously by single crystal X-ray studies had 4–13 gold atoms. In the late 1990s Teo reported the larger $[Au_{39}Cl_6(PPh_3)_{14}]$ cluster [20, 21] which had its origins in his related studies of gold–silver clusters based on linked icosahedra. In 1981 Schmid et al. reported [22–26] that if diborane B_2H_6 was used rather than $NaBH_4$ to reduce gold (I) triphenylphosphine complexes, then the resultant product was $[Au_{55}(PPh_3)_{12}Cl_6]$, which he formulated as a very stable closed shell and close-packed cluster with 42 surface atoms, 12 of which are stabilised by triphenylphosphine and 6 by chloro-ligands. Au_{55} is the second of the series of the so-called close-packed "full-shell clusters". Although definitive and evidence-based single crystal structural data were not obtained, it was proposed on the basis of spectroscopic data that it consists of a cuboctahedral fragment of the close-packed metallic structure of bulk gold.

As the result of the synthetic work described above, which led to well-defined molecular clusters, it was established that their nuclearities could be reduced or increased by the addition of specific reagents. Typical degradation reactions of gold cluster cations result in the loss of a few gold atoms and are induced by the addition of phosphines or soft anionic ligands which shift the equilibrium by coordinating to the outgoing $AuPR_3^+$ cation. Complementary aggregation reactions are encouraged either by the addition of mononuclear gold(I) complexes or the replacement of phosphine by smaller (and harder) anionic ligands. In general these degradation and aggregation reactions are faster than those observed for metal carbonyl clusters of the platinum metals. These studies also underlined the importance of the steric requirements of the ligands, and this led to the definition of a cluster cone angle which is closely related to the Tolman cone angle [27–29]. In recent years these procedures have been elegantly extended by Konishi and others, who have shown that growth and etching processes may be used to convert, for example, $[Au_6(dppp)_4](NO_3)_2$ into $[Au_8(dppp)_4Cl_2]^{2+}$ and $[Au_8(dppp)_4](NO_3)_2$ [30, 31].

The replacements of tertiary phosphine by organothiolates have led to an interesting range of molecular cluster compounds with 25–144 gold atoms, some of which have proved to be sufficiently crystalline for definitive structural analyses. The wide range of organothiolato-clusters results in part from their negatively charged sulphur ligand which leads to stronger gold–ligand bonding and their smaller cone angles which encourage larger cluster sizes. The syntheses and structures of organothiolato-clusters have been discussed in some detail in the previous volume [32, 33]. It should be noted that the syntheses of phosphine and organothiolato-clusters via reductive routes lead to mixtures and the use of chromatographic and size separation techniques has had a major impact on reducing the dispersities and the isolation of atomically precise compounds which are crystalline.

2   Single Crystal X-Ray Structural Determinations

The first single crystal X-ray structures of tertiary phosphine gold cluster compounds were reported in the 1950s by Mason and McPartlin [34]. At that time they represented the largest known cluster molecules of the transition metals. Since that date numerous X-ray structural determinations have been reported and the relevant literature up to 2000 was summarised by Mingos and Dyson [35]. The bond lengths in these clusters are longer than that reported for the Au_2 dimer (2.472 Å) and comparable to that which has been determined for the parent bulk metal which adopts a face-centred cubic lattice (Au–Au = 2.884 Å). These bond lengths may be related to a thermodynamic data, and it is noteworthy that the bond dissociation enthalpy in the gold dimer is 228 kJ mol^{-1}, which is intermediate between Cl_2 (243 kJ mol^{-1}) and Br_2 (193 kJ mol^{-1}). The enthalpy of vaporisation for the bulk metal is 343 kJ mol^{-1}, which is similar to that for copper 307 kJ mol^{-1} and greater than that for silver 258 kJ mol^{-1}. Therefore, the metal–metal bond strengths in gold clusters are anticipated to be reasonably strong and comparable with those in other related cluster compounds of the transition metals. This suggests that gold is capable of forming a wide range of cluster compounds particularly in low oxidation states. The ligands PR_3 and SR^- enhance the strengths of the metal–metal bond, and the linear moiety Au–Au–L enhances s/d$_{z^2}$ hybridisation and encourages strong radial bonding in clusters. This hybridisation and the relatively small contribution from the gold 6p orbitals is associated with the large relativistic effects associated with gold – an effect which has been reviewed extensively by Pyykkö [36–38]. The great majority of cluster compounds of gold have a formal average oxidation state which lies between 0 and +1 which corresponds to a partially filled 6s orbital in its valence shell of $\ldots\{5d^{10}\}\{6s\}^x$, where $x = 0.5$ in $[Au_4(PR_3)_4]^{2+}$ to 0.20 in $[Au_{39}(PPh_3)_{12}Cl_6]^{2+}$. Linear metal–ligand bonding is a characteristic of gold (I) compounds and is also a feature of gold cluster compounds. There are some rare examples of compounds with gold–gold bonds in the +2 oxidation state, but they are relatively rare. Gold–gold bonding is also a feature of alloys of gold with the alkali metals where the formal oxidation state of gold lies between −1 and 0 [39].

The majority of gold(I) complexes have a linear geometry and the hybridisation at the metal is approximately to sd_{z^2}. Many of these linear gold compounds show additional gold contacts to other gold atoms, which are significantly longer than the metal–metal distance in the metal (2.88 Å), but shorter than those expected from the sum of the van der Waals' radii (3.60 Å). It has been suggested that these aurophilic interactions may contribute as much as 46 kJ mol^{-1} [40], but generally are estimated to be approximately 20–30 kJ mol^{-1}. In smaller gold clusters which do not have a central gold atom the average gold–gold bond lengths ranges from 2.76 to 2.87 Å, but in larger clusters with an interstitial gold atom, the bond lengths involving the central atom are generally shorter than those on the periphery, e.g. in Au_8 clusters, 2.59–2.77 Å for the former and 2.83–2.96 Å for the latter. In icosahedral Au_{13} phosphine clusters the centre to periphery distances are 2.716 (2)–2.789(2) Å, whereas the peripheral gold–gold bond lengths lie between 2.852 (3) and 2.949(3) Å. The observation of relatively strong radial bonds in these compounds led to some confusion in the 1970s as to whether these compounds were best described as cluster compounds or coordination compounds, and the term "porcupine" clusters was coined in order to emphasise the anisotropy in the metal–metal bonding [41]. $[Au_{39}Cl_6(PPh_3)_{14}]Cl_2$ which has a structure based on hexagonal close packing the central interstitial 12 coordinate gold atom has gold–gold bond lengths which average 3.040 Å, and the gold–gold bond lengths in successive hexagonal layers range between 2.792 and 2.882 Å [21]. This pattern of bond lengths does not follow that noted above for simpler phosphine clusters and remains somewhat surprising.

Molecular phosphine gold cluster molecules and ions with 4–7 metal atoms have been structurally characterised and their structures are summarised in Fig. 1 [35]. The tetrahedron is a common building block and is observed in the 4, 5, and 6 atom clusters. The seven-atom cluster has a pentagonal bipyramidal structure. It is noteworthy that this structure has a very flat geometry leading to a short Au–Au interaction (2.58(2) Å) between the axial gold atoms. The equatorial–equatorial and axial-equatorial gold–gold bonds average 2.9 Å and 2.8 Å respectively. The absence of a symmetric octahedral Au_6 structure is noteworthy. The cavity at the centre of the tetrahedron, trigonal bipyramid and octahedron are sufficiently small

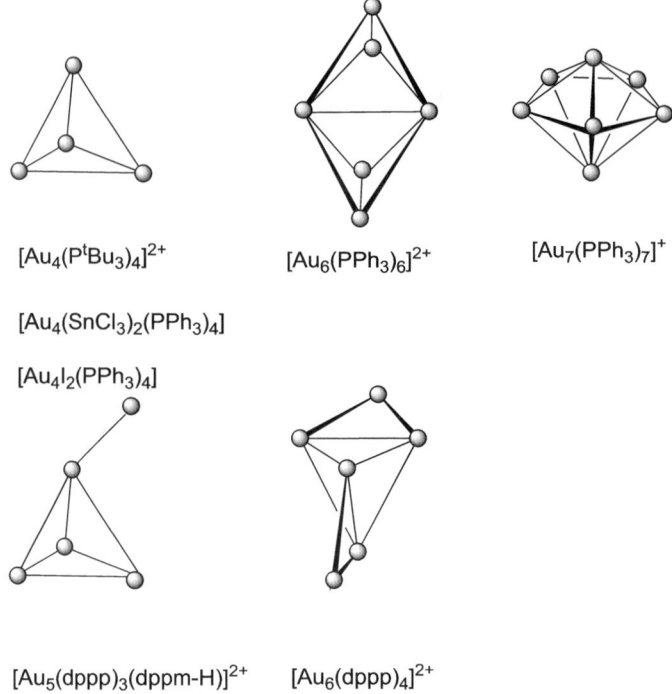

$[Au_4(P^tBu_3)_4]^{2+}$ $[Au_6(PPh_3)_6]^{2+}$ $[Au_7(PPh_3)_7]^+$

$[Au_4(SnCl_3)_2(PPh_3)_4]$

$[Au_4I_2(PPh_3)_4]$

$[Au_5(dppp)_3(dppm-H)]^{2+}$ $[Au_6(dppp)_4]^{2+}$

Fig. 1 Summary of structures of small gold phosphine clusters

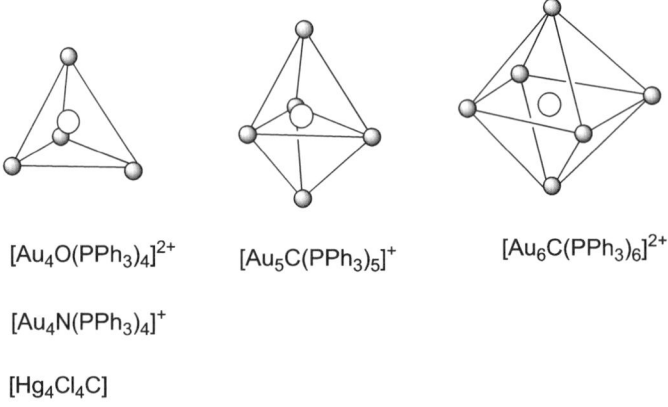

$[Au_4O(PPh_3)_4]^{2+}$ $[Au_5C(PPh_3)_5]^+$ $[Au_6C(PPh_3)_6]^{2+}$

$[Au_4N(PPh_3)_4]^+$

$[Hg_4Cl_4C]$

Fig. 2 Examples of gold clusters with interstitial main group atoms

to accommodate small main group atoms, and Schmidbaur's group have characterised the compounds illustrated in Fig. 2 [42–44].

Gold clusters with more than 7 metal atoms invariably have interstitial metal atoms and their structures are summarised in Fig. 3. The clusters fall into distinct

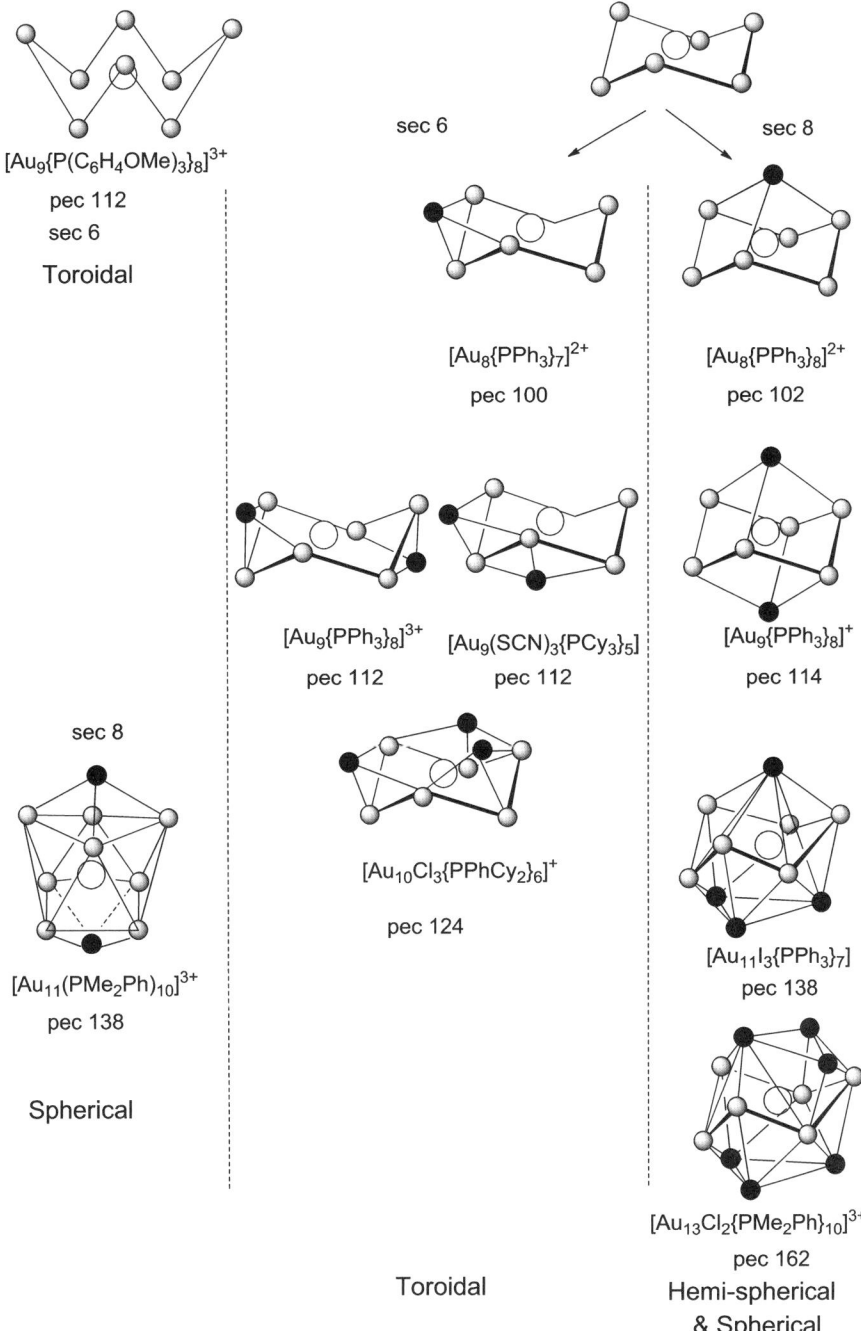

Fig. 3 Examples of centred gold clusters with phosphine ligands. The interstitial gold atom is shown as a *colourless sphere* and the added gold atoms by *black spheres*

classes based on either on a centred octagonal crown or a hexagonal chair shown at the top of the figure. Although the centred hexagonal ring is not known as a distinct species, it does give rise to the wide range of structures shown on the right hand side of the figure, which add gold atoms either to the edges of the molecule or along the symmetry axis. Adding single metal atoms or triangles of metal atoms along the threefold axis leads to hemi-spherical or spherical clusters depending on whether the addition occurs only on one side of the interstitial atom or both. Addition of a single metal atom along the rotation axis above and below the interstitial atom leads to a cube, and addition of triangles to these locations leads to an icosahedron. In contrast additions of bridging metal atoms around the chair lead to flatter structures which are described as toroidal or oblate. For the centred crown on the left hand side of the figure, the toroidal chair structure is converted into a spherical structure by squeezing together the four metal atoms which define planes above and below the interstitial atom to form the square antiprism shown at the bottom left of the figure. It is noteworthy that spherical and toroidal clusters with the same number of metal atoms have polyhedral electron counts (pec) which differ by 2, e.g. spherical $[Au_9(PPh_3)_8]^+$ has a pec of 114, whereas toroidal $[Au_9(PPh_3)_8]^{3+}$ has a pec of 112. The polyhedral electron count (pec) is defined as the total number of valence electrons donated by the gold atoms and the lone pairs donated to the surface gold atoms by the phosphine ligands minus the total positive charge on the cluster. The spherical clusters are therefore characterised by $12n + 18$ and the toroidal clusters by $12n + 16$ electrons where n is the number of non-interstitial atoms [6, 11, 13]. If the d shells of the gold atoms are excluded, the toroidal clusters have a sec count (*skeletal electron count*) of 6 and the spherical clusters a sec count of 8. A molecular orbital interpretation of this difference is discussed in some detail below in Sect. 3.

The single crystal X-ray determinations suggested that the potential surfaces connecting alternative toroidal or spherical structures are rather soft because for closely related compounds alternative structures were observed. For example, for $[Au_9\{P(C_6H_4OMe)_3\}_8]^{3+}$ careful crystallisation led to crystalline modifications and single crystal analyses showed one has the crown structure (shown on the top left hand side of Fig. 3) and the other a D_{2h} structure previously observed for $[Au_9\{PPh_3\}_8]^{3+}$ (shown in the third row of Fig. 3) [45, 46]. These isomers may be interconverted by applying high pressures [46]. $^{31}P\{^1H\}$ NMR experiments in the solid state and in solution have confirmed that these clusters are generally stereochemically non-rigid on the NMR time scale [47, 48]. The majority of transition metal carbonyl clusters of the later transition metals are stereochemically rigid, and therefore this represents an important difference between the two classes of compound which has to be accounted for by a reliable theoretical model [46]. Although the gold clusters shown in Fig. 3 are homonuclear, a wide range of cluster compounds with other interstitial metal atoms have been structurally characterised and shown to follow a structural paradigm identical to that shown in Fig. 3 [8, 49]. The toroidal and spherical classification introduced above is also applicable to these cluster complexes [49–57].

Higher nuclearity cluster compounds of the group 11 metals have also been studied, and their structures have shown that they do not necessarily follow the

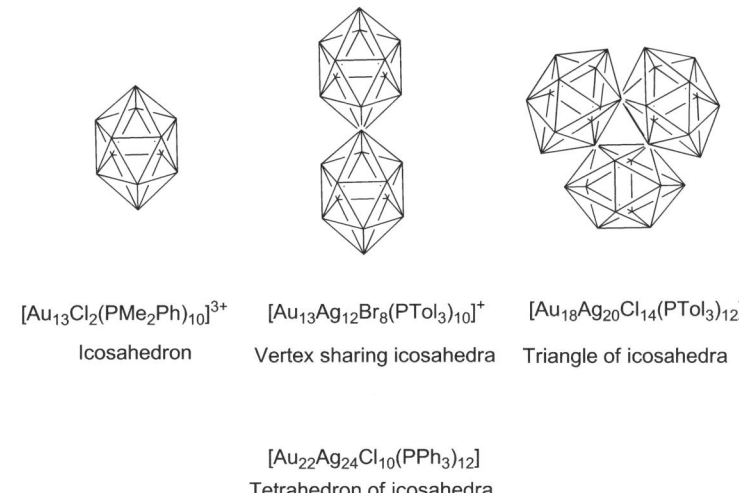

$[Au_{13}Cl_2(PMe_2Ph)_{10}]^{3+}$ $[Au_{13}Ag_{12}Br_8(PTol_3)_{10}]^+$ $[Au_{18}Ag_{20}Cl_{14}(PTol_3)_{12}]$

Icosahedron   Vertex sharing icosahedra   Triangle of icosahedra

$[Au_{22}Ag_{24}Cl_{10}(PPh_3)_{12}]$
Tetrahedron of icosahedra

Fig. 4 Linking of icosahedral polyhedral observed in higher nuclearity gold–silver clusters. The interstitial atoms at the centre of each icosahedron have been omitted

close-packed arrangement found in the bulk metal, but have structures which result from the sharing of vertices, edges and faces of simpler polyhedral moieties, e.g. octahedron and cuboctahedron. This general observation is consistent with the view that cluster growth is a kinetically controlled process which may result in the trapping of less symmetrical structures based on the condensation of simpler polyhedral species and particularly the cuboctahedron and the icosahedron. Teo and his co-workers have structurally characterised a number of vertex sharing icosahedral clusters containing silver and gold atoms [21, 58], which are illustrated in Fig. 4. The parent icosahedral gold cluster ion $[Au_{13}Cl_2(PMe_2Ph)_{10}]^{3+}$ [35] was predicted on the basis of theoretical calculations in 1976 and synthesised and structurally characterised in 1981. The related gold–silver clusters $[Au_{13}Ag_{12}Br_8\{P(p\text{-}tol)_3\}_{10}]^+$ and $[Au_{22}Ag_{24}Br_{10}\{PPh_3\}_{12}]$ and gold–silver–platinum clusters $[Pt_2Au_{10}Ag_{13}Cl_7\{P(p\text{-}tol)_3\}_{10}]^+$ have been structurally characterised and are illustrated in Fig. 4. Interestingly, and for reasons, which are discussed in more detail below, the clusters have skeletal electron requirements which represent a sum of those for the parent icosahedron. Icosahedral and cuboctahedral polyhedra are also important building blocks in organothiolato-gold clusters.

Recently Simon and his co-workers [59] have reported that Malaesta's original synthesis of $[Au_9(PPh_3)_8](NO_3)_3$ also provides, albeit in low yield a novel, but closely related, cluster with the formula $[Au_{14}(PPh_3)_8(NO_3)_4]$. It has two trigonal-bipyramidal Au_5 fragments directly connected by a short Au–Au bond (2.651 Å), and this bond also has a collar of 4 $AuNO_3$ fragments leading to an ellipsoidal shape and a polyhedron having triangular faces (see Ia). The radial Au–Au bond lengths are shorter than the surface gold bond lengths. Simon et al. have proposed that the structure of the complex may be interpreted in terms of an adaptation of a

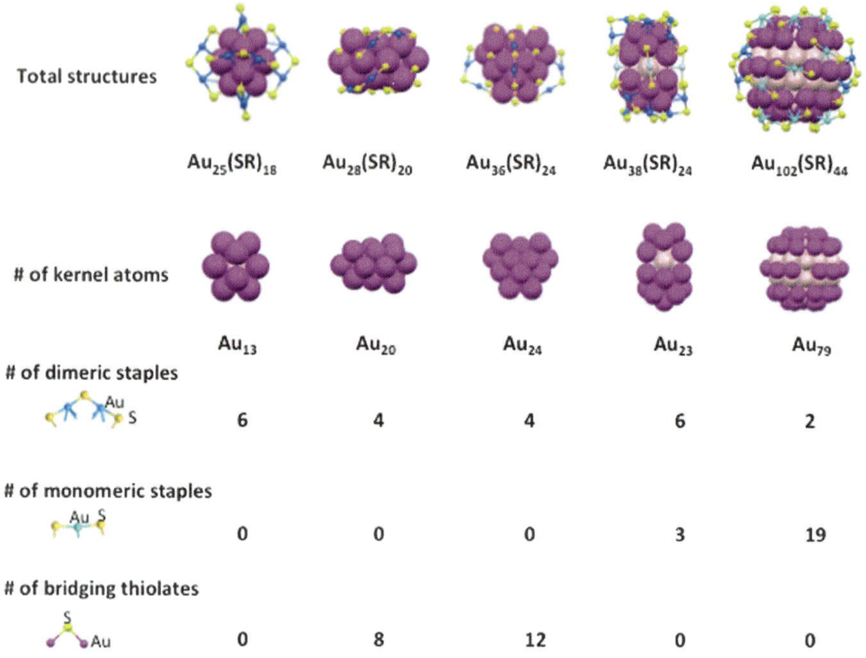
Total structures
$Au_{25}(SR)_{18}$
$Au_{28}(SR)_{20}$
$Au_{36}(SR)_{24}$
$Au_{38}(SR)_{24}$
$Au_{102}(SR)_{44}$
# of kernel atoms
Au_{13}
Au_{20}
Au_{24}
Au_{23}
Au_{79}
# of dimeric staples
Au
S
6
4
4
6
2
# of monomeric staples
Au S
0
0
0
3
19
# of bridging thiolates
S
Au
0
8
12
0
0

Fig. 5 Higher nuclearity organothiolato-gold cluster compounds. Reproduced with permission from Zeng and Jin [32]

superatom structure, and this aspect will be discussed in more detail in the bonding section below. Until 2007 $[Au_{39}(PPh_3)_{14}Cl_6]^{2+}$, which has a close-packed core of gold atoms stabilised by phosphine and chloro-ligands, represented the highest nuclearity gold cluster compound [21]. The metal framework comprises of layers of hexagonally close-packed gold atoms which down the threefold symmetry axis follow the sequence 1:9:9:1:9:9:1. Only one gold atom is completely encapsulated and resides at the centre of a hexagonal antiprism and has a coordination number of 12. The Au–Au bonds radiating from this gold average 3.04 Å, which surprisingly is not shorter than the tangential bonds Au–Au bonds. Eighteen of the surface gold atoms are sufficiently sheltered that they do nor form bonds to the 20 ligands.

The recent synthesis and characterisation of high nuclearity gold clusters stabilised by organothiolato-ligands (SR) has demonstrated that the central kernels consist of a close-packed polyhedron based on either icosahedra or cuboctahedra. $[Au_{25}(SR)_{18}]^-$ is based on an icosahedral kernel, $[Au_{38}(SR)_{24}]$ on a pair of icosahedra sharing a face and $[Au_{102}(SR)_{44}]$ on a decahedron of 79 atoms with fivefold symmetry [32, 33]. In contrast $[Au_{36}(SR)_{24}]$ has a cuboctahedral kernel and $[Au_{28}(SR)_{20}]$ is based on two cuboctahedra with shared faces (see Fig. 5).

A model which accounts for the structures and stoichiometries of these clusters presents a considerable challenge to theoreticians and clearly a model which works for both classes of clusters would provide an important unifying feature. It is also hoped that the resulting model would also provide insights into the relationship of these cluster compounds to gold colloids.

3 Theoretical Studies

3.1 Brief Historical Introduction to Theoretical Models

The early developments in cluster chemistry occurred at a time when computers were not readily available, and the initial theoretical models were based on semi-quantitative models which depended on symmetry arguments and semi-empirical estimates of the strengths of bonding interactions. Nevertheless, these primitive models greatly assisted the development of the subject and were able to account for the stabilities of those cluster molecules where the bonding cannot be described in terms of localised bonding schemes [60, 61]. In the 1950s Longuet-Higgins pioneered the combination of symmetry and semi-quantitative molecular orbital arguments to define the skeletal molecular orbitals in boron anions B_n^{2-} and rationalised the closed shell electronic requirements for octahedral B_6^{2-} and icosahedral B_{12}^{2-}. The borane anions $B_6H_6^{2-}$ and $B_{12}H_{12}^{2-}$ and the isoelectronic carboranes $C_2B_4H_6$ and $C_2B_{10}H_{12}$ were subsequently isolated and structurally characterised in the 1960s [62]. In this decade Cotton and Haas extended the basic molecular orbital analysis to octahedral and triangular transition metal clusters, e.g. $[Mo_6Cl_8]^{4+}$, $[Ta_6Cl_{12}]^{2+}$ and $[Re_3Cl_{12}]^{3-}$ [63], which required the calculation of the overlaps between the d orbitals of the transition metals. The growth in transition-organometallic chemistry in the 1960s led to a large range of polyhedral metal–metal compounds with carbonyl and acetylene ligands where the bonding could not be adequately described by localised bonding models.

The structural and electronic relationships connecting main group and transition metal polyhedral molecules were defined in the 1970s by Williams [64], Wade [65], Mingos [66] and Rudolph [67]. The ideas they introduced effectively broke down the conceptual barriers separating large areas of main group and transition metal polyhedral compounds. The *isolobal* relationship also formalised the connection between main group and transition metal chemistry and proved to be a valuable tool for synthetic chemists because it enabled them to inter-relate main group and transition metal fragments [68]. The capping and debor principles also enabled chemists to build interconnections between symmetrical polyhedral molecules and their less symmetrical derivatives with capping atoms or missing atoms [60, 61]. During the 1980s the theoretical basis of these generalisations was clearly exposed by Stone's Tensor Surface Harmonic theory, and this led to group theoretical algorithms for defining the symmetries of the radial and tangential molecular orbitals in cluster polyhedra. The tensor surface harmonic model (TSHM) is a free electron particle on a sphere model which utilises scalar, vector and tensor harmonics to describe the radial and tangential delocalised molecular orbitals in polyhedral molecules [69, 70]. The TSHM when combined with group theoretical analyses led to a clearer understanding of apparent exceptions to the electron counting rules which were formalised in the polyhedral skeletal electron pair theory (PSEPT) developed primarily by Wade and Mingos [10, 14, 65, 71–81]. This model was subsequently extended to condensed and high nuclearity clusters [71–73].

According to the PSEPT the polyhedral geometry is influenced primarily by the total number of valence electrons in the cluster molecule. This is determined by the number of bonding and non-bonding molecular skeletal molecular orbitals, which reflects the geometry of the cluster, and the number of molecular orbitals formed by the overlap of ligand and metal molecular orbitals. The Tensor Surface Harmonic Theory underlined the fact that the relationship between isostructural transition metal and main group clusters originates because they both share an identical set of antibonding molecular orbitals which are unavailable for bonding, because they are strongly metal–metal antibonding and are unable to accept electron pairs from the ligands [74–77]. For rings and three-connected polyhedral molecules, the results are exactly the same as those predicted by the noble gas rule. This connection is lost for four-connected and deltahedral polyhedral molecules because the relevant metal carbonyl and main group fragments use only three orbitals for skeletal bonding and the one-to-one relationship between edges and orbitals is lost. The majority of transition metal clusters are stereochemically rigid as far as their metal skeletons are concerned although the ligands on the surface of the cluster may migrate rapidly over the surface of the cluster. From an early stage it was apparent that the later transition metals and particularly palladium and platinum and silver and gold did not conform to the generalisation embodied in PSEPT. This led to specific calculations on platinum and gold clusters which accounted for their exceptional behaviour, and this led to the prediction of the electronic requirements for gold clusters with interstitial main group, e.g. octahedral $[Au_6C(PPh_3)_6]^{2+}$, and group 11 metal atoms [43, 44, 77], e.g. icosahedral $[Au_{13}(PR_3)_{12}]^{5+}$ [77]. As the number of structures of gold clusters increased and NMR studies were completed, it became apparent that many of the higher nuclearity gold clusters were stereochemically non-rigid in solution and displayed alternative isomeric structures in the solid state. The acceptance that these clusters were interconnected by soft potential energy surfaces suggested that it was not appropriate to classify these structures in a rigid manner which had been developed for clusters of the earlier transition metals, which conform to PSEPT, and it is more appropriate to classify them according to whether they adopt spherical, prolate or oblate (toroidal) topologies [10, 14, 78–84].

3.2 Bonding in Simple Gold Clusters

The bonding in molecular cluster compounds of gold was initially based on rather unsophisticated semi-empirical molecular orbital calculations. These calculations led to some interesting insights into the bonding in phosphine gold clusters. These initial and by today's standards primitive theoretical analyses worked remarkably well for gold because the bonding is dominated primarily by the 6s valence orbital

Fig. 6 The skeletal
molecular orbitals of a
tetrahedral gold cluster. The
molecular orbitals are
illustrated in terms of 6s
orbitals on the gold atoms,
but calculations indicate
significant $6s/5d_{z^2}$
hybridisation

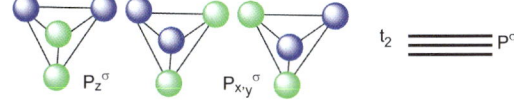

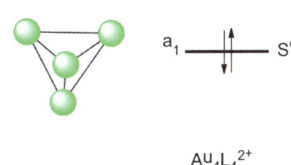

$Au_4L_4^{2+}$

which hybridise to some extent with the $5d_{z^2}$. The relative energies of the 5d, 6s and
6p valence orbitals are greatly affected by relativistic effects [36–38, 77] which
place the 6s and 5d orbitals at similar energies and thereby encourage effective
hybridisation. The relativistic effects create a large energy gap between these
orbitals and the 6p orbitals. Consequently the 6p orbitals do not participate greatly
in the metal–metal bonds in cluster compounds. This simplifications lead to the
following generalisations [77]:

1. The energies of the bonding skeletal molecular orbitals in single shell gold
   clusters are dominated primarily by radial bonding interactions and the number
   of nearest gold neighbours.
2. Low nuclearity $[Au_nL_n]$ ($n = 2$–4) are characterised by a single bonding molecular orbital resulting from the in-phase overlap of 6s orbitals on the individual
   gold atoms. This in-phase combination has no angular nodes and therefore
   resembles an s atomic orbital and is designated the symbol S^σ (see Fig. 6).
3. For larger clusters $[Au_nL_n]$ ($n > 4$) additional P^σ skeletal molecular orbitals,
   which are singly noded, become available and contribute significantly to the
   skeletal bonding.
4. The partial filling of these P^σ molecular orbital shell leads to prolate- and oblate-
   shaped clusters which maximise the occupation of the components of the P^σ
   shell which are bonding.
5. If the P^σ shell is completely filled, then a spherical cluster geometry which
   maximises the number of nearest neighbours is preferred.

Figure 6 uses the point group symmetry labels for the skeletal molecular orbitals,
and the pseudo-spherical symmetry labels which emphasises the number of nodes
associated with each orbital are also given. The number of nodes determines the
relative energies of the molecular orbitals, viz. S^σ which has no nodes along the

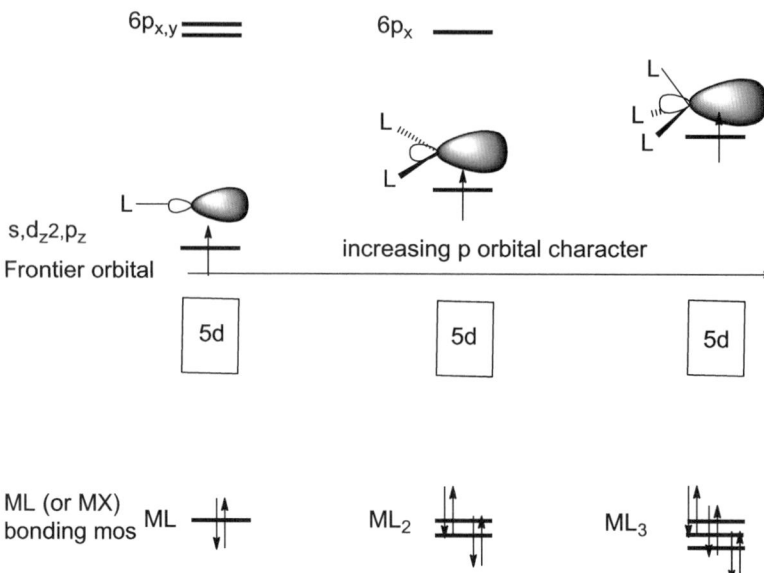

Fig. 7 Frontier orbitals of AuL$_n$ ($n = 1$–3) fragments – the introduction of more p character is introduced as the number of ligands increases, but the essential symmetry character of the frontier orbital is retained [78]

tetrahedral edges is strongly bonding and the P$^\sigma$, which is noded along half the edges, is somewhat antibonding. At its simplest level a Hückel [77] interpretation suggests that the bonding molecular S$^\sigma$ molecular orbital is stabilised by 3β (since it has three nearest neighbour gold atoms) and each component of the P$^\sigma$ molecular orbitals is destabilised by –β. The triple degeneracy of P$^\sigma$ leads to an equal spreading of the -3β antibonding character. Therefore a stable closed shell electronic configuration is associated with triangular [Au$_3$L$_3$]$^+$ and tetrahedral [Au$_4$L$_4$]$^{2+}$, which both have [S$^\sigma$]^2 configurations. Although triangular clusters of gold are not well established, there are several examples of tetrahedral gold clusters, viz. [Au$_4$(PBu$^t_3$)$_4$]$^{2+}$, [Au$_4$\{P(C$_6$H$_3$Me$_3$-1,3,5)$_4$\}]$^{2+}$, where each gold atom is associated with a single phosphine ligand [35, 42]. The stable [Au$_4$]$^{2+}$ entity may also be stabilised by a combination of phosphine ligands and anionic bridging ligands along two edges of the tetrahedron, namely [Au$_4$(μ-I)$_2$(PPh$_3$)$_4$] and [Au$_4$(μ-SnCl$_3$)$_2$(PPh$_3$)$_4$] [35].

Lower nuclearity clusters of gold are sufficiently flexible and have sufficient space to accommodate ML (or MX) and ML$_2$ (or MLX) ligand combinations, because each of them has similar frontier orbitals which are able to participate in metal–metal bonding (see Fig. 7). Gold may also form trigonal ML$_3$ fragments, but they are rarely observed within the clusters. The expansion of ligand sphere involves some interactions with the gold 6p orbitals although they are rather high lying, and this raises the energy of the frontier orbital but does not change the nodal characteristics of the frontier orbital, which remains essentially a 6s/5d$_{z^2}$ hybrid with small variations in 6p character. Therefore, gold clusters may be associated

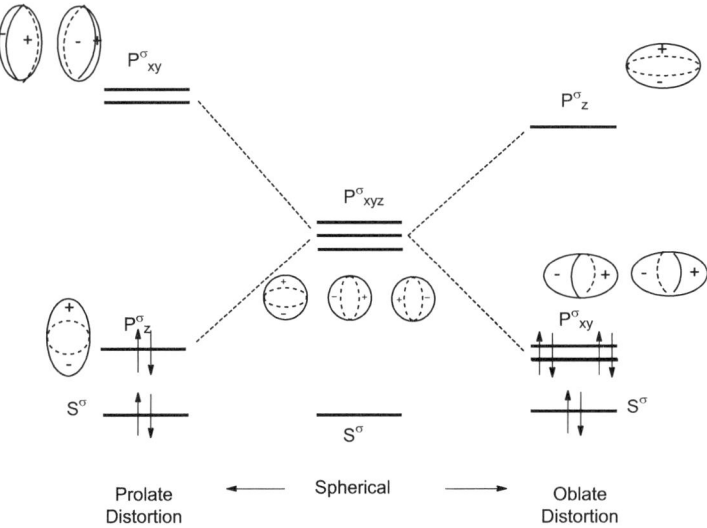

Fig. 8 Effect of prolate and oblate distortions on the energy levels of the particle in a sphere problem

with 0, 1 or 2 donors, and this affects the total cluster electron count although the number of skeletal bonding molecular orbitals remains the same. In such cases it is less confusing if the number of electron involved in skeletal bonding (*sec*) is identified, i.e. in these tetrahedral clusters, $[S^\sigma]^2$ defines the nodal characteristics of the molecular orbital and the number of electrons involved in metal–metal bonding [78]. As indicated above additional ligands may be involved in bridging situations, e.g. $[Au_4(\mu\text{-}I)_2(PPh_3)_4]$ and $[Au_4(\mu\text{-}SnCl_3)_2(PPh_3)_4]$, or may incorporate another gold atom, e.g. $[Au_5(dppm)_3(dppm\text{-}H)]^{2+}$, which has a gold containing molecule acting as a tridentate ligand (see Ia). This provides perhaps the simplest example of a "staple" ligand, which has proved to be important such a distinctive feature of the structures of organothiolato-gold cluster [32, 33].

[Au$_n$L$_n$]$^{m+}$ clusters ($n = 5\text{--}7$) have a set of four skeletal molecular orbitals which are potentially available for bonding, i.e. S^σ and P^σ and $(n-4)D^\sigma$ antibonding skeletal molecular orbitals. For the complete occupation of the S^σ and P^σ shells, 8 electrons are required, and this would lead to anionic gold phosphine clusters where the negative charges would be destabilising, e.g. $[Au_5(PPh_3)_5]^{3-}$, $[Au_6(PPh_3)_6]^{2-}$, $[Au_7(PPh_3)_7]^{-}$. Partial filling of the P^σ shell reduces the negative charge and can lead to a closed sub-shell (and a diamagnetic ground state) if a distortion occurs which removes the degeneracy of the P^σ shell. This can be achieved by undergoing prolate and oblate distortions as shown in Fig. 8. The relationship between the distortion and the orbital splitting may be analysed using perturbation theory methods based on either molecular orbital or crystal field models. The results are summarised in Fig. 8 [82, 84].

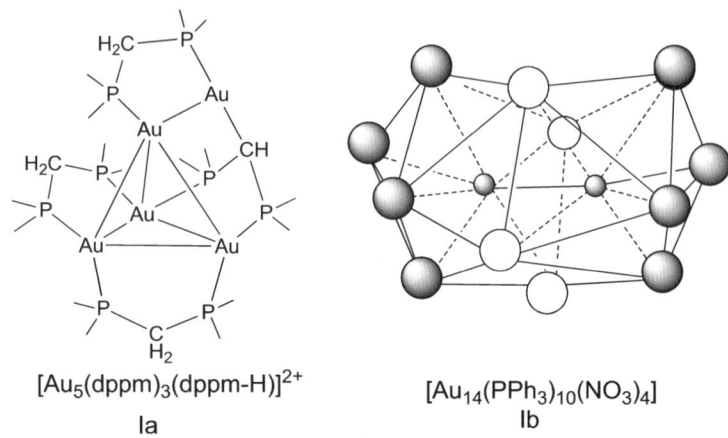

$[Au_5(dppm)_3(dppm-H)]^{2+}$ $[Au_{14}(PPh_3)_{10}(NO_3)_4]$

Ia   Ib

A prolate distortion is favoured for clusters with $[S^\sigma]^2$ $[P^\sigma]^2$ electronic configurations because the $P_{x,y}{}^\sigma$ components are destabilised relative to the $P_z{}^\sigma$ component as shown on the left hand side of Fig. 8. A prolate geometry is exemplified by the $[Au_6(PPh_3)_6]^{2+}$ (see II) cluster which has a pair of tetrahedra sharing a common edge and $[Au_6(dppp)_4]^{2+}$ (see III) which has a structure based on a di-edge bridged tetrahedron. The $P_{x,y}{}^\sigma$ components are localised on the atoms lying in the equatorial plane, and the nodal plane leads to antibonding next neighbour gold–gold interactions. A prolate distortion, which pushes the equatorial atoms closer amplifies the splitting energy. In contrast an oblate distortion which pushes the polar atoms closer together results in a greater destabilisation of the $P_z{}^\sigma$ component (see Fig. 8).

The pentagonal bipyramidal cluster $[Au_7(PPh_3)_7]^+$ has an oblate structure (see IV) and indeed the shortest bond in the cluster is between the two apical gold atoms. This structure is consistent with the closed sub-shell structure $[S^\sigma]^2$ $[P_{x,y}{}^\sigma]^4$ shown in Fig. 8 (right hand side). It is noteworthy that the distortions described above create a sufficiently large HOMO–LUMO gap to ensure that the clusters adopt a low spin state which is diamagnetic.

This analysis has proved sufficiently robust to account for the geometries of gold clusters stabilised by soft ligands such as phosphines, iodide and $SnCl_3$. It is important to recognise that the ligands play an important role in influencing the stabilities and geometries of the clusters. In the absence of such ligands the bare clusters either adopt rather different geometries or decompose to the bulk metal.

These structures of bare gold clusters have been established by careful spectroscopic studies in the gas phase and are supported by DFT calculations. The structures of these clusters are discussed in some detail by Woodham and Fielicke in the chapter "Gold Clusters in the Gas Phase" [85]. Their studies on clusters up to Au_{20} have shown that the smaller clusters generally favour close-packed planar structures and the 2D to 3D transition occurs at Au_{12} for anions. This has been a puzzle for theory, but has now been correctly described, thanks to more recent developments in density functional theory, which are able to handle dispersion interactions. For metal carbonyl clusters, where radial and tangential interactions are both significant, a larger number of bonding skeletal molecular orbitals result.

They have been analysed in detail using the symmetry implications of Tensor Surface Harmonic Theory. This has resulted in the development of the capping and condensation principles, which have shown that the skeletal molecular orbitals of the inner core of metal atoms play a dominant role [71–76]. According to the *capping principle* the molecular orbitals of a multicapped cluster retain a strong memory of the bonding molecular orbitals of the initial cluster and are characterised by the same number of skeletal MOs, unless the capping fragments introduce orbital combinations which did match those of the parent cluster [71, 72, 75, 76]. Since carbonyl clusters have many more bonding molecular orbitals, this mismatch rarely occurs, and two or three capped structures have the same number of bonding skeletal molecular orbitals as the parent cluster. For gold clusters, where radial bonding predominates and fewer bonding molecular orbitals are present, then this mismatch occurs more frequently observed. For example, if two edge bridging AuL_x fragments are added to a tetrahedral $[Au_4L_4]^{2+}$ cluster, then the in-phase $[S^\sigma]$ combination matches the tetrahedral $[S^\sigma]$ bonding skeletal molecular orbital leading to in-phase $[1S^\sigma]$ and out-of-phase $[2S^\sigma]$ molecular orbitals and the $[P^\sigma]$ combination interacts with the $[P_z^\sigma]$ component of the antibonding skeletal molecular orbitals of the parent tetrahedron, leading to $[1P_z^\sigma]$ and $[2P_z^\sigma]$ (see Fig. 9). Occupation of the bonding components of these interactions leads to a closed shell for $[1S^\sigma]^2 [1P^\sigma]^2$, which of course corresponds to the bonding orbitals associated with a prolate gold cluster (as shown in Fig. 9). For prolate and oblate clusters which are associated with the ground state configurations $[1S^\sigma]^2[1P^\sigma]^2$ or $[1S^\sigma]^2[1P^\sigma]^4$ respectively, the addition of a pair of additional gold atoms leads to symmetry matched interactions as long as the bridging metals occupy appropriate positions with respect to the nodal planes of the occupied P^σ orbitals. In these examples the capping principle holds and the resulting clusters have the same number of bonding skeletal molecular orbitals as the uncapped cluster. Examples of such clusters are to be found in the gold cluster compounds studied by Konishi and his group [86–89].

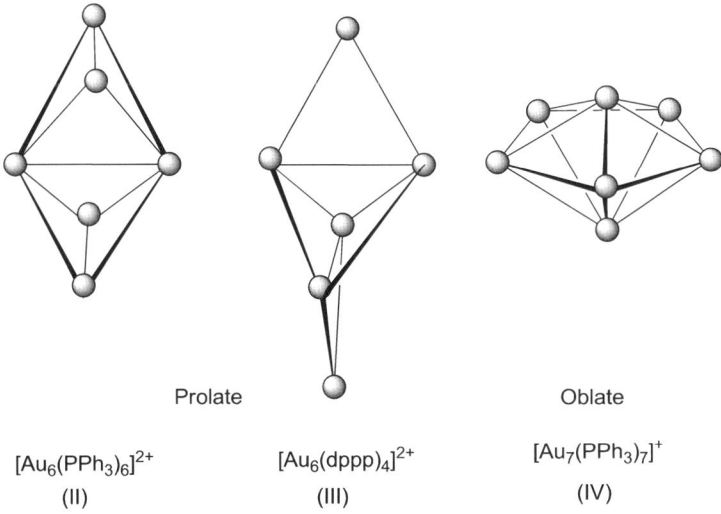

Prolate   Oblate

$[Au_6(PPh_3)_6]^{2+}$ $[Au_6(dppp)_4]^{2+}$ $[Au_7(PPh_3)_7]^+$

(II)   (III)   (IV)

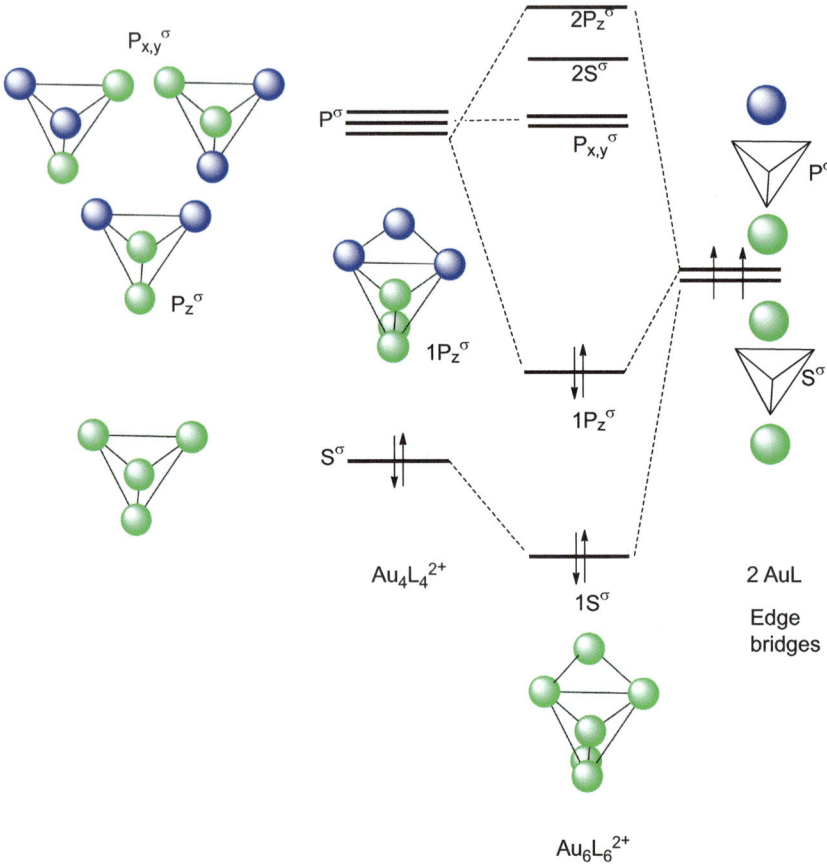

Fig. 9 Interaction diagram for the addition of two capping gold fragments to the central gold tetrahedron. This leads to a prolate edge-bridged structure

3.3 Introduction of Interstitial Atoms

Figure 10 illustrates the skeletal molecular orbitals for octahedral $[Au_6(PPh_3)_6]^2$ which has a closed shell electronic configuration $[S^\sigma]^2 [P^\sigma]^6$, but has a destabilising double negative charge. It is therefore not surprising that such a cluster has never been isolated. An alternative way of satisfying the closed shell electronic require-ments and retaining a positive charge on the cluster is possible if an interstitial atom is introduced. For an octahedral gold cluster an additional gold atom is too large to satisfy the steric requirements of the interstitial cavity, but a small main group atom, e.g. C, N or O, is able to satisfy these steric requirements. Figure 11 gives the relevant interaction diagram for octahedral $[Au_6C(PPh_3)_6]^{2+}$ and an interstitial carbon atom, which contributes just the right number of electrons to lead to a $[S^\sigma]^2[P^\sigma]^6$ pseudo-spherical ground state configuration. It is noteworthy that this

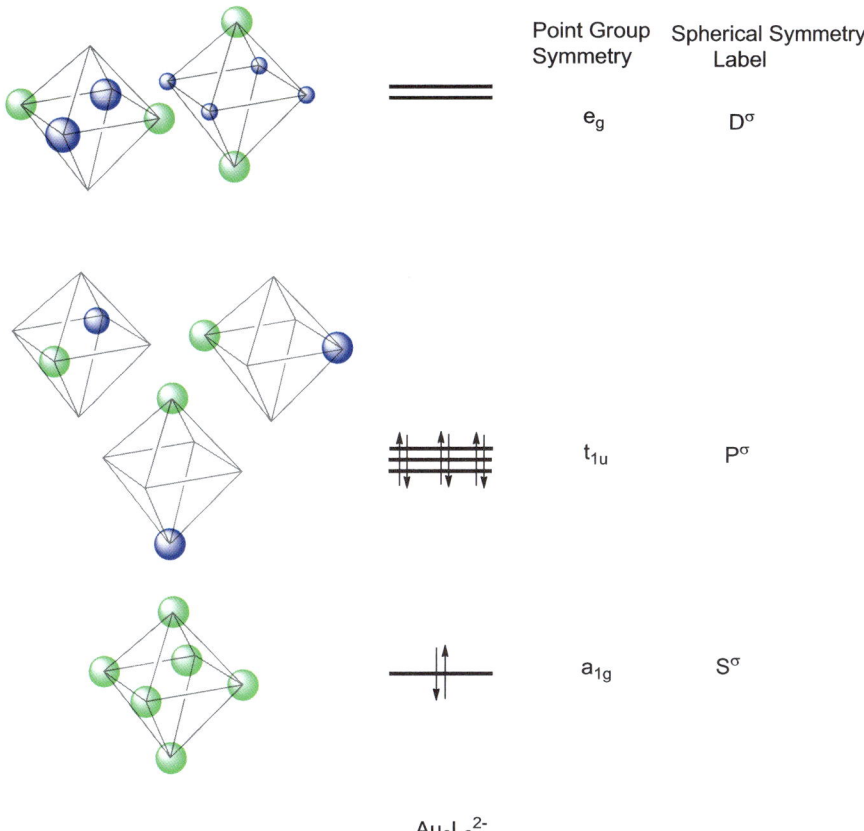

Point Group Symmetry	Spherical Symmetry Label
e_g	D^σ
t_{1u}	P^σ
a_{1g}	S^σ

$Au_6L_6^{2-}$

Fig. 10 Skeletal molecular orbitals for octahedral $[Au_6(PPh_3)_6]^{2-}$

simple theoretical analysis was first published in 1976 [77], and the existence of $[Au_6C(PPh_3)_6]^{2+}$ was confirmed by Schmidbaur and his co-workers in 1989 [42–44]. This compound had been isolated and structurally characterised previously, but the interstitial carbon had not been identified. The introduction of the interstitial atom strengthens the radial interactions significantly as a result of effective overlaps between the carbon 2s and 2p orbitals and the matching S^σ and P^σ cluster molecular orbitals. In broad brush terms the stabilisation of the valence orbitals of the central atom are stabilised by $n\beta^\sigma$/degeneracy of the molecular orbitals. If $\beta^\sigma(s) = \beta^\sigma(p) = \beta^\sigma(d)$, the relative stabilisations are $6\beta^\sigma(s)$, $2\beta^\sigma(p)$, and $3\beta^\sigma(d)$. Therefore, the greatest stabilisation involves the s orbitals of the central atom and increases as the number of metal atoms, n, increases. For filled shells the stabilisation energies are independent of geometry as long as the ligand polyhedron approximates to a sphere. It follows that gold clusters with main group interstitial atoms are characterised by a pec of $12n + 8$ (sec 8) valence electrons, since each $AuPPh_3$ fragment is associated with a filled d shell containing 10 electrons and a bonding Au–P bonding molecular orbital.

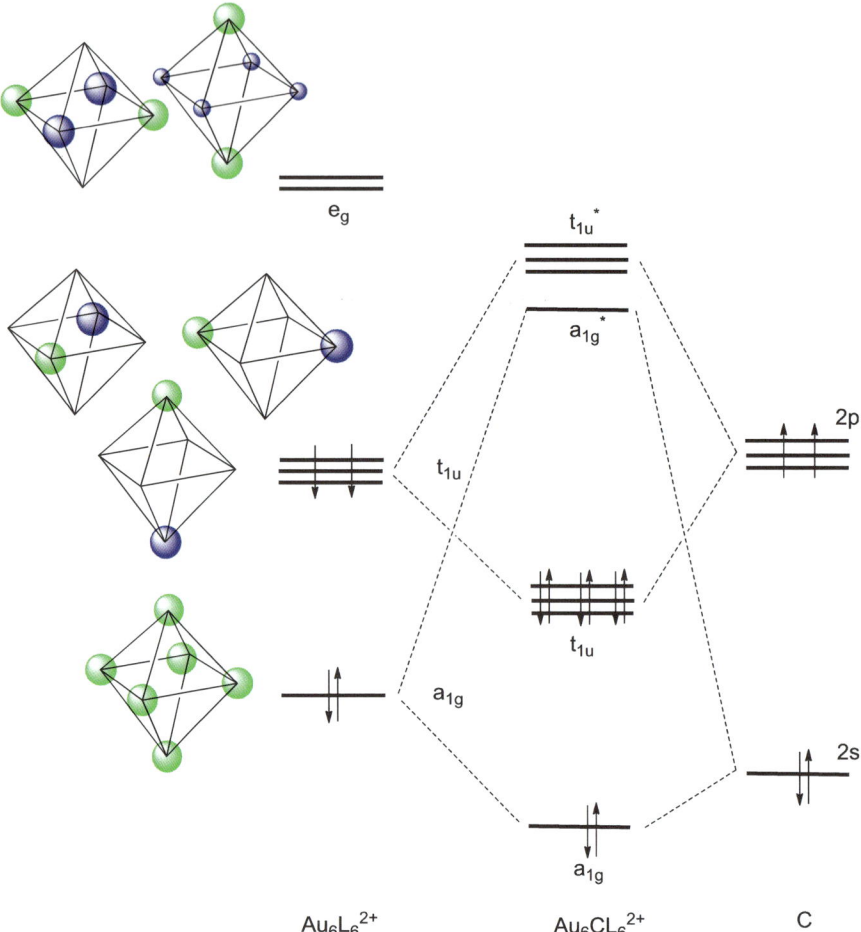

Fig. 11 Molecular orbital interaction diagram for $[Au_6C(PPh_3)_6]^{2+}$. Similar analyses may be constructed for trigonal bipyramidal $[Au_5N(PPh_3)_5]^{2+}$ and tetrahedral $[Au_4O(PPh_3)_4]^{2+}$

For gold clusters with 8 metal atoms or more metals the central cavity can accommodate a metal atom with a comparable radius. Therefore, a wide range of gold clusters with interstitial gold, rhodium, palladium and platinum interstitial atoms have been synthesised and characterised crystallographically. The introduction of an interstitial metal with nd, $(n + 1)$s and $(n + 1)$p valence orbitals for spherical clusters is represented schematically in Fig. 12 for an icosahedral cluster. The interstitial metal atom has d, s and p valence orbitals and the s and p valence orbitals symmetry match the skeletal molecular orbitals $[S^\sigma]$ and $[P^\sigma]$ in much the same way as that described above for carbon. The d orbitals of the gold atom are stabilised somewhat by a weaker interaction with the D^σ antibonding skeletal molecular orbitals, and therefore a spherical cluster is characterised by the

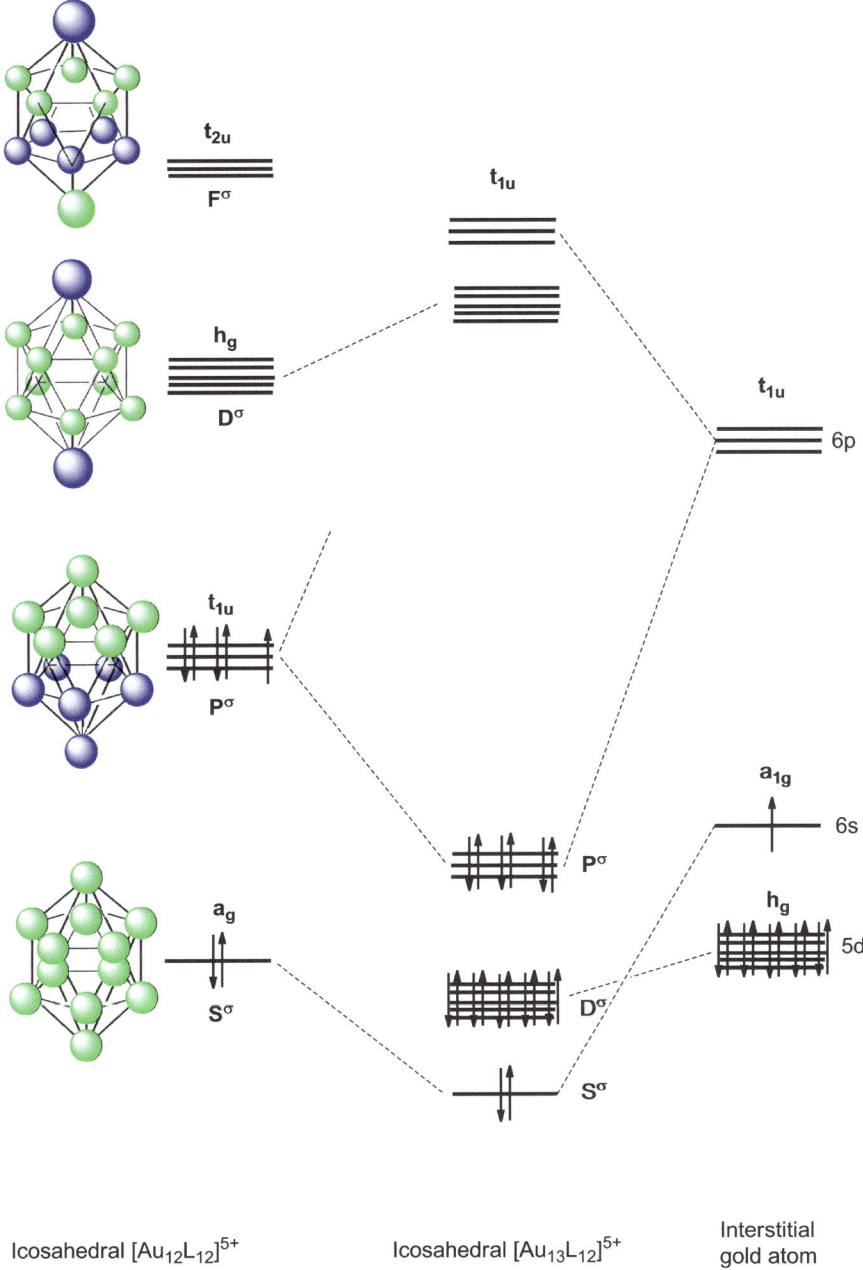

Fig. 12 Interaction molecular orbital diagram for metal-centred icosahedra

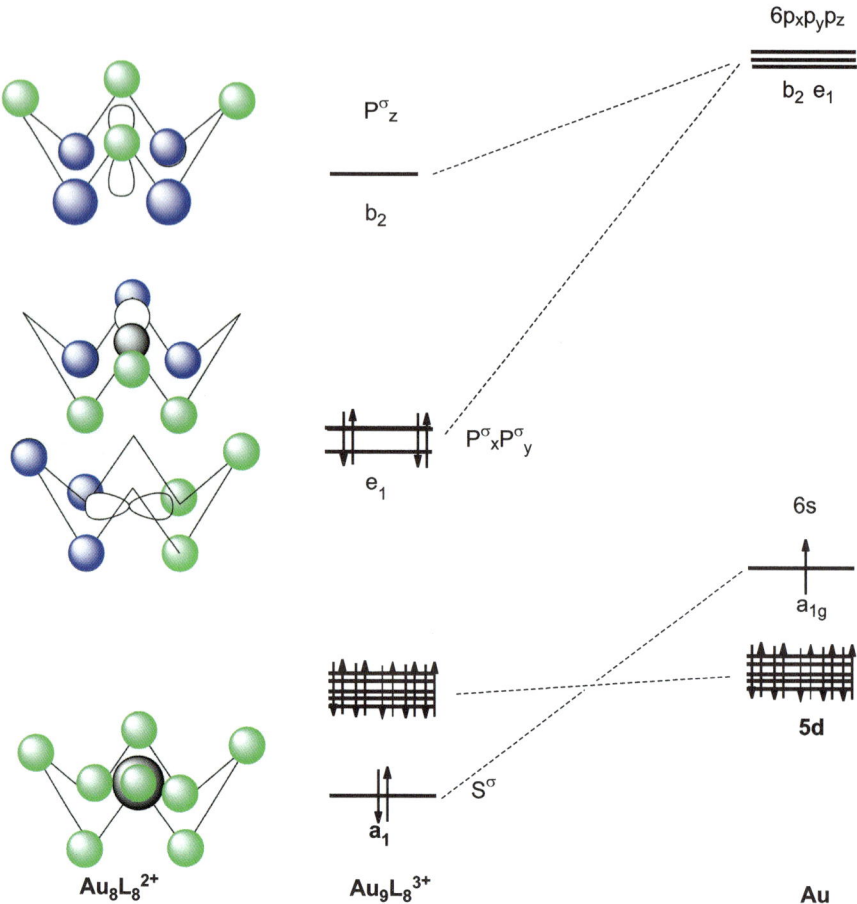

Fig. 13 Skeletal molecular orbitals for a centred crown of gold atoms

following filled molecular orbitals $[S^\sigma]^2$ $[d]^{10}[P^\sigma]^6$, i.e. when there is a total of 18 valence electrons, 10 of which are localised on the interstitial gold atom, i.e. these clusters are characterised by $12n + 18$ valence electrons (or $2n + 8$ if the d shells on the gold atoms are excluded because of their relatively small bonding contribution). If the filled d orbitals of gold and the Au–P bonds are ignored, then the skeletal electron count (sec) for these spherical clusters is 8. For centred gold clusters with 16 valence electrons (or 6 electrons if the d electrons on the interstitial atom are ignored), an oblate (or toroidal) topology is favoured, in much the same way as that described above for non-centred clusters. The relevant interaction diagram is shown in Fig. 13 for a crown geometry of gold atoms with an interstitial atom. The degeneracy of the $[P^\sigma]$ shell is removed in the oblate topology and one component $[P^\sigma_z]$ is no longer bonding, and consequently these clusters are characterised by the closed shell $[S^\sigma]^2[d]^{10}[P^\sigma_{xy}]^4$, i.e. 16 electrons. Figure 3 gives specific examples of spherical and toroidal clusters and their associated polyhedral and skeletal electron counts (pec and sec).

The strong radial interactions in such clusters result in a soft potential energy surface for skeletal rearrangements and consequently the broader classification based on topology is more appropriate than the designation of specific polyhedra. This difference distinguishes gold clusters from the carbonyl clusters of the earlier transition metals, which conform to the PSEPT. The presence of these soft potential energy surfaces also results in the observation of alternative skeletal geometries depending on the precise nature of the ligands, the counterions and the mode of crystallisation. Indeed for one cluster this has resulted in the structural character-isation of skeletal isomers [45, 46]. It also is manifested in geometry changes which are induced by high pressures. The bonding analysis developed above has been widely used to account for the structures of a wide range of homonuclear and heteronuclear cluster compounds of gold [49].

The soft potential energy surface separating alternative skeletal geometries has consequences also for understanding the geometries and bonding characteristics of higher nuclearity clusters. For example, if one compares the energies of icosahedral and cuboctahedral clusters, the energies are not greatly different. For both polyhe-dra there are 12 strong radial interactions, but the icosahedron has marginally stronger tangential interactions because it has exclusively triangular faces. The surface interactions for both polyhedra are weaker than the radial interactions, and the icosahedron has five (rather than four) nearest neighbour surface interactions leading to this small energy difference. The relevant skeletal molecular orbitals for a cuboctahedron are shown in Fig. 14.

3.4   Bonding in Condensed and Fused Clusters

3.4.1   Linking of Polyhedra

The structures of capped clusters of gold have been introduced and the bonding implications have been discussed above. Capping is just a specific example of a more general mechanism of cluster growth based on the condensation of polyhedra through sharing of vertices, edges or faces. It is noteworthy that higher-nuclearity ligand-stabilised clusters do not necessarily adopt the close-packed arrangements characteristic of the parent bulk metal, but frequently adopt less symmetrical structures based on the condensation of symmetric polyhedra [71–73]. This sug-gests that cluster growth is a kinetically controlled process where the activation energies for the condensation processes are smaller than those which are required for the rearrangement of the polyhedra into close-packed arrangements.

The closed shell requirements of these condensed clusters of the earlier transi-tion metals, especially with carbonyl ligands, have been brought within the frame-work of the PSEPT [77]. These principles are not directly applicable to cluster

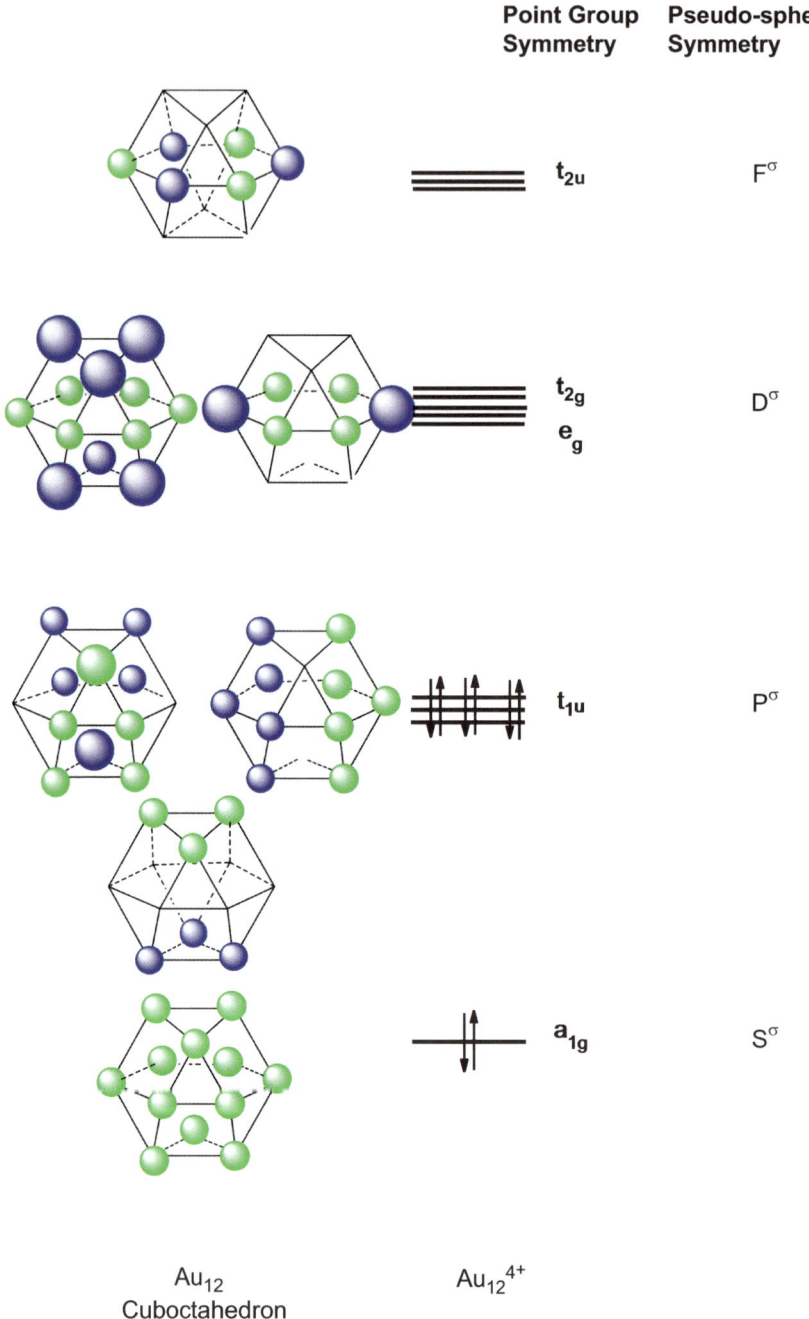

Fig. 14 Skeletal molecular orbitals for a non-centred cuboctahedral gold cluster. Note the similarity to that of icosahedrons

compounds of the group 11 metals, although similar conclusions may be derived for these metals by using a fragment molecular orbital mode of analysis, which utilises the fundamental results of perturbation theory.

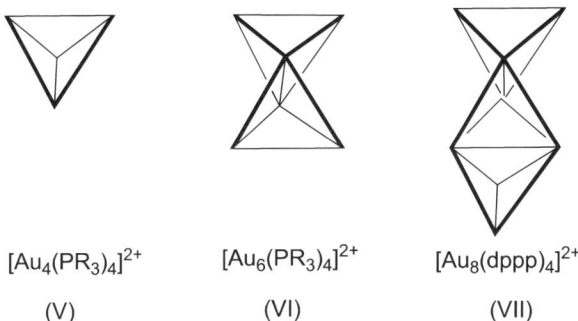

$[Au_4(PR_3)_4]^{2+}$ $[Au_6(PR_3)_4]^{2+}$ $[Au_8(dppp)_4]^{2+}$

(V)   (VI)   (VII)

For example, there are a series of gold clusters based on edge-linked tetrahedra, which are illustrated in (V)–(VII). The initial tetrahedral structure was established by Steggerda, Mingos and Schmidbaur, the bi-tetrahedron by Mingos and the tri-tetrahedron by Kominishi and his co-workers [86–89], who have also studied the UV-visible spectral properties of these ions. The series of clusters share a 2+ charge, and this suggests a 2e increment in the sec as pairs of gold atoms are added. Starting with the molecular orbitals of the tetrahedron which were introduced in Fig. 6, it is possible to evaluate the effect of adding successive tetrahedra by sharing edges and thereby increasing the number of metal atoms by 2. The problem has close similarities to the linking of hydrogen atoms in a linear chain since the parent tetrahedron is characterised by a single in-phase molecular orbital $[S^\sigma]$ (see Fig. 6). The problem is also analogous to the classical quantum mechanical solution to the particle in a box problem which results in the introduction of successive nodes as the principle quantum number is increased. The number of bonding skeletal molecular orbitals increases by one for each pair of gold atoms added, and the resultant molecular orbitals are illustrated in Fig. 15. In the notation which we have introduced above the relevant molecular orbitals are given the pseudo-atomic symbols $[S^\sigma]^2$ for Au_4, $[S^\sigma]^2[P^\sigma]^2$ for Au_6 and $[S^\sigma]^2[P^\sigma]^2[D^\sigma]^2$ for Au_8. This notation emphasises the number of nodal planes which result from the successive addition of pairs of gold atoms to generate the chain structures. The analysis is consistent with the observed electron counts and may also be used as a basis for interpreting their optical spectral characteristics [86–89]. The chain of tetrahedra has molecular orbitals which mimic those of alternant hydrocarbons which are characterised by a non-bonding molecular orbital when there are an odd number of carbon atoms. The occurrence of such non-bonding molecular orbitals results in related cationic, radical and anionic species, and it is possible that the gold chains may show similar redox properties for odd numbers of tetrahedra.

For the group 11 metals a number of structures have been determined which have several icosahedra either linked in a line or joined in a triangle or tetrahedron. Teo and his co-workers have been particularly active in developing this area and

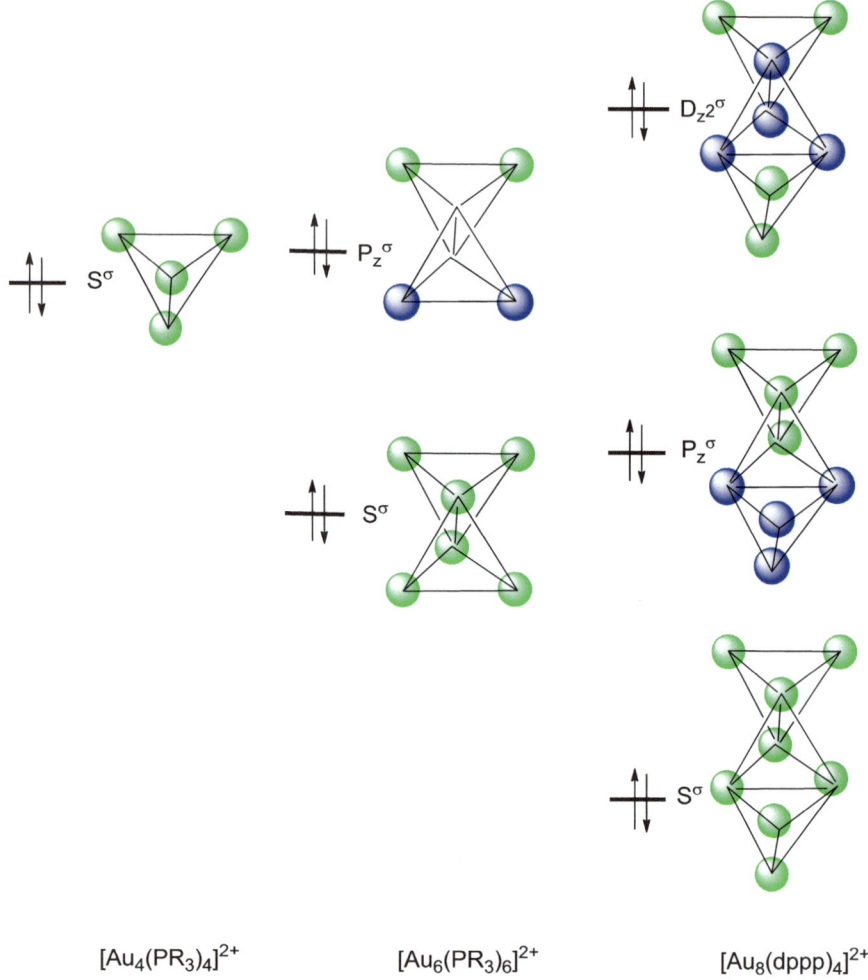

Fig. 15 Skeletal molecular orbitals for edge-linked tetrahedra

have reported the structures of several gold–silver clusters [59], and the electronic characteristics of these clusters have been analysed using the molecular orbital ideas developed above [84]. The parent icosahedron exemplified by $[Au_{13}X_m(PMe_2Ph)_{12-m}]^{x+}$ $(m = 2\text{--}4; x = 5 - m)$ has a pec of 162 electrons and a sec associated with 8 electrons occupying $[S^\sigma]^2[P^\sigma]^6$. For vertex sharing icosahedra which adopt a linear arrangement, the skeletal molecular orbitals are simple combinations of the skeletal molecular orbitals of the parent icosahedra. Furthermore, vertex sharing does not lead to strong inter-icosahedral interactions, because the overlap integrals are small [84]. Consequently these aggregates of n icosahedra are characterised by $[1S^\sigma][1P^\sigma]^6$, $[2S^\sigma][2P^\sigma]^6$,...... $[nS^\sigma]^2[nP^\sigma]^6$ occupied

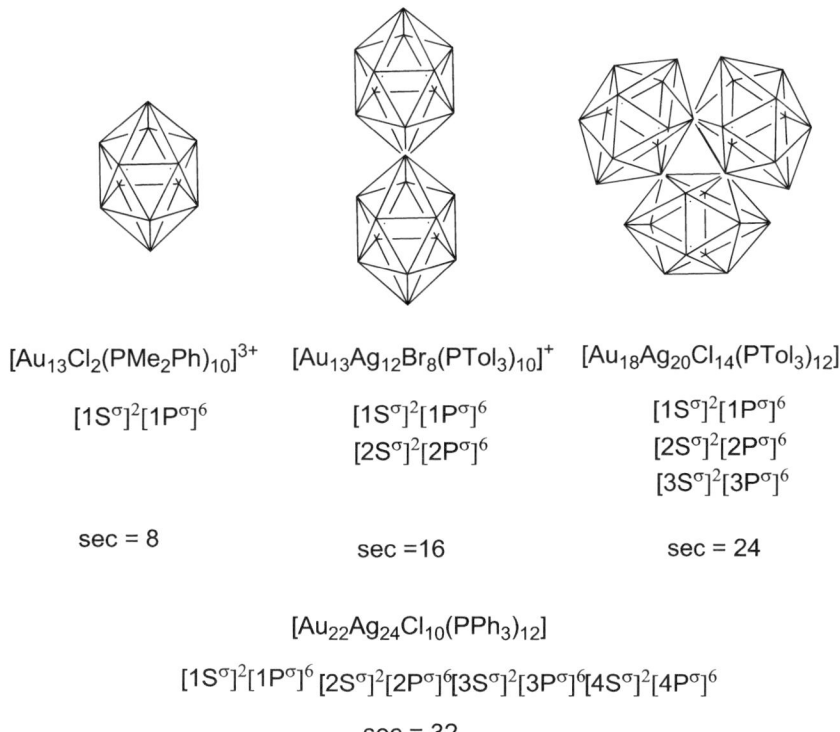

$[Au_{13}Cl_2(PMe_2Ph)_{10}]^{3+}$ $[Au_{13}Ag_{12}Br_8(PTol_3)_{10}]^+$ $[Au_{18}Ag_{20}Cl_{14}(PTol_3)_{12}]$

$[1S^\sigma]^2[1P^\sigma]^6$ $[1S^\sigma]^2[1P^\sigma]^6$ $[1S^\sigma]^2[1P^\sigma]^6$

$[2S^\sigma]^2[2P^\sigma]^6$ $[2S^\sigma]^2[2P^\sigma]^6$

$[3S^\sigma]^2[3P^\sigma]^6$

sec = 8   sec =16   sec = 24

$[Au_{22}Ag_{24}Cl_{10}(PPh_3)_{12}]$

$[1S^\sigma]^2[1P^\sigma]^6\,[2S^\sigma]^2[2P^\sigma]^6[3S^\sigma]^2[3P^\sigma]^6[4S^\sigma]^2[4P^\sigma]^6$

sec = 32

Fig. 16 Examples of vertex-linked icosahedral clusters and their closed shell electronic requirements

molecular orbitals. The total number of skeletal electrons (sec) is given by the simple formula sec = 8n, where n is the number of vertex-sharing icosahedra. Specific examples of these clusters are illustrated in Fig. 16.

Icosahedra may also condense to form more compact structures, and specific examples resulting from sharing a triangular face or by fusing an icosahedron and a pentagonal bipyramid are illustrated in Fig. 17 [90]. In the former example the resulting polyhedron has successive layers of atoms on the threefold rotations axis and in the latter along a fivefold axis. Figure 18 illustrates the skeletal molecular orbitals for a pair of icosahedra sharing a triangular face. The resulting fused polyhedron has D$_{3d}$ symmetry and is characterised by 7 skeletal bonding molecular orbitals. The molecular orbitals resemble those for a diatomic molecule such as F$_2$, where the 2s and 2p atomic orbitals combine to form (s + s), (s − s), (p$_z$ + p$_z$), (p$_z$ − p$_z$), (p$_{x,y}$ + p$_{x,y}$) and (p$_{x,y}$ − p$_{x,y}$) linear combinations. These molecular orbitals are given united atoms symmetry symbols in Fig. 6, i.e. $[1S^\sigma]^2[1P_z^\sigma]^2$, $[1P_{x,y}^\sigma]^4$, $[1D_{z^2}^\sigma]^2$, and $[1D_{xz,yz}^\sigma]^4$. The only linear combination which is too antibonding to be occupied is (p$_z$ − p$_z$) which has 1F$^\sigma$ symmetry in the united atom description. It follows that face rather than vertex sharing results in one molecular orbital becoming unavailable and consequently the total electron count

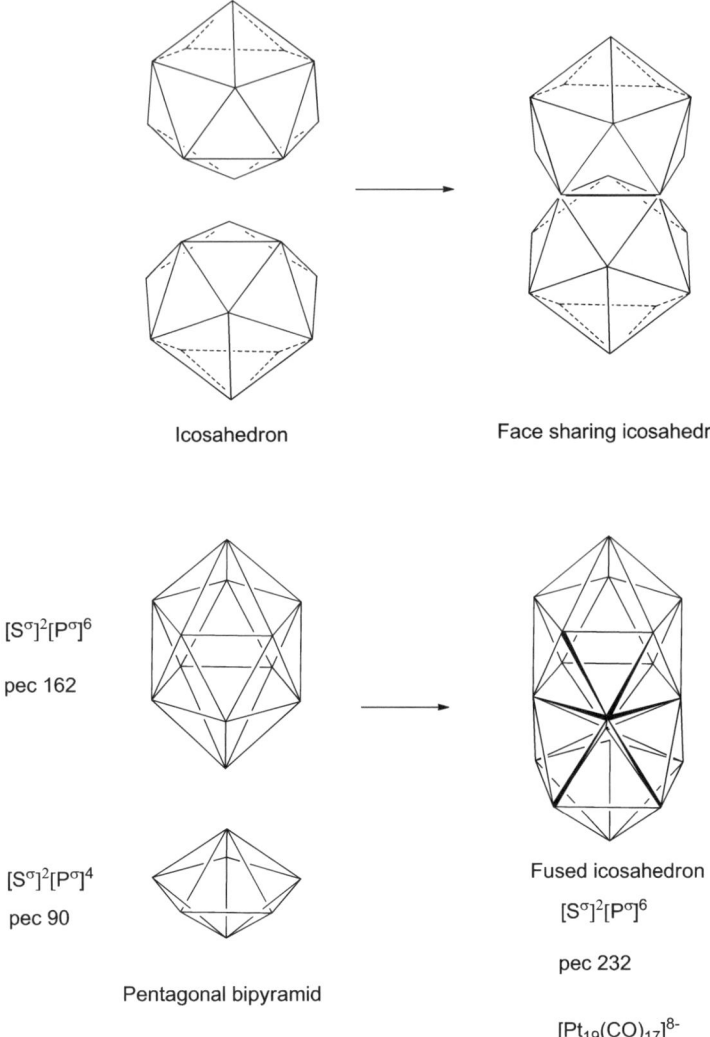

$[S^\sigma]^2[P^\sigma]^6$

pec 162

$[S^\sigma]^2[P^\sigma]^4$

pec 90

Icosahedron   Face sharing icosahedra

Fused icosahedron

$[S^\sigma]^2[P^\sigma]^6$

pec 232

$[Pt_{19}(CO)_{17}]^{8-}$

Pentagonal bipyramid

Fig. 17 Condensed clusters based on the fusion of icosahedra

(sec) for the dimer is 14 rather than 16 as observed for the vertex sharing pair of icosahedra.

The 19 atom fused icosahedron has the skeletal molecular orbitals shown in Fig. 19 and the number of bonding skeletal molecular orbitals is reduced to four: $[1S^\sigma]^2[1P_z^\sigma]^2$, $[1P_{x,y}^\sigma]^4$ because the doubly noded linear combinations which have D symmetry are too strongly antibonding to be accessible. The more condensed structure effectively makes the out-of-phase $(p_{x,y} - p_{x,y})$ linear combinations more strongly antibonding and therefore unavailable, i.e. it resembles the multiply bonded diatomics N_2 and C_2.

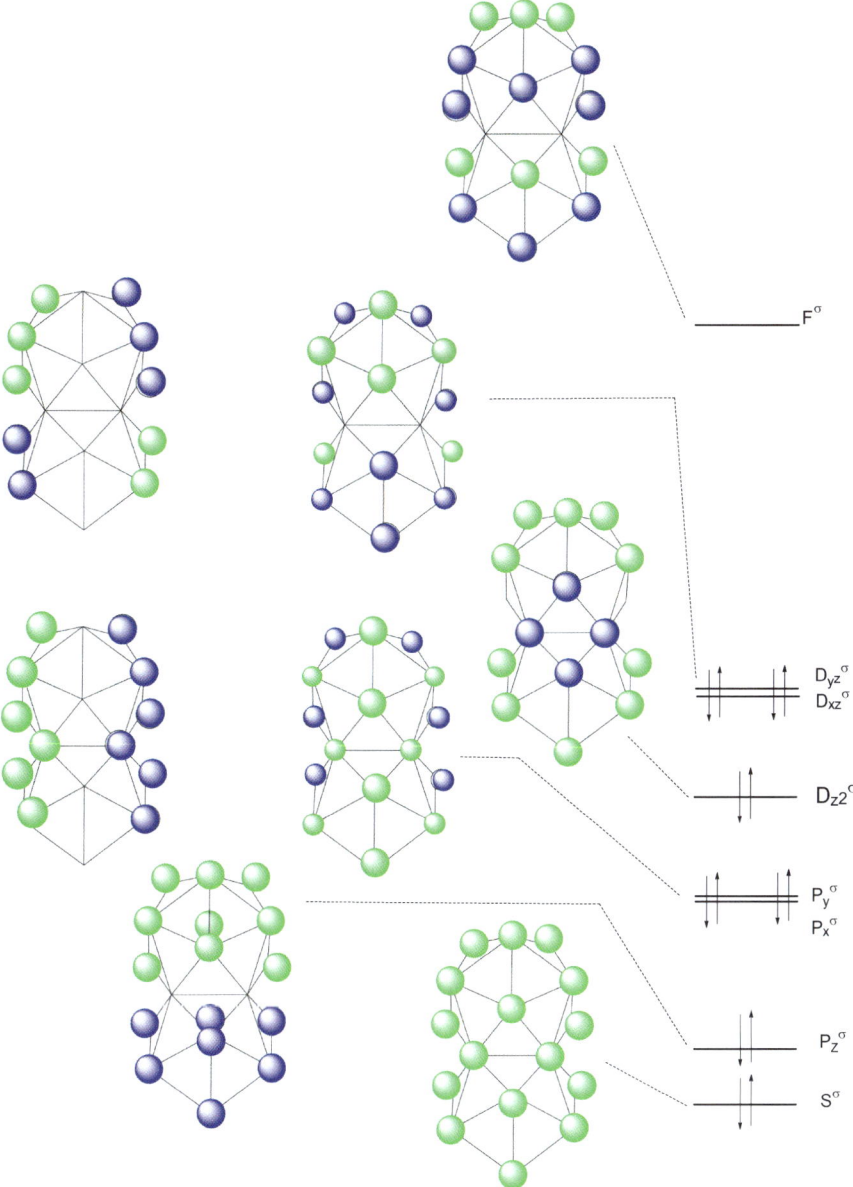

Fig. 18 Skeletal molecular orbitals of a pair of centred icosahedra sharing a triangular face. The total number of atoms is 23 (2 × 13 (centred icosahedron) − 3 for the shared triangular face)

This analysis of the nodal characteristics of the molecular orbitals of vertex, face and fused icosahedra centred clusters results in the number of bonding skeletal molecular orbitals being reduced from 8 to 7–4. The relevant skeletal molecular

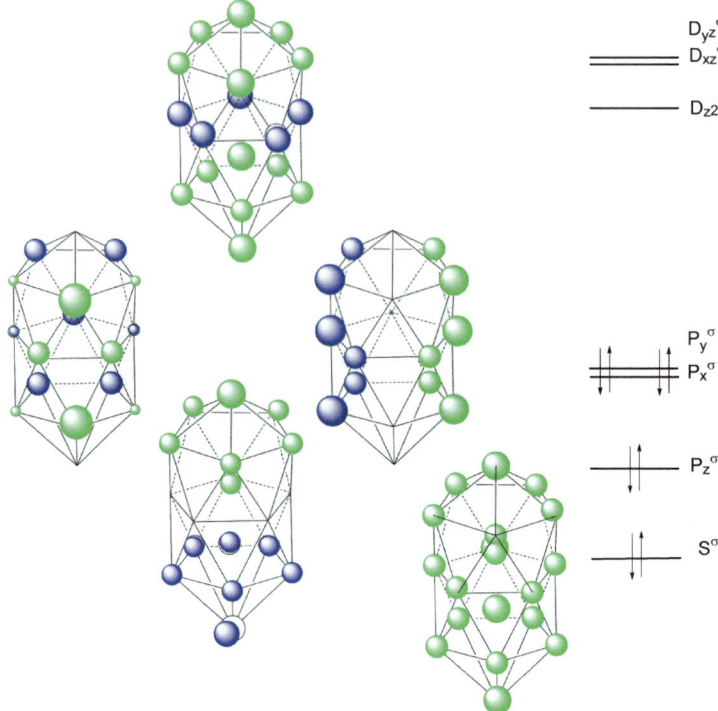

Fig. 19 Skeletal molecular orbitals for an icosahedron and a pentagonal bipyramid which has 19 metal atoms, 2 of which occupy interstitial sites

orbitals are illustrated in Figs. 18 and 19. The most spherical and most closely packed M_{19} cluster therefore behaves analogous to the centred icosahedron since both clusters are associated with four $[1S^{\sigma}]^2[1P_{z\sigma}]^2[1P_{x,y}^{\sigma}]^4$ skeletal molecular orbitals although the former has two interstitial atoms. This behaviour is reminiscent of the united atom approach which interrelated the molecular orbitals of diatomic molecules to the atomic orbitals of a single atom. The orbitals are correlated using symmetry arguments and are required to obey the non-crossing rule. The number of bonding skeletal molecular orbitals in carbonyl clusters of the earlier transition metals has been analysed using Tensor Surface Harmonic Theory [77], and they show a similar pattern of behaviour. For example, vertex, edge and face sharing octahedra have the following number of skeletal molecular orbitals: 11, 9 and 7 and the most condensed example has the same number of skeletal molecular orbitals as the parent octahedron [71, 72].

Although a gold phosphine cluster with face sharing or fused icosahedral structures have not been characterised to date, the corresponding and isostructural platinum carbonyl cluster $[Pt_{19}(CO)_{17}]^{8-}$ has been structurally determined [90]. It is noteworthy that gold clusters based either on a pair of face sharing icosahedra or the fused geometry have been stabilised by organothiolato-ligands recently. They have been structurally characterised and are discussed in more detail below.

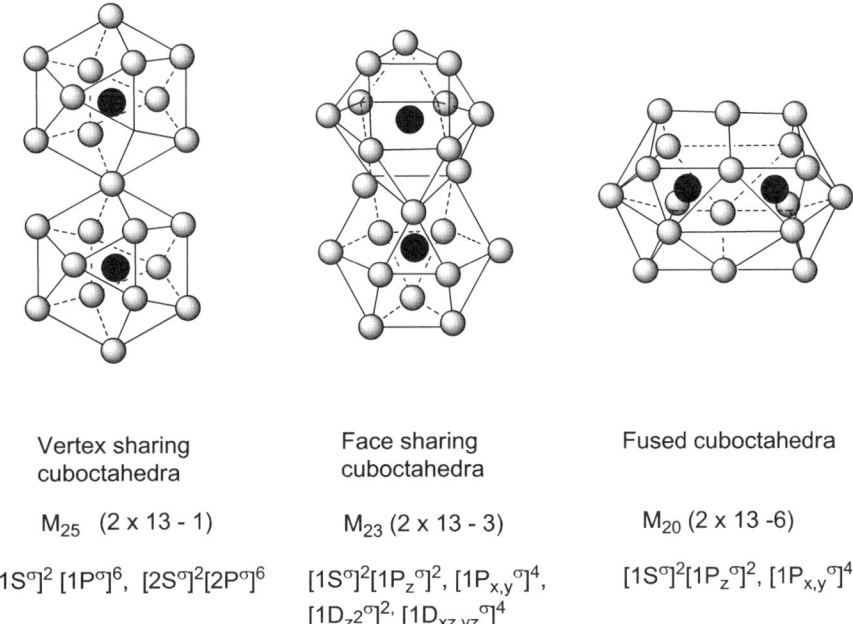

Vertex sharing cuboctahedra	Face sharing cuboctahedra	Fused cuboctahedra
M_{25} (2 x 13 - 1)	M_{23} (2 x 13 - 3)	M_{20} (2 x 13 -6)
$[1S^{\sigma}]^2 [1P^{\sigma}]^6$, $[2S^{\sigma}]^2[2P^{\sigma}]^6$	$[1S^{\sigma}]^2[1P_z{}^{\sigma}]^2$, $[1P_{x,y}{}^{\sigma}]^4$, $[1D_{z^2}{}^{\sigma}]^{2}$ $[1D_{xz,yz}{}^{\sigma}]^4$	$[1S^{\sigma}]^2[1P_z{}^{\sigma}]^2$, $[1P_{x,y}{}^{\sigma}]^4$

Fig. 20 Vertex, face and fusion of cuboctahedral centred clusters

It was noted above that the centred icosahedral and the cuboctahedral Au_{13} clusters have a very similar pattern of molecular orbitals, because of the predominance of the radial metal–metal bonding. The tangential bonding–bonding interactions are slightly larger for the icosahedron, because of the greater number of nearest neighbour interactions. They are both characterised by 4 very stable skeletal bonding molecular orbitals, which in the pseudo-spherical symmetry correspond to $[S^{\sigma}]^2[P^{\sigma}]^6$. The cuboctahedron may also form more complex cluster architectures by vertex, edge or face sharing as illustrated in Fig. 20. These condensation processes lead to a very similar pattern of molecular orbitals to that noted above for icosahedra, and the relevant occupied skeletal molecular orbitals are summarised at the bottom of Fig. 20. Most importantly both classes of polyhedral show a similar pattern of behaviour – the vertex sharing pair of clusters are analogous to a pair of isolated clusters, and as the degree of condensation increases, the number of skeletal molecular orbitals decreases until it reaches the limit for the spherical cluster. The relevance of these conclusions for organothiolato-gold clusters is discussed in more detail below.

The structure of the recently reported cluster $[Au_{14}(PPh_3)_8(NO_3)_4]$ (II) [59] has similarities to the cuboctahedral fused structure discussed above since it is based on two trigonal bipyramids joined by an Au–Au bond which is ringed by 4 gold atoms. The molecule has approximately D_{2h} symmetry which is the same point group as the fused cuboctahedron. The skeletal molecular orbitals would follow the patterns established in Figs. 18 and 19. The sec of 10 electrons is consistent with the occupation of the following molecular orbitals: $[1S^{\sigma}]^2[1P_z^{\sigma}]^2$, $[1P_{x,y}^{\sigma}]^4$, $[1D_{z^2}^{\sigma}]^2$. The occupation of $[1S^{\sigma}]^2[1P_{z\sigma}]^2$ and $[1D_{z^2}^{\sigma}]^2$ accounts for the short Au–Au distance between the interstitial gold atoms.

3.4.2 Summary

The skeletal bonding requirements of cluster compounds of gold, which have been developed since the 1980s and discussed and amplified in the previous section, are summarised in a graphical fashion in Fig. 21. The figure brings together the conclusions which have been derived from simple molecular orbital ideas for centred and non-centred clusters. It also highlights two important additional points – firstly the clusters do not adopt spherical structures exclusively and secondly the non-spherical structures arise either from shapes which result either from the elongation or compression of a sphere (i.e. prolate and oblate shapes) or from the condensation of spherical polyhedral via vertex, edge or face sharing.

The notation which has been used to describe the nodal characteristics of the skeletal molecular orbitals in these polyhedral gold clusters has its origins in the TSHM which is based on a free electron description for an electron constrained to a spherical surface [69–72, 84]. A closely related model is the jellium model which was developed originally to account for the relative abundance of alkali metal clusters generated in the gas phase in molecular beam experiments. The relative stabilities of the clusters was related to the quantum shell structure for the particle in a sphere problem [69, 70], i.e. the most stable structures were associated with molecules and ions which had complete electronic shells. The Tensor Surface Harmonic and jellium models both depend on expressing the solutions to the quantum mechanical problem in terms of spherical harmonics and both emphasise the importance of the number of angular nodes associated with the spherical harmonic s, p, d, f, etc. functions. However, they differ in several important respects – the TSH approach does not introduce radial nodes because the wave functions are restricted to a single spherical surface, whereas the jellium model requires radial and angular nodes to distinguish the spherical harmonic solutions. The TSH theory gives a more satisfactory account of single sphere clusters where tangential and radial bonding effects are both important because it introduces vector and tensor surface harmonics. Therefore, it has proved to be particularly important for describing boron and carbon clusters and metal cluster compounds of the transition elements where tangential and radial bonding interactions are both significant. The limitation of the single sphere TSH approach may be overcome by building up more complex polyhedral structures using capping and condensation principles and thereby creating a multilayered structure. Symmetry and perturbation theory considerations may be used to determine the number and nodal characteristics of the bonding skeletal molecular in the resulting multilayered structures [60, 61, 84].

The jellium model solves the Schrödinger equation for the particle in a sphere problem and the resulting wave function solutions are characterised by radial and angular nodes. The orbital energies depend somewhat on the potential used,

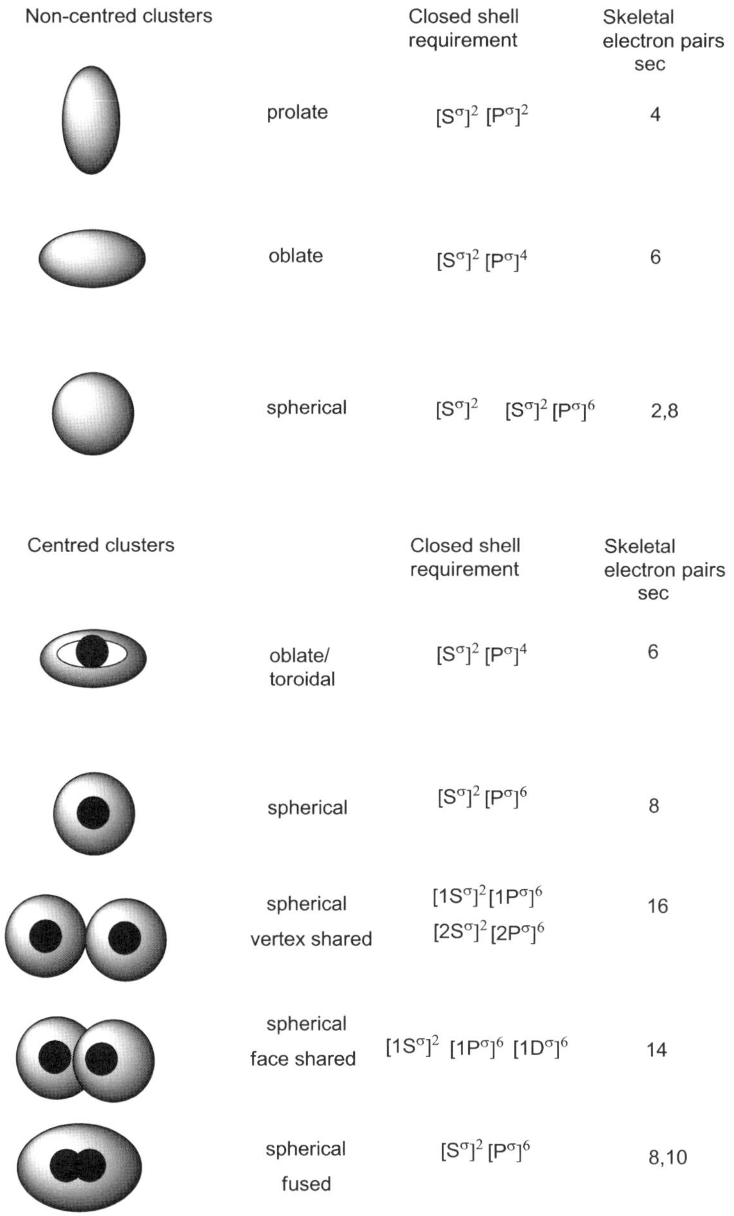

Non-centred clusters		Closed shell requirement	Skeletal electron pairs sec
	prolate	$[S^\sigma]^2 [P^\sigma]^2$	4
	oblate	$[S^\sigma]^2 [P^\sigma]^4$	6
	spherical	$[S^\sigma]^2 \quad [S^\sigma]^2 [P^\sigma]^6$	2,8

Centred clusters		Closed shell requirement	Skeletal electron pairs sec
	oblate/ toroidal	$[S^\sigma]^2 [P^\sigma]^4$	6
	spherical	$[S^\sigma]^2 [P^\sigma]^6$	8
	spherical vertex shared	$[1S^\sigma]^2 [1P^\sigma]^6$ $[2S^\sigma]^2 [2P^\sigma]^6$	16
	spherical face shared	$[1S^\sigma]^2 \ [1P^\sigma]^6 \ [1D^\sigma]^6$	14
	spherical fused	$[S^\sigma]^2 [P^\sigma]^6$	8,10

Fig. 21 Summary of bonding characteristics of gold clusters

i.e. 1s < 1p < 1d ~ 2s < 1f ~ 2p < 1g ~ 2d ~ 3s ... (using a 3-D harmonic potential), 1s < 1p < 1d < 2s < 1f < 2p < 1g ~ 2d (3-D square potential) and 1s < 1p < 1d < 2s < 1f < 2p < 1g ~ 2d (Woods–Saxon potential). When combined with the Aufbau principle, it suggests that for neutral alkali metal clusters the species M_2, M_8, M_{20}, M_{40}, M_{70}:M_2, M_8, M_{18}, M_{20}, M_{34}, M_{40}, M_{58}; M_2, M_8, M_{18}, M_{20}, M_{34}, M_{40}, M_{58} species should have stable closed shell configurations analogous to those of the noble gases. This approach was naturally extended to Cu, Ag and Au, which also have ns^1 ($n = 4, 5, 6$) outer configurations, making them analogous to the alkali metals. The complementary aspects of the TSH and jellium models to gold was discussed by Lin et al. [82, 84], and Lin and Mingos [84] first explored the consequences of satisfying both the closed shell electronic requirements of the jellium model and the geometric consequences of having close-packed and spherical arrangements of atoms within the cluster. The shell-correction method, which had its origins in the physics of shell effects in atomic nuclei, has been widely used for describing shell effects, the energetics and decay pathways of metal fragmentation processes. In early uses of the jellium model, the shape of the cluster was assumed in all instances to be spherical, but it was soon recognised that the constraint of spherical symmetry was too restrictive. For gold cluster molecules the geometric consequences of oblate and prolate distortions on the shell structures were recognised early on and then applied to those clusters with packed and spherical structures using a perturbation theory analysis of the jellium model [82, 84].

The Tensor Surface Harmonic Theory developed by Stone [69, 70] also takes into consideration the tangential skeletal molecular orbitals which originate from the transition metal np and ($n - 1$)d orbitals, which have π-local symmetries with respect to the radial coordinate. They require vector and tensor expansions of the basis p and d functions, but for the gold clusters such tangential interactions may be neglected. The two models come to very similar conclusions for metal atoms where the s orbital or radial interactions predominate. The TSH model which was developed for gold clusters almost 40 years ago has proved to be particularly useful for accounting for the structures gold phosphine clusters and was notable for the successful predictions of the structures and stoichiometries of octahedral [Au_6C $(PPh_3)_6$]$^{2+}$ and icosahedral [$Au_{13}Cl_2(PMe_2Ph)_{10}$]$^{3+}$. More generally the methodology and its group theoretical implications have proved useful for understanding the structures of condensed and fused clusters via the capping principle and the condensation rules [71, 72].

3.5 Introduction of Interstitial Atoms

The structures of high nuclearity clusters result from vertex, edge and face sharing of polyhedral units. The electron counting rules developed for metal carbonyl clusters were extended to condensed clusters by Mingos [71–73]. For more condensed clusters the observed structures are more conveniently described in terms of those close-packed arrangements which are characteristic of metals or crystallites of the metal which may have fivefold symmetry. As the nuclearity of transition metal clusters and the number of layers of close-packed atoms increases, the radial bonding interactions increases in importance compared to the tangential interactions and the bonding approaches the electron count $12n_S + \Delta_i$ where n_S is the number of surface atoms and Δ_i reflects the closed shell requirements of the interstitial group of atoms, i.e. 34 for M_2, 48 for M_3, 60 for a tetrahedron, 86 for octahedron and 162 for a centred icosahedron or cuboctahedron.

The model has important implications for the insulator to conductor transition for metal clusters and nano-particles. The recognition that even high nuclearity clusters conform to electron counting rules suggests that they are attaining well-defined closed shells. This behaviour is different from that which is characteristic of the bulk metal, which has partially filled bands. As the nuclearity of the cluster increases, the HOMO–LUMO gaps which define the closed shells decreases and the electron counting rules will become less valid.

3.6   Bonding Interrelationships Between Organothiolato- and Gold Phosphine Clusters

Following the development of flexible and convenient syntheses of organothiolato-gold cluster in the 1990s, a wide range of cluster compounds with radii less than 2 nm have been isolated and studied. The synthesis, separation and crystallisation of these new cluster compounds and their detailed structural analysis by single crystal X-ray crystallographic techniques presented a tremendous challenge to experimentalists, but their efforts have been rewarded by the determination of some key clusters structures, which have provided a profound insight into the structures of gold clusters, nanoparticles and colloids [32, 33, 91–96].

Until 2007 $[Au_{39}(PPh_3)_{14}Cl_6]^{2+}$, which has a close-packed core of gold atoms stabilised by phosphine and chloro-ligands, represented the highest nuclearity gold cluster compound [21]. It was initially assumed that organothiolato-gold cluster, which were isolated from the reductive routes developed initially for making gold colloids, also had structures based on a central close-packed core of gold atoms stabilised by SR ligands. It was appreciated, however, that SR ligands are more flexible than phosphines because they are able to bridge edges and faces of a close-packed gold polyhedron of metal atoms in a way which is not accessible to a simple Lewis base such as PR_3. The structural determinations of $[Au_{102}(p-MBA)_{44}]$ and $[Au_{25}(SCH_2CH_2Ph)_{18}]^-$ represented major breakthroughs [91–94], because they demonstrated that these clusters have several interesting features which differentiated them from the previously studied gold phosphine clusters. The structures revealed that in addition to an approximately spherical close-packed core of gold atoms, some of the gold atoms had separated from the core and combined with the SR ligands to form novel bidentate ligands, which are bound to the central metal core by dative S–Au bonds. Figure 22 illustrates the $[Au(SR)_2]$ and $[Au_2(SR)_3]$ metallothiolato-ligands, which have been revealed by recent crystallographic studies, and shows how these ligands bond to the central cluster core. The adoption of two distinct roles by the gold atoms in these gold cluster molecules has been described as the "divide and protect" principle. The oligomeric ligands $[Au(SR)_2]$

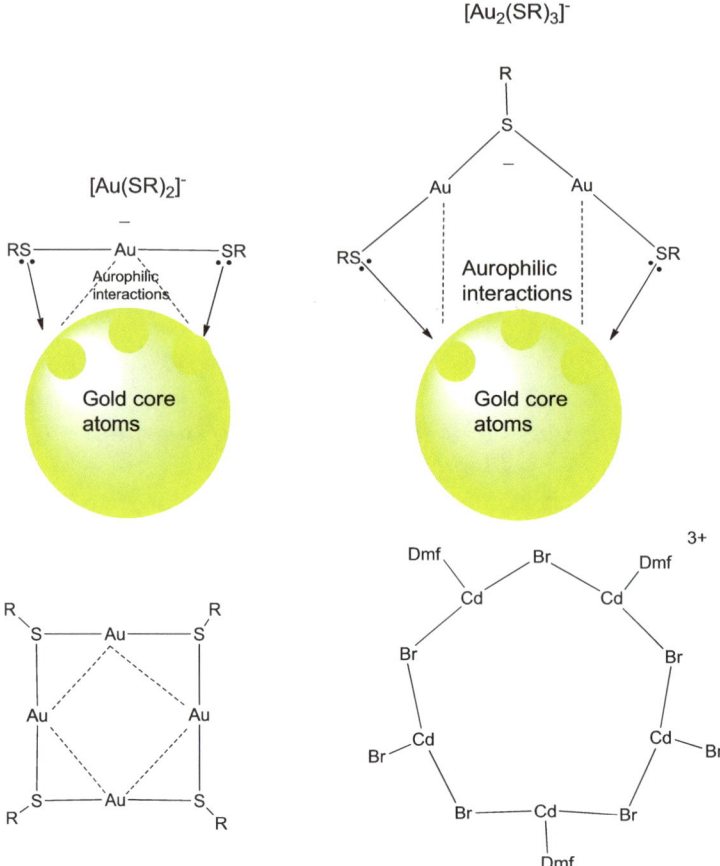

Fig. 22 Examples of gold organothiolato-and cadmium bromide staple ligands [97, 98]

and $[Au_2(SR)_3]$ have been given the portmanteau description "staple motifs", in order to emphasise their important role in protecting and stabilising the central gold core, by forming bridging bidentate ligands, with the appropriate bite angles. These two examples are the first three members of the series of $[Au_x(SR)_{x+1}]$ ligands ($m = 0, 1, 2, 3$, etc.), which have been studied theoretically and may be observed in future structure determinations. Related cyclic organothiolato-ligands, e.g. $[Au_4(SR)_4]$, also shown in Fig. 22, have also been proposed as potential ligating systems and studied extensively by DFT molecular orbital calculations by Häkkinen and co-workers [97, 99–105]. These calculations have provided interesting insights into the bonding capabilities of these ligands and higher oligomers of $[Au_x(SR)_{x+1}]$ as well as neutral cyclic $[Au_x(SR)_x]$ ligands. The strength of the gold–sulphur bond in organometallic clusters and the bonding preferences of "staple" ligands have been addressed by Häkkinen and concluded that the M–SR bond is strong and comparable to the metal–metal bond, i.e. Au–SR has been calculated to have a dissociation energy of 2.5 eV (241 kJ mol^{-1}) which is slightly larger than 2.3 eV (228 kJ mol^{-1}) dissociation energy of Au_2. They have also explored the conformations of the complexes and shown that these conformations may influence the S–Au–S bond angle adopted when the ligands are stapled to the gold core. The wider (124°) angle is observed in larger clusters and a smaller angle of 100° is observed in smaller clusters. The series of staple ligands $[Au_x(SR)_{x+1}]$ are probably in equilibrium in the solution surrounding the growing gold cluster molecule and provide a flexible source of those ligands which have just the right bite angle and spanning dimension for a specific gold cluster. Their presence therefore reduces the dispersity of the mixture of clusters present at the end of the reaction. The dative bonds between the sulphur atoms of the gold thiolato-oligomers and the central gold core may be supplemented by aurophilic interactions which are represented schematically in Fig. 22 by dotted lines.

If formal oxidation formalisms are used (i.e. Au^{+1} and SR^-) to assign electrons to the staple ligands, then the cluster molecules may be assigned the general formula $\{[Au]_{a+a'}\}^{x+}\{[Au(SR)_2]_b[Au_2(SR)_3]_c(SR)_d\}^{x-}$: $a + a'$ represents the total number of core gold atoms, a represents those gold atoms which are unavailable for bonding to the S ligands either because they are interstitial atoms or a surface atoms which lie in concave surface regions making them inaccessible for gold–sulphur bonding and a' represents the number of gold atoms of the central core which are capable of forming dative bonds with the staple ligands [106–112]. Table 1 gives specific examples of this notation. The formalism represents a reasonable formal partitioning of charges between the positively charged central core and the negatively charged surface staple motifs. Since the Au^{+1} ion has an empty 6s shell, it does contribute to the skeletal molecular orbitals of the central gold kernel. Recently a reversal of this formal charge separation has been proposed for cationic staple cadmium bromide ligands (see, e.g. Fig. 22), which have been observed to stabilise anionic platinum metal carbonyl anionic clusters, e.g. $[Pt_{19}(CO)_{17}]^{8-}$ stabilised by $\{[Cd_5(\mu-Br)_5Br_2(dmf)_3]^{3+}$. In this case the staple ligands are cyclic [90].

In the last decade, DFT calculations have been used increasingly to rationalise the observed structures of clusters and to use the information obtained to develop models to predict the stoichiometries and structures of new clusters with a good degree of accuracy. Years of chemical experience suggests that extra computing power is most effectively used when combined with intelligent qualitative and semi-quantitative models, which can provide an interpretation which may be used imaginatively by synthetic chemists.

The key structural determinations of $[Au_{25}(SCH_2CH_2Ph)_{18}]^-$ and $[Au_{102}(p\text{-}MBA)_{44}]$ and subsequent studies have led to the following generalisations [91, 96]:

1. The central gold core is approximately spherical and approximates to close packing, and cuboctahedral and icosahedral fragments are common components, e.g. an icosahedral Au_{13} core in the former and a decahedral D_{5h} Au_{79} core in the latter.
2. The staple gold thiolato-ligands are bonded to the central core using Au–S bonds and in general each atom on the outer face of the core is bonded to a single S

atom of the $[Au(SR)_2]$ and $[Au_2(SR)_3]$ staple ligands. The higher curvature of the core surface in the lower nuclearity clusters results in a preference for $[Au_2(SR)_3]$ staples and $[Au(SR)_2]$ staples in higher nuclearity clusters. Theoretical calculations have also indicated that the $[Au_3(SR)_4]$ and $[Au_4(SR)_5]$ ligands may also act as staples.
3. Bridging SR ligands have only been observed to date in lower nuclearity clusters.

Knowledge of these structural generalisations combined with DFT calculations have led Pei and Zeng [106–111] to propose a more general "inherent structural rule", which highlights principles based on the constraints which accompany the general formula of the subset of clusters, which do not have SR ligands, i.e. $[Au]_{a+a'}[Au(SR)_2]_b[Au_2(SR)_3]_c$, where $a, a', b, c, \ldots$ are integers. This analysis provides a more efficient starting point for DFT calculations by limiting the number of initial structures which have to be considered. Häkkinen has proposed an alternative "superatom model" for these gold clusters which draws heavily on the jellium model described above and to a lesser extent on earlier theoretical studies on gold phosphine clusters [97, 99–105]. This superatom model which is based on the Aufbau filling of the electronic shells $1s < 1p \ll 1d < 2s < 1f < 2p < 1g < 2d < 3s < 1h\ldots$. The superatom approach partitions the cluster so that the total number of free valence electrons in a cluster $[Au_m(SR)_n]^q$ is $m-n-q$ where m is the number of gold atoms. This corresponds to sec defined above. It follows that $[Au_{25}(SCH_2CH_2Ph)_{18}]^-$ and $[Au_{102}(p\text{-}MBA)_{44}]$ have 8 and 58 free valence electrons which correspond to the following shell closings in the jellium model: $1s^2\,1p^6$ and $1s^2\,1p^6\,1d^{10}\,2s^2\,1f^{14}\,2p^6\,1g^{18}$. Häkkinen has promoted the extension of the jellium model to these "superatom" clusters with closed shells for 2, 8, 20, 34, 40, 58, 92, 138 and 198 electrons. Häkkinen has also developed his model to understand the reactivities of gold cluster in catalysis. Both approaches described above have provided valuable insights into the bonding in organothiolato-gold clusters, although they have not fully united the phosphine and organothiolato-subdisciplines. The following section provides an analysis which encourages the attainment of this goal.

3.7 Construction of Organothiolato-Clusters from Phosphine Cluster Building Blocks

The bonding paradigm for phosphine clusters summarised in Fig. 21 may be used as a convenient starting point for understanding the structures and stoichiometries of organothiolate clusters and also serves a useful role in reducing the number of options required for more detailed DFT calculations [106]. Initially one may imagine that for $[Au_m(SR)_n]^q$ the DFT calculations may be used to minimise the energy until a global minimum is identified, but the process is complicated by the "divide and protect" principle which has the gold atoms alternating between ligand

and core locations. The interconversion of staple ligands requires some of the gold atoms to be sequestered from or added to the core. If the stoichiometry of the $[Au_m(SR)_n]^q$ cluster is to be maintained, then this requires the loss of 2 $[Au_2(SR)_3]$ staple ligands to create 3 $[Au(SR)_2]$ staple ligands, an increase in the number of gold atoms in the central core by one and the movement of two gold atoms from the core to the surface to form the new S–Au dative bonds (see Fig. 23). This creates a problem for bonding models which relate the geometry of the core with a specific number of metal atoms to the total number of electrons donated by the metal atom and the ligands. DFT calculations also face some difficulties since they have to model the migration of gold atoms to very different chemical environments, i.e. from core to metallo-ligand.

These conceptual and practical difficulties may be circumvented by an alternative procedure which has as its starting point a specific example of a phosphine gold cluster which has a well-defined geometry, i.e. an example drawn from Fig. 21. The chosen cluster, which has a well-defined stoichiometry and skeletal electron count, is used to establish an isoelectronic series which is derived by replacing the phosphines initially by thiolates (SR) and subsequently by $[Au(SR)_2]$ and $[Au_2(SR)_3]$ metallothiolato-ligands. Specifically the core cluster maintains its skeletal geometry by involving the same number of skeletal electrons (sec) across the series, and it maintains the contribution from the ligating ligands constant by replacing each phosphine ligand by an electron pair from the organothiolato-ligand, e.g. two phosphine ligands are replaced by a bridging SR^- ligand, which donates an electron pair to each of the gold atoms it bridges (see Fig. 24). The development of an isoelectronic series is assisted because SR, $[Au(SR)_2]$ and $[Au_2(SR)_3]$ are all bidentate ligands and consequently all bond to two metal atoms and do so by donating the same number of electrons. It is based on the following transformation of SR^- into the staple ligands shown in Fig. 25:

$$SR^- + AuSR \rightarrow \left[Au(SR)_2\right]^- + AuSR \rightarrow \left[Au_2(SR)_3\right]^-. \qquad (1)$$

Therefore, the following general substitutional procedure may be proposed, whereby the central core structure is retained throughout the series. The successive addition of Au(SR) fragments shown above permits the development of a series of clusters which have an identical Au_m core with the same number of skeletal electrons and the same number of gold atoms coordinated either to SR or the staple ligands. The process is illustrated in Fig. 25.

It may also be generalised as follows for homoleptic examples:

$$[Au_m(PR_3)_{2n}]^{q+}$$

$$[Au_{m+n}(SR)_{2n}]^{(q-n)+} \qquad\qquad [Au_{m+2n}(SR)_{3n}]^{(q-n)+}$$

$$\downarrow n(SR^-)$$

$$\|\| \qquad\qquad \|\|$$

$$[Au_m(\mu\text{-}SR)_n]^{(q-n)+} \xrightarrow{\;nAu(SR)\;} [Au_m\{Au(SR)_2\}_n]^{(q-n)+} \xrightarrow{\;nAu(SR)\;} [Au_m\{Au_2(SR)_3\}_n]^{(q-n)+}$$

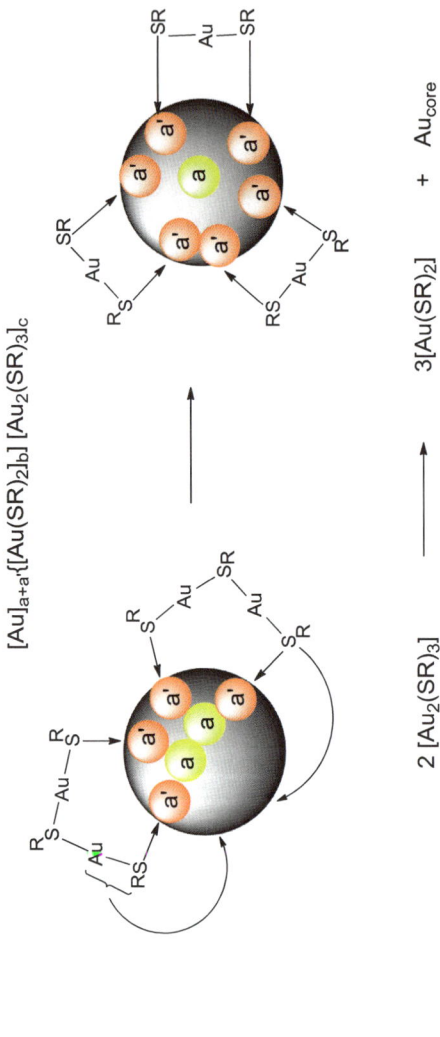

Fig. 23 Schematic illustration of the conversion of $[Au_2(SR)_3]$ to $[Au(SR)_2]$ staple ligands. In order to maintain the cluster stoichiometry 3 $[Au(SR)_2]$ staples are created for the loss of 2 $[Au_2(SR)_3]$ staples, and the number of available core atoms is increased by 2 in order to accommodate the extra bidentate ligand, and this means that the number of interstitial atoms has to be increased by one

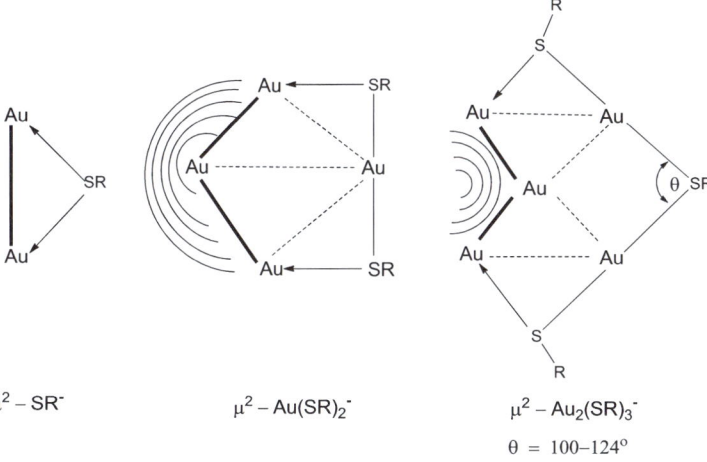

$\mu^2 - SR^-$ $\mu^2 - Au(SR)_2^-$ $\mu^2 - Au_2(SR)_3^-$

$\theta = 100\text{--}124^\circ$

Fig. 24 Examples of bridging thiolato- and metallothiolato-ligands. The dative bonds are indicated by *arrows* and aurophilic interactions by *dotted lines*

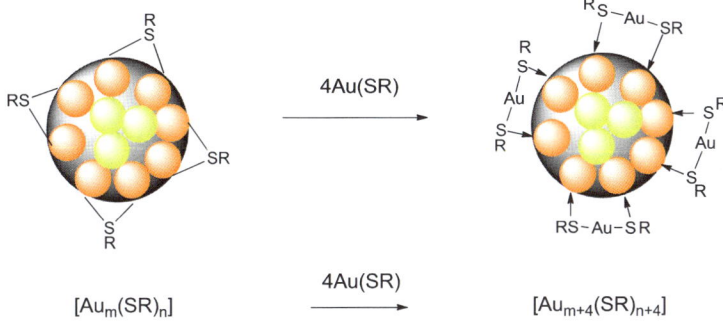

$[Au_m(SR)_n]$ $\xrightarrow{4Au(SR)}$ $[Au_{m+4}(SR)_{n+4}]$

Fig. 25 Schematic illustration of the building up of the surface thiolato-gold ligand layer by the addition of neutral AuSR molecules which do not affect the core geometry or electron count and maintain equivalent surface-core metal ligand bonding

The series $[Au_m(PR_3)_{2n}]^{q+}$: $[Au_m(SR)_n]^{(q-n)+}$: $[Au_{m+n}(SR)_{2n}]^{(q-n)+}$: $[Au_{m+2n}(SR)_{3n}]^{(q-n)+}$ have Au_m central cores which are isostructural and n or $2n$ gold atoms involved in the staple ligands. Some specific examples will serve to introduce the procedure.

3.7.1 Icosahedral-Based Structures [32, 33]

A *centred icosahedron*, which has a skeletal electron count (sec) of 8, is exemplified by $[Au_{13}Cl_2(PMe_2Ph)_{10}]^{3+}$ and may related to an isoelectronic organothiolato-cluster as follows:

$$\left[Au_{13}Cl_2(PR_3)_{10}\right]^{3+} \equiv \left[Au_{13}(PR_3)_{12}\right]^{5+} \equiv \left[Au_{13}(SR)_6\right]^-.$$

Since SR^- is capable of donating 4 electrons to gold atoms using 2 dative bonds (e.g. the bridging mode illustrated in Fig. 24), the 6(SR) ligands are electronically equivalent to 12 two-electron phosphine ligands in the original cluster. The charge difference between PR_3 and SR^- is reflected in the overall cluster which changes from 5+ to -1. The surface staple gold organothiolato-ligands in the resulting cluster may be formally created by adding neutral AuSR moieties to SR ligands according to Eq. (1) above.

This addition of Au(SR) molecules has no effect on the total charge on the cluster and does not alter the skeletal electron count. Therefore, the SR ligands may be replaced completely or partially by up to $6[Au(SR)_2]$ or $[Au_2(SR)_3]$ staple ligands, i.e. leading to $[Au_{13}\{Au(SR)_2\}_6]^-$ or $[Au_{13}\{Au_2(SR)_3\}_6]^-$. However, geometrically the ligands have the preferred bridging modes shown in Fig. 26. Specifically SR favours bonding to an adjacent pair of gold atoms, whereas the metallothiolates prefer non-adjacent pairs of metal atoms. The formulation $[Au_{13}\{Au_2(SR)_3\}_6]^-$ is preferred to $[Au_{13}\{Au(SR)_2\}_6]^-$ because $Au_2(SR)_3$ spans two non-adjacent gold atoms of the icosahedron more effectively. This simple analysis therefore not only reproduces the observed centred icosahedral structure for $[Au_{25}(SR)_{18}]^-$ illustrated schematically in Fig. 27 and clearly underlines the relationship with the parent phosphine cluster.

A pair of centred icosahedra sharing a triangular face is exemplified by $[Au_{23}L_{18}]^{9+}$ (see Fig. 17), with 18 of the 21 metal vertices available for bonding to phosphine-like ligands. The three vertices at the intersection are unavailable because of the concavity of the surface at this plane. This leads to the following isoelectronic relationships:

$$\left[Au_{23}L_{19}\right]^{9+} \equiv \left[Au_{23}(SR)_9\right] \equiv \left[Au_{23}\{Au(SR)_2\}_9\right] \equiv \left[Au_{32}(SR)_{18}\right]$$
$$\equiv \left[Au_{23}\{Au_2(SR)_3\}_9\right] \equiv \left[Au_{41}(SR)_{27}\right].$$

The homoleptic possibilities $[Au_{23}\{Au(SR)_2\}_9] \equiv [Au_{32}(SR)_{18}]$ and $[Au_{23}\{Au_2(SR)_3\}_9] \equiv [Au_{41}(SR)_{27}]$ both satisfy the closed shell electronic requirements, but the observed structure has a 3:6 combination of the two staples probably because the concave surface noted above is more effectively bridged by $3[Au(SR)_2]$ ligands for the geometric reasons illustrated in Fig. 26. Single crystal X-ray studies have confirmed that the $[Au_{38}(SR)_{24}]$ cluster indeed has the $[Au_{23}\{Au(SR)_2\}_3\{Au_2(SR)_3\}_6]$ structure.

A pair of fused icosahedra have not been structurally characterised for simple phosphine gold clusters, but the related platinum carbonyl cluster has been shown to have the structure shown in Fig. 17, e.g. $[Pt_{19}(CO)_{17}]^{8-}$ has two interstitial atoms in a structure with fivefold symmetry and successive layers containing 1:5:5:5:1 surface platinum atoms. This leads to the following isoelectronic relationships:

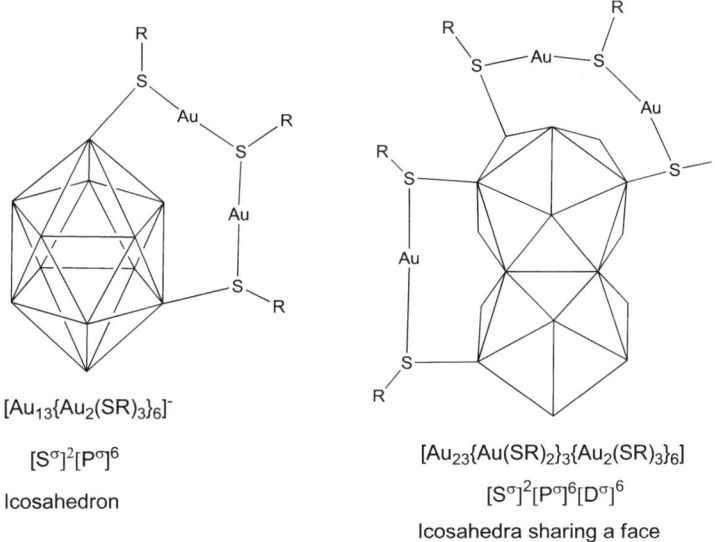

$[Au_{13}\{Au_2(SR)_3\}_6]^-$

$[S^{\sigma}]^2[P^{\sigma}]^6$

Icosahedron

$[Au_{23}\{Au(SR)_2\}_3\{Au_2(SR)_3\}_6]$

$[S^{\sigma}]^2[P^{\sigma}]^6[D^{\sigma}]^6$

Icosahedra sharing a face

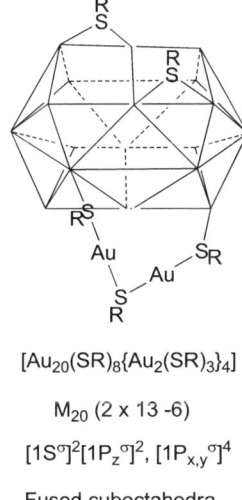

$[Au_{20}(SR)_8\{Au_2(SR)_3\}_4]$

M_{20} (2 x 13 -6)

$[1S^{\sigma}]^2[1P_z^{\sigma}]^2, [1P_{x,y}^{\sigma}]^4$

Fused cuboctahedra

Fig. 26 Illustration of skeletal geometries and staple binding modes in organothiolato-clusters. For reasons of clarity only representative examples of the staple ligands are shown

$$\left[Pt_{19}(Co)_{19}\right]^{8-} \equiv [Au_{19}L_{19}]^{11+} \equiv [Au_{19}(SR)_{11}] \equiv \left[Au_{19}\{Au(SR)_2\}_{11}\right].$$
$$\left[Au_{19}\{Au(SR)_2\}_{11}\right] \equiv \left[Au_{30}\{(SR)_{22}\}\right] \equiv \left[Au_{19}\{Au_2(SR)_3\}_{11}\right] \equiv \left[Au_{41}(SR)_{33}\right].$$

Consequently the cluster analogy predicts the possibility of gold thiolato-clusters based on a Au_{19} pair of fused icosahedra with $[S^{\sigma}]^2[P^{\sigma}]^6$ filled skeletal

Fig. 27 Molecular orbital
for triangular Au_6

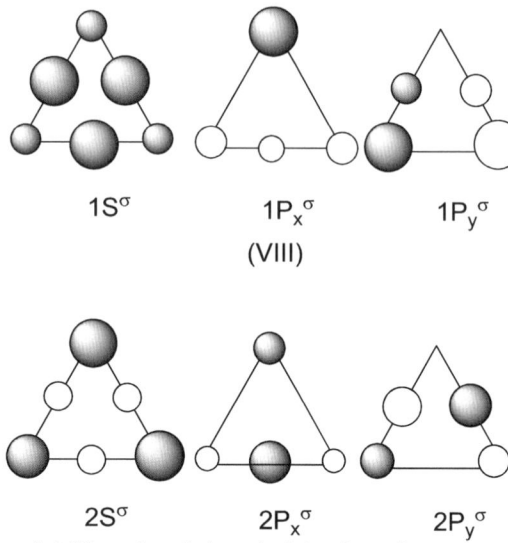

$$1S^\sigma \qquad 1P_x^\sigma \qquad 1P_y^\sigma$$

(VIII)

$$2S^\sigma \qquad 2P_x^\sigma \qquad 2P_y^\sigma$$

Additional radial node introduced

molecular orbitals with the stoichiometries $[Au_{30}\{(SR)_{22}]$ based on $[Au(SR)_2]$ and $[Au_{41}(SR)_{33}]$ based on $[Au_2(SR)_3]$ staple motifs. Of course intermediate combinations of these ligands may also be present in the isolated cluster.

3.7.2 Cuboctahedral-Based Structures

The previous section suggested that a centred cuboctahedron of gold atoms has similar bonding characteristics to the icosahedron because they are both essentially spherical. The centred Au_{13} cuboctahedron could also form the basis of a similar series of conjoined clusters – starting with $[Au_{25}(SR)_{18}]$ which is predicted to have a centred cuboctahedron inside a staple ligand shell based on $6[Au_2(SR)_3]$. $[Au_{38}(SR)_{24}]$ would be based on a pair of cuboctahedra sharing a triangular face. Finally a pair of fused cuboctahedra has 20 surface atoms and the skeletal bonding requirement of $[S^\sigma]^2[P^\sigma]^6$ suggests the following homoleptic relationships:

$$[Au_{20}L_{24}]^{12+} \equiv [Au_{20}(SR)_{12}] \equiv [Au_{20}\{Au(SR)_2\}_{12}].$$

$$[Au_{20}\{Au(SR)_2\}_{12}] \equiv [Au_{32}\{(SR)_{24}] \equiv [Au_{20}\{Au_2(SR)_3\}_{12}] \equiv [Au_{44}(SR)_{36}].$$

The single crystal structure determination of $[Au_{28}(SR)_{20}]$ has confirmed that it has a fused bi-cuboctahedral core with 8 bridging SR and 4 $[Au_2(SR)_3]$ staples and corresponds to the following intermediate substitution:

$$\left[\mathrm{Au}_{20}(\mathrm{SR})_{12}\right] + 8\mathrm{AuSR} \rightarrow \left[\mathrm{Au}_{20}(\mathrm{SR})_8\left\{\mathrm{Au}_2(\mathrm{SR})_3\right\}_4\right] \equiv \left[\mathrm{Au}_{28}(\mathrm{SR})_{20}\right].$$

The SR ligands bridge pairs of metal atoms and the $[\mathrm{Au}_2(\mathrm{SR})_3]$ ligands bridge non-adjacent metal atoms on folded diamonds of Au_4 made from two adjacent faces as shown in Fig. 26. The stoichiometric and structural relationships for these clusters are summarised in Table 1.

3.7.3 Alternative Geometries

The same principles may be applied to the recently reported $[\mathrm{Au}_{14}(\mathrm{PPh}_3)_8(\mathrm{NO}_3)_4]$, which has two trigonal-bipyramidal Au_5 fragments directly connected by a short Au–Au bond (2.651 Å), and this bond also has a collar of 4 AuNO_3 leading to an ellipsoidal shape and a polyhedron having triangular faces (see II).

$$\left[\mathrm{Au}_{14}(\mathrm{PPh}_3)_8(\mathrm{NO}_3)_4\right] \equiv \left[\mathrm{Au}_{14}\mathrm{L}_{12}\right]^{4+} \equiv \left[\mathrm{Au}_{14}(\mathrm{SR})_6\right]^{2-} \equiv \left[\mathrm{Au}_{14}\left\{\mathrm{Au}(\mathrm{SR})_2\right\}_6\right]^{2-}$$
$$\equiv \left[\mathrm{Au}_{14}\left\{\mathrm{Au}_2(\mathrm{SR})_3\right\}_6\right]^{2-}.$$

This leads to the prediction of three homoleptic possibilities, viz. $[\mathrm{Au}_{14}(\mathrm{SR})_6]^{2-}$, $[\mathrm{Au}_{20}(\mathrm{SR})_{12}]^{2-}$ or $[\mathrm{Au}_{26}(\mathrm{SR})_{18}]^{2-}$ or mixed species. All would share in common a pair of interstitial atoms inside a 12-atom polyhedron.

The same principles may be applied to other classes of clusters in Fig. 22. For example, toroidal or oblate clusters are characterised by $[\mathrm{S}^\sigma]^2[\mathrm{P}^\sigma]^4$ and are exemplified by $[\mathrm{Au}_9(\mathrm{PPh}_3)_8]^{3+}$ and $[\mathrm{Au}_{10}\mathrm{Cl}_3(\mathrm{PCy}_2\mathrm{Ph})_6]^+$ and lead to the following homoleptic replacement series:

$$\left[\mathrm{Au}_9(\mathrm{PPh}_3)_8\right]^{3+} \equiv \left[\mathrm{Au}_9(\mathrm{SR})_4\right]^- \equiv \left[\mathrm{Au}_9\left\{\mathrm{Au}(\mathrm{SR})_2\right\}_4\right]^-$$
$$\equiv \left[\mathrm{Au}_9\left\{\mathrm{Au}_2(\mathrm{SR})_3\right\}_4\right]^- \left[\mathrm{Au}_{13}(\mathrm{SR})_8\right]^- \left[\mathrm{Au}_{17}(\mathrm{SR})_{12}\right]^-.$$
$$\left[\mathrm{Au}_{10}\mathrm{Cl}_3(\mathrm{PCy}_2\mathrm{Ph})_6\right]^+ \equiv \left[\mathrm{Au}_{10}(\mathrm{SR})_5\right]^- \equiv \left[\mathrm{Au}_{10}\left\{\mathrm{Au}(\mathrm{SR})_2\right\}_5\right]^-$$
$$\equiv \left[\mathrm{Au}_{10}\left\{\mathrm{Au}_2(\mathrm{SR})_3\right\}_5\right]^- \left[\mathrm{Au}_5(\mathrm{SR})_{10}\right]^- \left[\mathrm{Au}_{20}(\mathrm{SR})_{15}\right]^-.$$

The analysis may also be used to provide an insight into structures which have been calculated from DFT calculations. For example, $[\mathrm{Au}_{18}(\mathrm{SR})_{14}]$ has two tetrahedral Au_4 moieties (which are characterised by $[\mathrm{S}^\sigma]^2$) which are well separated and with two pairs of $[\mathrm{Au}_2(\mathrm{SR})_4]$ and $[\mathrm{Au}_2(\mathrm{SR})_3]$ staple ligands [109]:

$$2\left[\mathrm{Au}_4(\mathrm{PPh}_3)_4\right]^{2+} \equiv 2\left[\mathrm{Au}_4(\mathrm{SR})_2\right]$$
$$\equiv 2\left[\mathrm{Au}_4\left\{\mathrm{Au}_2(\mathrm{SR})_3\right\}\left\{\mathrm{Au}_3(\mathrm{SR})_4\right\}\right]\left[\mathrm{Au}_{18}(\mathrm{SR})_{14}\right].$$

The analysis above has underlined in a very transparent manner the structural relationships between phosphine and organothiolato-cluster compounds, which has been proposed previously by Häkkinen.

Table 1 Summary of stoichiometries, staple types, charges skeletal electrons in gold and platinum clusters

Formulae	Structure description	m n	a a'	b c d	s x
Icosahedra based					
$[Au_{13}Cl_{12}(PPh_2Me)_{10}]^{3+}$	Centred icosahedron	13	1 12	0 0 0	8 5
$[Au_{13}(SR)_6]^-$	Centred icosahedron	13 6	1 12	0 0 6	8 5
$[Au_{19}(SR)_{12}]^-$	Centred icosahedron	19 12	1 12	6 0 0	8 5
$[Au_{25}(SR)_{18}]^-$	**Centred icosahedron**	**25 18**	**1 12**	**0 6 0**	**8 5**
$[Au_{25}Cl_2(SR)_5(PPh_3)_{10}]^{2+}$	Vertex sharing pair of icosahedron	25	3 22	0 0 0	16 9
$[Au_{23}(PR_3)_{19}]^{9+}$	Two icosahedra sharing a triangular face	23	5 18		14 9
$[Au_{23}(SR)_9]$		23	5 18	0 0 9	14 9
$[Au_{32}(SR)_{18}] \equiv [Au_{23}\{Au(SR)_2\}_9]$	9 type b staples	23	5 18	9 0 0	14 9
$[Au_{41}(SR)_{27}] \equiv [Au_{23}\{Au_2(SR)_3\}_9]$	9 type c staples	23	5 18	0 9 0	14 9
$[Au_{5+18}\{[Au(SR)_2]_3\,[Au_2(SR)_3]_6\}$	**3 b and 6 c type staples**	**23**	**5 18**	**3 6 0**	**14 9**
$[Pt_{19}(CO)_{19}]^{8-} \equiv [Au_{19}L_{19}]^{11+}$	Fused icosahedron	19	2 17		8 11
$[Au_{19}(SR)_{11}]$	Fused icosahedron	19 11	2 17	0 0 11	8 11
$[Au_{19}\{Au(SR)_2\}_{11}]$	Fused icosahedron	19 22	2 17	11 0 0	8 11
$[Au_{19}\{Au_2(SR)_3\}_{11}]$	Fused icosahedron	19 33	2 17	0 11 0	8 11
$[Au_{102}(SR)_{44}]\,[Au]_{79}[\{Au(SR)_2\}_{19}$ **$\{Au_2(SR)_3\}_2\}$**	**Marks dodecahedron and 19 b type and 2 c type staples**	**102 44**	**39 40**	**19 2 0**	**58 21**
Cuboctahedra based					
$[Au_{13}(SR)_6]^-$	Centred cuboctahedron	13 6	1 12	0 0 6	8 5
$[Au_{19}(SR)_{12}]^-$	Centred cuboctahedron	19 12	1 12	6 0 0	8 5
$[Au_{25}(SR)_{18}]^-$	Centred cuboctahedron	25 18	1 12	0 6 0	8 5
$[Au_{23}(PR_3)_{19}]^{9+}$	Two cuboctahedra sharing a triangular face	23	2 21		14 9
$[Au_{23}(SR)_9]$		23	2 21	0 0 9	14 9
$[Au_{32}(SR)_{18}] \equiv [Au_{23}\{Au(SR)_2\}_9]$	9 type b staples	23	2 21	9 0 0	14 9
$[Au_{41}(SR)_{27}] \equiv [Au_{23}\{Au_2(SR)_3\}_9]$	9 type c staples	23	2 21	0 9 0	14 9
$[Au_{20}L_{24}]^{12+}$	Fused cuboctahedra	20	2 18		8 12
$[Au_{20}(SR)_{12}]$		20 12	2 18	0 0 12	8 12
$[Au_{20}\{Au(SR)_2\}_{12}] \equiv [Au_{32}(SR)_{24}]$		32 24	2 18	12 0 0	8 12
$[Au_{20}\{Au_2(SR)_3\}_{12}] \equiv [Au_{44}(SR)_{36}]$		44 36	2 18	0 12 0	8 12
$[Au_{20}(SR)_8\{Au_2(SR)_3\}_4] \equiv [Au_{28}(SR)_{20}]$	**Fused cuboctahedra 8SR and 4 type c staples**	**28 20**	**2 18**	**0 4 8**	**8 12**
$[Au_{36}(SR)_{24}]\,[Au]_{4+24}[\{Au_2(SR)_3\}_4(\mu^2\text{-}SR)_{12}]$	**Cuboctahedral with central tetrahedron and 4 type c staples $12(\mu^2\text{-}SR)$**	**36 24**	**4 24**	**0 4 12**	**12 16**

The gold clusters described above are built up from the smaller polyhedra, which suggests kinetic control of the cluster growth process which favours processes which freezes in structures which are not necessarily spherical or close packed. The kinetics of the growth process presumably does not favour an annealing process which enables the cluster to rearrange to a traditional close-packed structure. It follows that triangles and tetrahedra may also get trapped in an interstitial site of a close-packed polyhedron of gold atoms. The equilibria connecting SR^- and the oligomeric gold staples may play a significant role in the kinetic trapping of specific clusters because they provide a soup of alternative ligands from which the appropriate ligand for stabilising a particular structure may be extracted.

The structure of $[Au_{28}(\mu\text{-}SR)_{12}\{Au_2(SR)_3\}_4]$ may be described in terms of a fusion of four centred cuboctahedra to give a 28-atom truncated polyhedron with a central tetrahedron of gold atoms made up from the interstitial atoms of the four fused cuboctahedra. According to the bonding model developed above, a tetrahedron is associated with a sec of 2 corresponding to the occupation of $[S^\sigma]$. In the Au_{28} core each of the tetrahedral faces has located above it a triangle of 6 gold atoms (i.e. a total of 24 gold atoms). The interaction between these triangles and the central tetrahedron may be analysed using the capping principle described above. The molecular orbitals generated from the 6 gold 6s orbitals are illustrated in Fig. 28 – the three most stable orbitals may be classified as S^σ and P^σ by virtue of the nodal properties, the higher lying orbitals have additional radial nodes which makes them antibonding and they do not interact strongly with the frontier orbital of the central tetrahedron. The triangles behave like a main group bridging atom with the S^σ combination stabilising S^σ of the interstitial tetrahedron, but P^σ_x and P^σ_y do not match and contribute 5 additional skeletal molecular orbitals ($e + t_2$ in the tetrahedral point group) resulting in a total of 6 skeletal molecular orbitals for the Au_{28} kernel. In a localised description this corresponds to the edges of the tetrahedron, or in the spherical harmonic description, this corresponds to $[S^\sigma]^2[P^\sigma]^6[D^\sigma]^4$. Filling these molecular orbitals leads to a central tetrahedral kernel of 28 atoms characterised by a sec of 12 electrons. The following phosphine substitutional series connects phosphine and organothiolato-stabilised clusters:

$$\left[Au_{28}(PR_3)_4(\mu - X)_4\right]^{16+} \equiv \left[Au_{28}(SR)_{16}\right] \equiv \left[Au_{28}(\mu - SR)_{12}\{Au_2(SR)_3\}_4\right].$$

The single crystal X-ray analysis has revealed the structure described above with 4 $[Au_2(SR)_3]$ staples bridging (111) facets of the polyhedron and 12 SR ligands bridging square (100) facets (see Fig. 26).

Finally the super cluster $[Au_{102}(SR)_{44}]$ has no direct analogue in phosphine cluster chemistry, but it may be constructed from a five Au_{20} tetrahedra coming together along their four-atomed edges to give a central core based on a Marks dodecahedron with fivefold symmetry. This 79-atom polyhedron would be associated with the formulation $[Au_{79}L_{42}]^{21+}$ corresponding to the filling of the following jellium sub-shells with 58 electrons, $1s < 1p < 1d < 2s < 1f < 2p < 1g$:

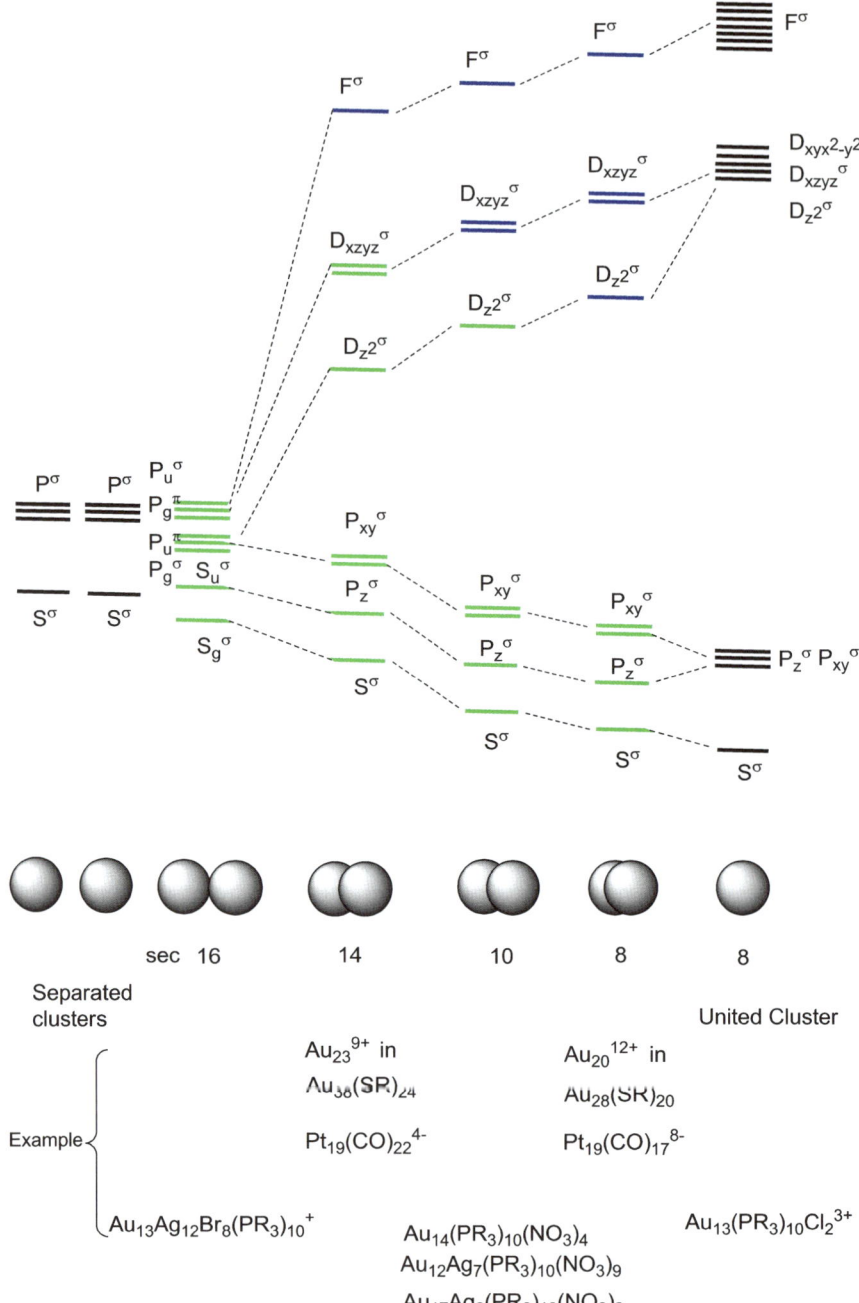

Fig. 28 Illustration of the *united cluster model*. The *right hand side* illustrates the molecular orbitals of two well-separated spherical clusters and the *left hand side* the united cluster which results from progressively squeezing the clusters together. For icosahedral clusters the interstitial metal atoms diminish from approximately 5.5 Å for a pair of icosahedra sharing a vertex to 2.64 Å in [$Au_{17}Ag_2(PR3)_{10}(NO_3)_9$], 2.55 Å in [$Pt_{19}(CO)_{17}$]$^{8-}$ and 2.58 Å in [$Au_{14}(PPh_3)_{10}(NO_3)_4$]

$$[Au_{79}L_{42}]^{21+} \equiv [Au_{79}(SR)_{21}] \equiv [Au_{79}\{Au(SR)_2\}_{21}] \equiv [Au_{79}\{Au_2(SR)_3\}_{21}]$$
$$\equiv [Au_{79}\{Au(SR)_2\}_{19}\{Au_2(SR)_3\}_3].$$

The latter is the structurally determined structure with the $Au_2(SR)_3$ ligands and the $Au(SR)_2$ capping opposite vertices of square faces.

3.8   Inherent Structure Rule and the Superatom Model [97, 99–112]

In the initial stages of characterisation when the formula of $[Au_m(SR)_n]$ has been established, but the single crystal analysis has not defined the structure, it is helpful to reduce the number of structural possibilities. The previous section has suggested a method for reducing the possibilities based on phosphine analogies. Tsukuda [112] and Pei and Zeng [106–111] have proposed an alternative algorithm for a subset of gold clusters with the general formula $[Au]_{a+a'}[Au(SR)_2]_b[Au_2(SR)_3]_c$. They have set up a set of equations which limit the number of structural possibilities, and the following section amplifies their analyses and relates it to the bonding analyses described in the previous section. The problem is mathematically underdetermined and consequently multiple integer solutions result. The relationships between the integer variables combine mathematically and chemically based information in order to limit the number of structural possibilities. Taking as a specific example $[Au_{25}(SR)_{18}]^-$, which may be represented as $[Au]_{a+a'} = [Au]_{1+12}$, where $a = 1$ (representing the interstitial atom) and $a' = 12$ (the surface atoms of the core which bond to SR), and $c = 6$; $[Au_{102}(SR)_{44}]$ is represented as $[Au]_{39+40}$ $[Au(SR)_2]_{19}[Au_2(SR)_3]_2$, with $b = 19$ and $c = 4$. More generally for $[Au]_{a+a'}$ $[Au(SR)_2]_b[Au_2(SR)_3]_c$ the following mathematical relationships follow directly from the general formula by connecting the variables which contain the number of SR ligands, n, the number of metal atoms, m, and the number of metal sites which form metal sulphur bonds to the bidentate gold organothiolato ligands:

$$n = 2b + 3c. \qquad (2)$$

$$m = a + a' + b + 2c. \qquad (3)$$

$$a' = 2b + 2c. \qquad (4)$$

Equation (3) follows from the recognition that each of the a' outer surface gold atoms are coordinated to one sulphur atom and both $[Au(SR)_2]$ and $[Au_2(SR)_3]$ have two sulphur donors. This limits a' to an even integer. This conclusion agrees with that proposed in the previous section, namely SR, $Au(SR)_2$ or $Au_2(SR)_3$ replace two phosphine ligands in $[A_{a+a'}(PR_3)_{a'}]^{x+}$ clusters. Some of the clusters with SR ligands have two SR ligands coordinated to a single gold atom, and therefore Eq. (4) is no longer valid. These clusters resemble those phosphine clusters which do not conform to the simple formula $[Au_n(PR_3)_n]^{x+}$ (see Fig. 7).

Accepting this caveat it follows that $b \leq n/2$ and $c \leq n/3$ and that $[Au(SR)_2]$ and $[Au_2(SR)_3]$ will only coexist when n is divisible by 3 and 4, i.e. $n = 12, 24, 36, 48 \ldots$, and if it is only divisible by 3, then only $[Au_2(SR)_3]$ will be present with $c = n/3$ and $b = 0$. $[Au(SR)_2]$ will be exclusively present when n is even and not divisible by 2, i.e. when $n = 10, 14, 16, 20, 22, 26, 28, 32\ldots$, and will equal to $n/2$. Since the total number of SR groups must equal those in the two ligand classes in the ratio 2:3, $b_{max} = a'_{max}/2$ and $c_{max} = a'_{max}/3$. The number of solutions to Eqs. (2)–(4) is $\{(b_{max}-b_{min})/3 + 1\}$ and $\{(c_{max}-c_{min})/2 + 1\}$ (see Fig. 23). Pei and Zeng [109] have confirmed that for $[Au_{38}(SR)_{24}]$ the equations lead to the following 5 possibilities:

$$[Au]_{2+24}\big\{[Au(SR)_2]_{12}\big\} \quad [Au]_{3+22}\big\{[Au(SR)_2]_9[Au_2(SR)_3]_2\big\}$$

$$[Au]_{4+20}\big\{[Au(SR)_2]_6[Au_2(SR)_3]_4\big\}\mathbf{[Au]_{5+18}\big\{[Au(SR)_2]_3[Au_2(SR)_3]_6\big\}}$$

$$[Au]_{6+16}\big\{[Au(SR)_2]_0\big\}[Au_2(SR)_3]_8$$

Pei and Zeng [109] explored these alternative possibilities using DFT calculations and concluded that $[Au]_{5+18}\{[Au(SR)_2]_3[Au_2(SR)_3]_6\}$ (in bold above) was the most stable by a large margin. The structure which is based on a bi-icosahedral structure with two icosahedra fused subunits forming the core and the staples distributed evenly over this core has been described above and illustrated in Fig. 26. Their recent review [109] has expanded on the consequences of "the inherent structural rule" and DFT calculations to other gold thiolato-clusters.

The above solutions form a series which may be understood by recognising that addition of an interstitial atom to the core ($a \rightarrow a + 1$) removes a pair of donor S sites from b and c and therefore a' must be reduced by 2. Furthermore, the loss of a gold atom from the staples must be compensated by replacing $[Au(SR)_2]$ ligands by $[Au_2(SR)_3]$ in the ratio of 2:3, i.e. $a \rightarrow (a + 1)$ leads $b \rightarrow (b - 2)$ and $c \rightarrow (c + 2)$, thereby satisfying the following equation:

$$Au + bAu(SR)_2 + cAu_2(SR)_3 \rightarrow (b - 3)Au(SR)_2 + (c + 2)Au_2(SR)_3.$$

The limited restraints on a, a' and $a + a'$ makes it impossible to reduce the number of possibilities further unless additional constraints are introduced.

The five solutions enumerated above have been based exclusively on the formulae of the compounds and the valency requirements of sulphur and gold(I). Additional relationships may be derived if the conclusions of the molecular orbital analysis and electron counting rules discussed in the previous section are incorporated. Specifically the sec count of cluster s is equal to $m - n$, because the total number of gold atoms contributes m 6s valence electrons, but n gold atoms bare a charge of $+1$ because of coordination to SR ligands to make up the staple ligands.

Therefore the number of electrons which remain for skeletal bonding in the gold core is $(m - n)$:

$$s = m - n. \tag{5}$$

It follows from this that s is a constant for the series of isomers which are consistent with the equations. Therefore m, n and s are constant for $[Au_m(SR)_n]$ and the equations may be used to define a, a', b, c and the charges on the central core and the surface ligands.

Häkkinen [97, 98, 102–112] has used this relationship to develop his superatom model which has been discussed above. This analysis if used in conjunction with the jellium model does not provide a specific geometric description of the cluster or the number and type of staple ligands since it can only conclude that the cluster has a spherical close-packed structure with a closed shell electronic shell with 8, 18, 20, ... valence electrons.

The charge on the core gold cluster (x) and the staple ligands are equal and of opposite sign. In a neutral $[Au_m(SR)_n]$ and it follows that:

$$m = s + x. \tag{6}$$

$$x = b + c. \tag{7}$$

$$x = a + a' - s. \tag{8}$$

The total number of valence electrons for sulphur ($3n$) and gold (m) leads to the following equalities:

$$3n + m = 2a' + 4b + 8c + s. \tag{9}$$

$$x = n - b - 2c. \tag{10}$$

The maximum value of $x = b_{max}$ and the minimum value $= c_{max} = 8$, whereas s is independent of b and c and a and a' and is equal to 14 for all possible structures in the series.

$[Au_{38}(SR)_{24}]$	a	a'	b	c	x	s
$[Au]_{2+24}\{[Au(SR)_2]_{12}\}$	2	24	12	0	12	14
$[Au]_{3+22}\{[Au(SR)_2]_9[Au_2(SR)_3]_2\}$	3	22	9	2	11	14
$[Au]_{4+20}\{[Au(SR)_2]_6[Au_2(SR)_3]_4\}$ 4	4	20	6	4	10	14
$[Au]_{5+18}\{[Au(SR)_2]_3 [Au_2(SR)_3]_6\}$ observed	**5**	**18**	**3**	**6**	**9**	**14**
$[Au]_{6+16}\{[Au(SR)_2]_0] [Au_2(SR)_3]_8$	6	16	0	8	8	14

s carries the information regarding the relationship between the number of the electrons in skeletal molecular orbitals and the polyhedral geometry of the core. This links this approach to the polyhedral electron counting paradigm for gold phosphine clusters summarised in Fig. 22. Although s is constant within the series, the other variables vary as shown above. Chemical considerations may be used to identify the more favourable structures. For example, the sequence in *b* suggests a threefold axis and the total number of gold atoms in the kernel ($a + a'$) suggests one or two interstitial gold atoms. The methodology developed above may be extended to other organothiolato-clusters and the results are summarised in Table 2.

3.9 Spherical and Close-Packed Arrangements

The discussion of high nuclearity clusters above has stressed the electronic factors which favour specific nuclearities and geometries and the purely geometric aspects, which become increasingly important as the clusters increase in size, have not been discussed in detail. The terms spherical and close packed have been mentioned, but not defined very precisely. Ignoring possible ligand affects, a cluster which is spherical in shape, has a close-packed arrangement of atoms and the requisite number of valence electrons to complete an electronic shell is very likely to be very stable. However, if it is not possible to satisfy these criteria simultaneously and none of the criteria predominates, then a balance has to be struck. Ab initio, crystal field perturbation analyses and the TSHMs all suggest that stable alkali metal and gold clusters are associated with closed electronic shells with 2, 8, 18, 20, 34, ... electrons, i.e. the closed shells which Häkkinen has emphasised in his superatom analyses [84, 97, 99–105]. There have also been a number of theoretical studies on non-spherical clusters and their occurrence when the shells are partially filled [98, 113–118]. The sphericity of a metal cluster may be defined by doing a moment of inertia analysis or using the definition originally proposed by Wadell in 1935 [117, 118]. He defined sphericity Ψ as the ratio of surface area of a sphere with the same volume as the given polyhedron to the surface area of the particle:

$$\Psi = \frac{\pi^{\frac{1}{3}}\left(6V_\mathrm{p}\right)^{\frac{2}{3}}}{A_\mathrm{p}},$$

where V_p is volume of the particle and A_p is the surface area of the particle. The sphericity of a sphere is 1, and by the isoperimetric inequality, any particle which is not a sphere will have sphericity less than 1. Table 3 summarises the sphericities of some common high symmetry polyhedra which are relevant to the current review and emphasises that for clusters with up to 13 atoms the icosahedron is the most and the tetrahedron the least spherical. As the sphericities of the polyhedra approach 1, their description using particle in a sphere free electron models become more appropriate. It is also noteworthy that by this criterion the icosahedron is significantly more spherical than the cuboctahedron although they have the same number of vertices. Condensed polyhedra become more spherical as they progress from vertex and edge sharing to face sharing, and this transformation underlines the united atom approach which has been discussed above. Since structures with fivefold symmetry are incompatible with translational symmetry, they cannot

Table 2 Summary of the conclusions of the inherent structure rule for organothiolato-gold cluster

		m n	a d'	b c d	s x
Icosahedra based					
$[Au_{25}(SR)_{18}]^{-}$ $[Au]_{1+12}\{[Au_2(SR)_3]_6\}$	Centred icosahedron with 6 c type staples	25 18	1 12	0 6 0	8 5(+1)
$[Au_{38}(SR)_{24}]$ $[Au]_{5+18}\{[Au(SR)_2]_3 [Au_2(SR)_3]_6\}$	Two icosahedra sharing a triangular face with 3 b and 6 c type staples	38 24	5 18	3 6 0	14 9
$[Au_{102}(SR)_{44}]$ $[Au]_{79}\{[Au(SR)_2]_{19} [Au_2(SR)_3]_2\}$	Marks dodecahedron and 19 b type and 2 c type staples	102 44	39 40	19 2 0	58 21
Cuboctahedra based					
$[Au_{28}(SR)_{20}]$ $[Au]_{2+18}\{[Au_2(SR)_3]_4(\mu^2\text{-}SR)_8\}$	Two cuboctahedra sharing six faces 4 type c staples	28 20	2 18	0 4 8	8 12
$[Au_{36}(SR)_{24}]$ $[Au]_{4+24}\{[Au_2(SR)_3]_4(\mu^2\text{-}SR)_{12}\}$	Cuboctahedral with central tetrahedron and 2 type c staples	36 24	4 24	0 4 12	12 16

Table 3 Sphericities of common polyhedral

	Volume	Surface area	Sphericity
Tetrahedron	$\sqrt{2}/12s^3$	$\sqrt{3}s^2$	0.671
Cube	s^3	$6s^2$	0.806
Octahedron	$\sqrt{2}/3s^3$	$2\sqrt{3}s^2$	0.846
Trigonal prism	$\sqrt{3}/4s^3$	$\sqrt{3}/2s^2 + 3s^2$	0.716
Dodecahedron	$(15 + 7\sqrt{5})s^3/4$	$3\sqrt{(25 + 10\sqrt{5})}s^2$	0.910
Cuboctahedron	$(5\sqrt{2})s^3/3$	$(6 + \sqrt{3})s^2/12$	0.877
Icosahedron	$5(3 + \sqrt{5})s^3/12$	$5\sqrt{3}s^2$	0.939
Icosahedra sharing vertex	$10(3 + \sqrt{5})s^3/12$	$10\sqrt{3}s^2$	0.745
Icosahedra sharing face	$10(3 + \sqrt{5})s^3/12$	$380/40\sqrt{3}s^2$	0.890
Fused icosahedra	$3.760s^3$	$12.99s^2$	0.900
Bicapped pentagonal prism	$2.324s^3$	$9.330s^2$	0.909
Fused bicapped pentagonal prisms	$4.044s^3$	$14.330s^2$	0.857
Sphere			1.00

The calculations are based on polyhedra constructed from regular polygons with sides $= s$

form the basis of infinite bulk structures, which are characterised by fcc, hexagonal or bcc packing arrangements. Therefore, larger clusters with fivefold symmetry must at some stage undergo a phase transition when they reach a certain size in order to achieve infinite close packing. The precise size at which this transformation occurs remains the subject of intense study and debate [98]. The sphericity index may also be used to quantify the distortions in oblate, prolate and toroidal gold clusters, which have been discussed above.

Close-packed structures with high symmetries, i.e. T_d, O_h, and I_h [84], may be generated, but once again they do not necessarily have high sphericities. For example, tetrahedral close-packed arrangements of metals with $1/6\{k(k + 1)(k + 2)\}$ ($k = 2, 3, 4$, etc. = number of atoms on equivalent edges) atoms may be constructed, but their sphericities (0.671) deviate greatly from the spherical ideal. The icosahedron and cuboctahedron which have higher sphericity indices provide a better basis for constructing clusters which are simultaneously close packed and approximately spherical. Related formulae for the cube and octahedron are summarised below. Close-packed structures based on fcc packing with high symmetry polyhedra depends on the number of atoms on equivalent edges (k)

k	2	3	4	5	
$N_{\text{Tetrahedron}}$	4	10	20	35	$1/6\{k(k + 1)(k + 2)\}$
$N_{\text{Octahedron}}$	6	19	44	85	$1/3\{k(2k^2 + 1)\}$
N_{Cube}	14	63	172	465	$k\{4k^2 - 6k + 3\}$

For 12 vertex polyhedra, e.g. icosahedra, decahedra and cuboctahedra, with K concentric shells, the total number of atoms N is given by

$$N(K) = 1/3\{10K^3 + 15K^2 + 11K + 3\}.$$

This corresponds to $N = 13, 55, 147, 309, 561, 923, 1{,}415, 2{,}057$, etc.

An analogous formula exists for bbc close-packed structures based on the 14 vertex rhombic dodecahedron:

$$N(K) = 4K^3 + 6K^2 + 4K + 1.$$

This corresponds to $N = 15, 65, 175, 369, 671, 1{,}105, 1{,}695, 2{,}465$, etc.

The number of atoms in the resultant close-packed arrangement may be increased by adding capping atoms or decreased by truncation. However, if the symmetry of the central polyhedron is to be maintained, the resulting number of capping or truncated atoms has to reflect their positions relative to the symmetry elements.

If attention is focussed on T_d, O_h and I_h structures, then the point group symmetries may be used to establish the number of atoms either in general or special positions lying on symmetry elements. For example, for M_{20} the following permutations are the only ones allowed. The local symmetries of the atoms are indicated in brackets:

$$T_d \quad 4(C_{3v}) + 4(C_{3v}) + 12(Cs)$$
$$O_h \quad 6(C_{4v}) + 6(C_{4v}) + 8(C_{3v})$$
$$12(C_{2v}) + 8(C_{3v})$$
$$I_h \quad 20(C_{3v})$$

The resultant arrangements are composite structures based on tetrahedra (C_{3v}), octahedra (C_{4v}), cubes $8(C_{3v})$ and cuboctahedra $12(C_{2v})$. The most stable close-packed structure is one with the maximum number of nearest neighbours and has successive layers of atoms in complementary positions. On the other hand the most spherical structure is that which has the maximum number of atoms on the surface layer, and this is achieved for evenly distributed arrangements. Of the M_{20} structures, the first choice is preferred since it satisfies the requirements of complementary angular coordinates on successive layers. In the second possibility, the location of atoms on two sets of C_{3v} special positions leads to a structure which has atoms placed above each other. The third and fourth structures cannot lead to close-packed structures because there are too many atoms on one layer (12 or 20) or a cube is generated as the central moiety. These geometric constraints have been discussed in more detail by Lin and Mingos, who also noted coincidences between the geometric and jellium electronic shell structures [84].

In recent years the determination of an increasing number of clusters by crystallographic techniques has provided additional experimental evidence regarding the preferred close-packed and spherical arrangements, and Dahl and his co-workers and Longoni and Iapalucci have provided detailed reviews and discussions of cluster structures based on icosahedral packing modes [119–124].

3.10 Summary

This review has provided an introduction to the bonding in gold clusters based on relatively simple molecular orbital models based on free electron models and interpreted with the assistance of perturbation theory concepts. More detailed computational studies on these systems present significant challenges. The precise calculation of their physical and chemical properties requires the incorporation of relativistic effects in an accurate fashion. Pykkö has emphasised that relativistic effects become more important as the atomic number increases, but have particularly dramatic consequences for gold. The consequences of relativistic effects in bare gold clusters have been hinted at above and have been more fully discussed in Woodham and Fielicke's chapter [85]. The large number of gold atoms in their clusters and the need to accurately model the ligands which at times also incorporate gold atoms impose further burdens on molecular orbital calculations even when density functional calculations are employed. Many of these difficulties have been overcome by the spectacular advances in computer technology and software. The "divide and protect" nature of the metallo-organothiolato-ligands provides additional challenges, because the theoretical models have to satisfactorily account for the "metallic" properties of the gold atoms in the central core, as well as the "insulator" properties of the surface ligands where the bonding is essentially covalent although it contains gold atoms. Recent reviews particularly by Häkkinen, Zeng and Fielicke have provided an accurate summary of the important recent developments in this area [85, 97, 99–111].

These detailed calculations have provided important insights into the preferences of staple ligands for binding to the metal core. Since SR, $Au(SR)_2$ and $Au_2(SR)_3$ are all bidentate, form two donor bonds and donate equal numbers of electrons, then their bonding preferences cannot be interpreted simply in terms of qualitative bonding models, or electron counting procedures, although differences in their bite angles may give some preliminary indications of their bonding preferences (see Sect. 3.6). These problems may be addressed very satisfactorily using DFT molecular orbital calculations. Furthermore, these calculations have provided important insights into the HOMO–LUMO gaps in these clusters. For example, for small spherical clusters such as $[Au_{11}Cl_3(PR_3)_7]$ the HOMO–LUMO gap is 1.5–2.1 eV, but this falls to 0.8 eV for Au_{39} phosphine clusters and dropping to 0.5 eV for Au_{102} cluster. Related calculations on ground and excited states may be used to interpret the spectral characteristics of these clusters. The calculations also provide a more accurate description of the charge distribution within the cluster and highlight the important differences between core and surface gold atoms. The former are close to having a neutral charge, whereas the latter are more positively charged. The determination of accurate structural parameters for gold clusters based on DFT calculations has also been used to interpret X-ray powder diffraction radial distribution functions.

Traditionally phosphine and thiolato-clusters have been regarded as distinct branches of gold cluster chemistry. Recent theoretical and experimental work has provided guiding principles for unifying these areas, and this review has summarised these developments and extended the analysis to a wider range of clusters. This will perhaps emphasise that the two areas have many common features and may both be understood within a common framework which is based on free electron models. This review has suggested that although closed spherical shells derived from the free electron model provide important milestones, the kinetic control of the cluster growth sequence and the stabilising effects of the ligands can lead to a wide range of cluster compounds which have partially filled shells. In the context of cluster chemistry these partially filled shells have important geometric consequences. Specifically, smaller clusters adopt spherical structures when they have closed S^σ, P^σ or D^σ shells, but those which have partially filled shells adopt prolate and oblate structures when the P^σ shell is partially filled. Larger clusters may be described by a "superatom" spherical model, but only if the cluster has a high sphericity index. For condensed clusters a polyspherical free electron model which models the effect of spherical clusters fusing together through vertex, edge and face sharing is more appropriate. Figure 28 suggests that the cluster condensation process for two clusters is akin to two drops progressively coming together. Initially the two drops retain their initial characteristics, but as they touch they progress through a series of stages where the diameter of the newly formed double drop decreases. Initially it has some concave surfaces at the interface and then forms an oblate spheroid before finally forming a new sphere. The quantum mechanical analogue of this is the formation of diatomic molecule from two separated atoms and the final molecule which has been described for more than 80 years as the *united atom model* [125]. Therefore, there are a series of clusters with 13–25 metal atoms which may be described using the *united cluster model* illustrated in Fig. 28. The molecular orbitals of the condensed clusters are represented by linear combinations of the molecular orbitals of the isolated clusters in the same way that linear combinations of atomic orbitals are taken together in diatomic molecules [125]. As the two initial clusters are squeezed together, the number of available bonding molecular orbitals decreases until a super cluster is formed which is characterised by a sec of 8 corresponding to occupation of S^σ and P^σ. The figure also provides specific examples of clusters, which span the spectrum of electron counts from 16 to 8. The examples are drawn from phosphine and organothiolato-cluster areas. The sphericity and in particular the distance between the two interstitial gold atoms play a very important role in determining the specific position of a cluster along the united atom coordinate. Small changes in the ligands or metals may influence the observed electron count. For example, the central fused cuboctahedral kernel in $[Au]_{2+18}\{[Au_2(SR)_3]_4(\mu^2\text{-}SR)_8\}$ has a sec electron count of 8 making it analogous to $[Pt_{19}(CO)_{17}]^{8-}$, which has a fused icosahedral structure. However, $[Au_{17}Ag_2(PR_3)_{10}(NO_3)_9]$ [126], which has a fused icosahedral structure, has a sec of 10, which suggests that $D_{z^2}^\sigma$ is occupied in the former cluster. Comparing $[Pt_{19}(CO)_{17}]^{8-}$ and $[Pt_{19}(CO)_{22}]^{4-}$ [127], the sec count increases from 8 to 14 as a result in the change of geometry of the central moiety from pentagonal antiprismatic to pentagonal prismatic, which results in an increased separation between the interstitial atoms and a more prolate geometry for the cluster

Table 4 Summary of the electron counting analysis for organothiolato-gold cluster

Formula	Description	a	d'	$10a + 12d'_{ML} + 14d'_{ML.2} + \mathbf{sec}$	$\mathbf{pec} = 11(a + d') + 3$ $(b + c + d)$
Icosahedra based					
$[Au_{25}(SR)_{18}]^-$ $[Au]_{1+12}\{[Au_2(SR)_3]_6\}$	Centred icosahedron with 6 c type staples	1	12	$10 \times 1 + 12 \times 12 + \mathbf{8} = 162$	162
$[Au_{38}(SR)_{24}]$ $[Au]_{5+18}\{[Au(SR)_2]_3$ $[Au_2(SR)_3]_6\}$	Two icosahedra sharing a triangular face with 3b and 6 c type staples	5	18	$10 \times 5 + 18 \times 12 + \mathbf{14} = 280$	$253 + 27 = 280$
$[Au_{102}(SR)_{44}]$ $[Au]_{79}\{[Au(SR)_2]_{19}$ $[Au_2(SR)_3]_2\}$	Marks dodecahedron and 19 b type and 2 c type staples	39	42	$42 \times 12 + 370 + \mathbf{58} = 932$	$79 \times 11 + 21 \times 3 = 932$
Cuboctahedra based					
$[Au_{28}(SR)_{20}]$ $[Au]_{2+18}\{[Au_2(SR)_3]_4(\boldsymbol{\mu}^2\text{-}\mathbf{SR})_8\}$	Two cuboctahedra sharing six faces 4 type c staples	2	18	$2 \times 10 + 12 \times 12 + 6 \times 14 + \mathbf{8} = 256$	$20 \times 11 + 12 \times 3 = 256$
$[Au_{36}(SR)_{24}]$ $[Au]_{4+24}\{[Au_2(SR)_3]_4$ $(\boldsymbol{\mu}^2\text{-}\mathbf{SR})_{12}\}$	Cuboctahedral with central tetrahedron and 2 type c staples	4	24	$4 \times 10 + 16 \times 12 + 8 \times 14 + \mathbf{12} = 356$	$28 \times 11 + 16 \times 3 = 356$

as a whole. The united atom approach developed above is not limited to the condensation of pairs of clusters, but may also be extended to collections of three, four or more atoms [125]. It may also be used to interpret the structures of heteronuclear clusters [128], where the site preferences provide an interesting additional theoretical problem which has been analysed in general terms.

The construction of high nuclearity cluster from vertex, edge and face sharing of smaller polyhedral units is not limited to gold clusters, but is a characteristic of metal carbonyl clusters of the later transition metals. The bonding models developed for metal carbonyl clusters were extended to their condensed clusters by Mingos [71–73], and the pecs of condensed clusters were related to those of the parent polyhedron via simple relationships. It was also noted that as the clusters became larger, radial bonding takes on a more important role than tangential bonding and the closed shell electronic structures may be conveniently described in terms of the simple formula $12n_S + \Delta_i$, where n_S is the number of surface atoms and Δ_i reflects the closed shell requirements of the interstitial group of atoms, i.e. 34 for M_2, 48 for M_3, 60 for a tetrahedron, 86 for octahedron and 162 for a centred icosahedron or cuboctahedron. The $12n_S$ arises from a filled d shell and a single terminal metal ligand bond.

For gold clusters a similar formula may be proposed and related to the sec which has been introduced above. In the absence of bridging carbonyls, the thiolato- or phosphine ligands do not form exclusively ML fragments and ML_2 fragments are also present (see Fig. 7). Their metal–ligand requirements lead to $12a'_{ML}$ and $14a'_{ML_2}$ components to the electron count. The a interstitial atoms and those surface gold atoms which are not able to bind to ligands because of the concavity of the surface contribute 10 electrons. The electron counts for organothiolato-clusters are summarised in Table 4. The pec of the clusters is given by $11(a + a') + 3$ $(b + c + d)$ since each of the staple ligands donates 3 electrons to the gold kernel. It is noteworthy that the electron counts in the final two columns of the table agree for all these examples. The sec is indicated in bold in column 3. It needs to be emphasised that if formula 4 ($a' = 2b + 2c$) which relates the number of Au–S bonds to the number of gold atoms capable of forming dative bonds is not valid, then the total polyhedral electron count will not prove to be a good indicator of the total pec. The calculation of pec needs to be corrected for the number of 14 electron centres. These compounds are shown in bold in column 1 of Table 4. The table illustrates the good agreement between observed and calculated pecs when this correction is made.

Acknowledgments I would like to thank Professors Roy Johnston, Zhenyang Lin, Rongchao Jin, Larry Dahl and Dr Evgueni Mednikov for reading and providing helpful comments on the drafts of this chapter. I also like to thank Man Sing Cheung and Zhenyang Lin for their assistance with the molecular orbital calculations.

References

1. Naldini L, Cariati F, Simonetta G, Malatesta L (1965) Gold tertiary phosphine derivatives with intermetallic bonds. J Chem Soc Chem Commun 212–213
2. Malatesta L, Naldini L, Simonetta G, Cariati F (1966) Triphenylphosphine gold(0)-gold(I) compounds. Coord Chem Rev 1:255–262
3. Naldini L, Cariati F, Simonetta G, Malatesta L (1967) Ethyldiphenylphosphine-gold derivatives with intermetallic bonds. Inorg Chim Acta 1:24–26
4. Naldini L, Cariati F, Simonetta G, Malatesta L (1967) Clusters of gold compounds with 1,2-bis(diphenylphosphino)ethane. Inorg Chim Acta 1:315–318
5. Malatesta L (1975) Cluster compounds of gold. Gold Bull 8:48–52
6. Mingos DMP (1993) Recent developments in the cluster chemistry of gold. Roy Soc Chem Spec Publ [Chemistry of the Copper and Zinc Triads] 131:189–197
7. Mingos DMP (1992) High-nuclearity clusters of the transition metals and a re-evaluation of the cluster surface analogy. J Cluster Sci 3:397–409
8. Mingos DMP, Watson MJ (1992) Heteronuclear gold cluster compounds. Adv Inorg Chem 39:327–399
9. Mingos DMP, Watson MJ (1991) TMC literature highlights – 27. Recent developments in the homo- and hetero-metallic cluster compounds of gold. Transition Met Chem 16:285–287
10. Mingos DMP (1984) Structure and bonding in cluster compounds of gold. Polyhedron 3:1289–1297
11. Hall KP, Mingos DMP (1984) Homo- and heteronuclear cluster compounds of gold. Progr Inorg Chem 32:237–325
12. Mingos DMP (1984) Gold cluster compounds. Are they metals in miniature? Gold Bull 17:5–12
13. Mingos DMP (1982) Some theoretical and structural aspects of gold cluster chemistry. Phil Trans R Soc (London) 308:75–83
14. Mingos DMP (1980) Theoretical and structural studies on organometallic cluster molecules. Pure Appl Chem 52:705–712
15. Mingos DMP (1996) Gold: a flexible friend in cluster chemistry. J Chem Soc Dalton Trans 561–566
16. Steggerda JJ, Bour JJ, van der Velden JWA (1982) Preparation and properties of gold cluster compounds. Recl Trav Chim Pays Bas 101:164–170
17. Vollenbrook FA, Bouten DCP, Trooster JM, van der Berg JP, Bour JJ (1978) Inorg Chem 17:1345–1347
18. van der Velden JWA, Bour JJ, Steggerda JJ, Beurskens PT, Roseboom M, Noordik JH (1982) Inorg Chem 21:4321–4324
19. Kanters RF, Steggerda JF (1992) Spherical and toroidal clusters of gold. J Cluster Sci 19:229–239
20. Teo BK (1988) Cluster of clusters: a new series of high nuclearity gold–silver clusters. Polyhedron 1:2311–2320
21. Teo BK, Shi Z, Zhang H (1992) Pure gold cluster of 1:9:9:1:9:9:1 layered structure: a novel 39-metal-atom cluster $[(Ph_3P)_{14}Au_{39}Cl_6]Cl_2$ with an interstitial gold atom in a hexagonal antiprismatic cage. J Am Chem Soc 114:2473–2745
22. Schmid G, Pfeil R, Boese R, BandermannF MS, Galis GHM, van der Velden JWA (1981) $Au_{55}(PPh_3)_{12}Cl_6$ – a gold cluster of unusual size. Chem Ber 114:3634–3642
23. Wallenberg LR, Bovin JO, Schmid G (1985) Au55(PPh3)12Cl6 – TEM study of a gold cluster of unusual size. Surf Sci 156:256–264
24. Schmid G (1985) Developments in transition metal cluster chemistry, the way to large clusters. Struct Bond 62:52–85
25. Schmid G (1988) Large transition metal clusters-VI. Ligand exchange reactions on the gold triphenylphosphine chloro cluster, $Au_{55}(PPh_3)_{12}Cl_6$ – the formation of a water soluble gold (Au_{55}) cluster. Polyhedron 7:605–608

26. Schmid G, Lehnert A (1989) Complexation of gold coloids. Angew Chem Int Ed 28:773–774
27. Mingos DMP (1982) Steric effects in metal cluster chemistry. Inorg Chem 21:466–488
28. Vollenbroek FA (1979) Ph. D. Thesis, University of Nijmegen
29. Tolman CA (1977) Steric and electronic effects, steric effects of phosphorus ligands in organometallic chemistry and homogeneous catalysis. Chem Rev 77:313–348
30. Kobayashi N, Kamei Y, Shichibu Y, Konishi K (2013) Electronic properties of [Core+exo]-type gold clusters: factors affecting the unique optical transitions. J Am Chem Soc 135:16078–16081
31. Yukatsu SY, Kai S, Konishi K (2012) Unique [core+2] structure and optical property of dodeca-ligated undecagold cluster: critical contribution of the exo- gold atoms to the electronic structure. Nanoscale 4:4125–4129
32. Zeng C, Jin R (2014) Gold nanoclusters: size-controlled synthesis and crystal structures. Struct Bond. DOI: 10.1007/430_2014_146
33. Lu Y, Chen W (2014) Progress in the synthesis and characterization of gold nanoclusters. Struct Bond. DOI: 10.1007/430_2013_126
34. McPartlin M, Mason R, Malatesta L (1969) Custer compounds of gold(0)-gold(I). J Chem Soc Chem Commun 334–335
35. Dyson PJ, Mingos DMP (1999) Homonuclear clusters and colloids of gold: synthesis, reactivity, structural and theoretical considerations. In: Schmidbaur H (ed) Gold progress in chemistry, biochemistry and technology. Wiley, Chichester, pp 512–513
36. Pyykkö P, Zhao Y-F (1991) The ab initio calculations for the dimer chloro(phosphine)gold [ClAuPH$_3$]$_2$ with relativistic pseudopotential. Is the aurophile attraction a correlation effect? Angew Chem Int Ed 103:622–623
37. Li J, Pyykkö P (1993) The ab initio calculations for the dimer chloro(phosphine) gold [ClAuPH$_3$]$_2$ with relativistic pseudopotential. Is the aurophile attraction a correlation effect? Inorg Chem 32:2630–2634
38. Pyykko P, Runeberg N (1993) J Chem Soc Chem Commun 1812–1814
39. Zachwieja U (1999) The structural chemistry of alkali metal aurides M$_n$Au$_m$ with M=Na, K, Rb and Cs. In: Schmidbaur H (ed) Gold progress in chemistry, biochemistry and technology. Wiley, Chichester, pp 496–508
40. Harwell DE, Mortimer MD, Knobler CB, Anet FAL, Hawthorne MF (1996) Auracarboranes with and without Au–Au interactions: an unusually strong aurophilic interaction. J Am Chem Soc 118:2679–2685
41. King RB (1972) Transition metal cluster compounds. Prog Inorg Chem 15:287–473
42. Schmidbaur H, Grohmann A, Olmos ME (1999) Organogold chemistry. In: Schmidbaur H (ed) Gold progress in chemistry, biochemistry and technology. Wiley, Chichester, pp 648–731
43. Steigelmann O, Bissinger P, Schmidbaur H (1990) Angew Chem Int Ed 27:1544
44. Steigelmann O, Bissinger P, Schmidbaur H (1991) Angew Chem Int Ed 30:1488
45. Briant CE, Hall KP, Mingos DMP (1984) Structural characterisation of two crystalline modifications of the [Au$_9${P(p-C$_6$H$_4$OMe)$_3$}$_8$](NO$_3$)$_3$ – the first example of skeletal isomerism in cluster chemistry. J Chem Soc Chem Commun 290–292
46. Coffer JL, Shapley JR, Drickamer HG (1990) Pressure-induced skeletal isomerization of octakis-(triphenylphosphine)nonagold(3+) hexafluorophosphate in the solid state. Inorg Chem 29:3000–3001
47. Clayden NJ, Dobson CM, Hall KP, Mingos DMP, Smith DJ (1985) Studies of gold cluster compounds using high-resolution phosphorus-31 solid-state nuclear magnetic resonance spectroscopy. Inorg Chem 25:1811–1814
48. Vollenbroek FA, Bouten PCP, Trooster JM, van der Berg JP, Bour JJ (1978) Mössbauer investigation and novel synthesis of gold cluster compounds. Inorg Chem 17:1345
49. Pignolet LH, Krogstad DA (1999) Molecular compounds of gold with main group and transition metals. In: Schmidbaur H (ed) Gold progress in chemistry, biochemistry and technology. Wiley, Chichester, pp 430–465

50. Ito LN, Sweet JD, Mueting AM, Pignolet LH, Schoondergang MFJ, Steggerda JJ (1989) Heterobimetallic gold–platinum phosphine complexes. X-ray crystal and molecular structures of [(PPh$_3$)Pt(AuPPh$_3$)$_6$](NO$_3$)$_2$ and [(PPh$_3$)(CO)Pt(AuPPh$_3$)$_6$](PF$_6$)$_2$. Inorg Chem 28:3696–37001
51. McNair RJ, Pignolet LH (1985) Heterobimetallic complexes of rhodium with platinum, silver, and gold containing bridging 2-[bis(diphenylphosphino)methyl]pyridine (PNP) ligands. Inorg Chem 24:4717–4723
52. Bour JJ, Kanters RPF, Schlebos PPJ, Steggerda JJ (1988) Mixed platinum–gold clusters. Synthesis, structure, and properties of PtAu$_8$(PPh$_3$)$_8$(NO$_3$)$_2$. Recl Trav Chim Pays Bas 107:211–215
53. Kanters RPF, Schlebos PPJ, Bour JJ, Bosman WP, Behm HJ, Steggerda JJ (1988) Reaction of carbon monoxide and an isonitrile with octakis(triphenylphosphine)octagoldplatinum(3+). Crystal structure of Pt(CO)(AuPPh$_3$)$_8$(NO$_3$)$_2$.2C$_4$H$_{10}$O.2CH$_2$Cl$_2$. Inorg Chem 27:4034–4037
54. Kanters RPF, Bour JJ, Schlebos PPJ, Bosman WP, Behm H, Steggerda JJ, Ito LN, Pignolet LH (1989) Structure and properties of the hydride-containing cluster ion Pt(H)(PPh$_3$)(AuPPh$_3$)$_7$$^{2+}$. Inorg Chem 28:2591–2594
55. Bour JJ, Schlebos PPJ, Kanters RPF, Bosman WP, Smits JMM, Beurskens PT, Steggerda JJ (1990) Synthesis and characterization of [Pt(CN)(AuCN)(AuPPh$_3$)$_8$](NO$_3$) and Pt(CO)(AuPPh$_3$)$_6$(AuCN)$_2$. Crystal structure of [Pt(CN)(AuCN)(AuPPh$_3$)$_8$](NO$_3$). Inorg Chim Acta 171:177–181
56. Kanters RPF, Schlebos PPJ, Bou J Jr, Bosman WP, Smits JMM, Beurskens PT, Steggerda JJ (1990) Electrophilic addition reactions of Ag$^+$ with Pt(AuPPh$_3$)$_8$$^{2+}$ and Pt(CO)(AuPPh$_3$)$_8$$^{2+}$. CrystalStructures of [Pt(AgNO$_3$)(AuPPh$_3$)$_8$](NO$_3$)$_2$ and [Pt(CO)(AgNO$_3$)(AuPPh$_3$)$_8$](NO$_3$)$_2$. Inorg Chem 29:324–328
57. Bour JJ, WVD B, Schlebos PPJ, Kanters RPF, Schoondergang MFJ, Bosman WP, Smits JMM, Beurskens PT, Steggerda JJ, van der Sluis P (1990) PtAu$_8$Hg$_2$ and PtAu$_7$Hg$_2$ cluster compounds. X-ray structure of [Pt(AuPPh$_3$)$_8$)Hg$_2$](NO$_3$)4.3CH$_2$Cl$_2$. Inorg Chem 29:2971–2975
58. Teo BK, Zhang H (1995) Polyicosahedracity: icosahedraon to icosahedrons of icosahedral growth pathway to bimetallic and trimetallic Au, Ag, M (M=Ni, Pd, Pt) supraclusters–synthetic strategies and stereochemical principles. Coord Chem Rev 143:611–636
59. Gutrath BS, Oppel IM, Presly O, Beljakov I, Meded V, Wenzel W, Simon U (2013) [Au$_{14}$(PPh$_3$)$_{10}$(NO$_3$)$_4$]: an example of a new class of Au(NO)$_3$ ligated superatom cluster. Angew Chem Int Ed 52:3529–3532
60. Mingos DMP, Johnston RL (1987) Theoretical models of cluster bonding. Struct Bond 68:31–82
61. Mingos DMP, Johnston RL (1987) Group theoretical paradigms for describing the skeletal molecular orbitals of cluster compounds. J Chem Soc Dalton Trans 647–656:1445–1456
62. Longuet-Higgins HC, de V Roberts M (1955) The electronic structure of an icosahedron of boron atoms. Proc R Soc London Ser A 230:110–119
63. Cotton FA, Haas TE (1964) A molecular orbital treatment of the bonding in certain metal atom clusters. Inorg Chem 3:10–17
64. Williams RE (1971) Carboranes and boranes; polyhedra and polyhedral fragments. Inorg Chem 10:210–214
65. Wade K (1971) Structural significance of the number of skeletal bonding electron-pairs in carboranes, the higher boranes, and borane anions, and various transition metal carbonyl cluster compounds. J Chem Soc Chem Commun 792–793
66. Mingos DMP (1972) A general theory for cluster and ring compounds of the main group and transition elements. Nature Phys Sci 236:99–102
67. Rudolph RW (1976) Boranes and heteroboranes: a paradigm for the electron requirements of clusters? Acc Chem Res 9:446–452
68. Elian M, Chen MML, Mingos DMP, Hoffmann R (1976) A comparative study of conical fragments. Inorg Chem 15:1148–1155

69. Stone AJ (1981) New approach to bonding in transition-metal clusters and related compounds. Inorg Chem 20:563–571
70. Stone AJ (1981) The bonding in boron and transition-metal cluster compounds. Polyhedron 3:1299–1306
71. Mingos DMP (1984) Polyhedral skeletal electron pair approach. Acc Chem Res 17:311–319
72. Mingos DMP (1983) Polyhedral skeletal electron pair theory – a generalised principle for condensed clusters. J Chem Soc Chem Commun 706–708
73. Mingos DMP (1985) Theoretical analysis and electron counting rules for high nuclearity clusters. J Chem Soc Chem Commun 1352–1354
74. Mingos DMP (1974) Molecular orbital calculations for an octahedral cobalt cluster complex $Co_6(CO)_{14}^{4-}$. J Chem Soc Dalton Trans 124–138
75. Mingos DMP, Forsyth MI (1977) Molecular orbital calculations on transition metal clusters containing six metal atoms. J Chem Soc Dalton Trans 610–616
76. Thomas KM, Mason R, Mingos DMP (1973) Stereochemistry of octadecacarbonyl-hexaosmium(0), a novel hexanuclear complex based on a bicapped tetrahedron of metal atoms. J Am Chem Soc 95:3802–3804
77. Mingos DMP (1974) Molecular orbital calculations on cluster compounds of gold. J Chem Soc Dalton Trans 133–116
78. Evans DG, Mingos DMP (1982) Molecular orbital analysis of the bonding in low nuclearity gold and platinum tertiary phosphine complexes and the development of isolobal analogies for the $M(PR_3)$ fragment. J Organomet Chem 232:171–179
79. Hall KP, Gilmour DI, Mingos DMP (1984) Molecular orbital analysis of the bonding in high nuclearity gold cluster compounds. J Organomet Chem 268:275–293
80. Gilmour GI, Mingos DMP (1986) Molecular orbital calculations on platinum–gold heterometallic clusters. J Organomet Chem 302:127–146
81. Evans DG, Mingos DMP (1985) Molecular orbital analysis of the bonding in penta- and heptanuclear gold tertiary phosphine clusters. J Organomet Chem 295:389–400
82. Wales DJ, Mingos DMP (1989) Splitting of cluster orbitals. Inorg Chem 28:2748–2754
83. Mingos DMP, Kanters RPF (1991) Molecular-orbital analysis of mono- and dicarbidogold cluster compounds. J Organomet Chem 384:405–415
84. Lin Z, Kanters RPF, Mingos DMP (1991) Closed-shell electronic requirements for condensed clusters of the group 11 elements. Inorg Chem 30:91–95
85. Woodham AP, Fielicke A (2014) Gold clusters in the gas phase. Struct Bond. DOI: 10.1007/430_2013_136
86. Yukatsu SY, Katsuaki K, Konishi K (2013) Facile synthesis and optical properties of magic-number Au_{13} clusters. Inorg Chem 52:6570–6575
87. Shichibu Y, Kamei Y, Konishi K (2012) Generation of small gold clusters with unique geometries through cluster-to-cluster transformations: octanuclear clusters with edge sharing gold tetrahedra motifs. J Chem Soc Chem Commun 7559–7561:1602
88. Kamei Y, Shichibu Y, Konishi K (2011) Generation of small gold clusters with unique geometries through cluster-to-cluster transformations: octanuclear clusters with edge sharing gold tetrahedra motifs. Angew Chem Int Ed 50:7442–7445
89. Jaw H-RC, Mason WR (1991) Electronic absorption and MCD spectra for the $Au_8 (PPh_3)_8^{3+}$ ion. Inorg Chem 30:3552–3555
90. Femoni C, Iapalucci MC, Longoni G, Zacchini S, Zarra S (2011) Icosahedral Pt-centered Pt_{13} and Pt_{19} carbonyl clusters decorated by $[Cd_5(\mu-Br)_5Br_5\text{-}x(solvent)_x]^{x+}$ rings reminiscent of the decoration of Au-Fe-CO and Au-thiolate nanoclusters: a unifying approach to their electron counts. J Am Chem Soc 133:2406–2409
91. Murray RW (2008) Nanoelectrochemistry: metal nanoparticles, nanoelectrodes, and nanopores. Chem Rev 108:2688–2720
92. Jadzinsky PD, Calero G, Ackerson CJ, Bushnell DA, Kornberg RD (2007) Structure of a thiol monolayer protected gold nanoparticle at 1.1A resolution. Science 318:430–433
93. Price R, Whetten RL (2007) Nano-golden order. Science 318:407–408

94. Zhu M, Aikens CM, Hollander FJ, Schatz GC, Jin R (2008) Correlating the crystal structure of a thiol-protected Au_{25} cluster and optical properties. J Am Chem Soc 130:3754–3755

95. Heaven MW, Dass A, White PS, Holt KM, Murray RW (2008) Crystal structure of the gold nanoparticle $[N(C_8H_{17})_4][Au_{25}(SCH_2CH_2Ph)_{18}]$. J Am Chem Soc 130:3754–3755

96. Mednikov EG, Dahl LF (2008) Crystallographically proven nanometer-sized gold thiolato-cluster $Au_{102}(SR)_{44}$: its unexpected molecular anatomy and resulting stereochemical and bonding consequences. Small 4:534–537

97. Häkkinen H (2008) Atomic and electronic structure of gold clusters: understanding flakes, cages, superatoms from simple concepts. Chem Soc Rev 37:1847–1859

98. Johnston RL (2002) Atomic and molecular clusters. Taylor and Francis, New York

99. Häkkinen H, Walter M, Grönbeck H (2006) Divide and protect: capping gold nanoclusters with molecular gold-thiolate rings. J Phys Chem B 110:9927–9931

100. Walter M, Akola J, Lopez-Acevedo O, Jadinsky PD, Calero G, Ackerson CJ, Whetten RL, Grönbeck H, Häkkinen H (2008) A unified view of ligand protected gold clusters as a super atom complexes. Proc Natl Acad Sci U S A 105:9157–9162

101. Häkkinen H, Barnett RN, Landman U (1999) Electronic structure of passivated $[Au_{38}(SCH_3)_{24}]$ nanocrystal. Phys Rev Lett 82:3264, Science 318:430–433

102. Salorinne K, Lahtinen T, Koivisto J, Kalenius E, Nissinen M, Pettersson M, Häkkinen H (2013) Non-destructive size determination of thiol-stabilised gold nanoclusters in solution by diffusion ordered NMR spectroscopy. Anal Chem 85:3489–3492

103. Häkkinen H (2012) Ligand protected gold nanoclusters as superatoms – insights from theory and computations. In: Johnston RL, Wilcoxon JP (eds) Metal nanoparticles and nanoalloys, Frontiers of nanoscience [Series Editor Palmer RE], vol 3. Elsevier, Amsterdam, pp 129–154

104. Häkkinen H (2012) Theoretical studies of gold nanoclusters in various chemical environ-ments: when the size matters. In: Louis C, Pluchery O (eds) Gold particles in physics, biology and chemistry. Chap 9. Imperial College Press, Amsterdam

105. Häkkinen H (2012) The gold–sulfur interface at the nanoscale. Nat Chem 4:443–455

106. Peng Y, Zeng XC (2012) Investigating the structural evolution of thiolate protected gold clusters from first principles. Nanoscience 4:4054–4072

107. Huang W, Pal R, Wang L-M, Zeng XC, Wang L-S (2010) Isomer identification and resolution in small gold clusters. J Chem Phys 132:054305

108. Wang L-M, Pal R, Huang W, Zeng XC, Wang L-S (2010) Observation of earlier two-to-three dimensional structural transition in gold cluster anions by isoelectronic substitution: MAun- (n=8–11; M=Ag, Cu). J Chem Phys 132:114306

109. Pei Y, Pal R, Liu C, Gao Y, Zhang Z, Zeng XC (2012) Interlocked catenane-like structure predicted in $Au_{24}(SR)_{20}$: implication to structural evolution of thiolated gold clusters from homoleptic gold(I) thiolates to core-stacked nanoparticles. J Am Chem Soc 134:3015–3024

110. Zhou R, Wei X, He K, Shield JE, Sellmyer DJ, Zeng XC (2011) Theoretical and experimental characterization of structures of MnAu nanoclusters in the size range of 1–3 nm. ACS Nano 5:9966–9976

111. Gao Y, Shao N, Pei Y, Chen ZF, Zeng XC (2011) Catalytic activities of subnanometer gold clusters (Au_{16}, Au_{18}, Au_{20}, and Au_{27} - Au_{35}) for CO oxidation. ACS Nano 5:7818–7829, Science 360 2 Feb 2012

112. Chaki NK, Negishi Y, Tsunoyama H, Shichibu Y, Tsukada T (2008) J Am Chem Soc 130:8608–8610

113. Ekardt W (1999) Metal clusters. Wiley, New York

114. Knight WD, Clemenger K, De Heer WA, Saunders WA, Winston A, Chou MY, Cohen ML (1984) Electronic shell structure and abundances of sodium clusters. Phys Rev Lett 52:2141–2143

115. Knight WD, Clemenger K, De Heer WA, Saunders WA, Winston A (1985) Electronic shell structure in potassium clusters. Solid State Commun 53:445–446

116. Clemenger K (1985) Ellipsoidal structure in free-electron metal clusters. Phys Rev B Condens Matter Mater Phys 32:1359–1362

117. Mingos DMP, McGrady JE, Rohl AL (1992) Moments of inertia in cluster and C-ordination compounds. Struct Bond 79:1–54
118. Wadell H (1935) Volume, shape and roundness of quartz particles. J Geol 43:250–280
119. Tran NT, Kawano M, Dahl L (2001) High-nuclearity palladium carbonyl trimethylphosphine clusters containing unprecedented face-condensed icosahedral-based transition-metal core geometries: proposed growth patterns from a centered Pd_{13} icosahedron. J Chem Soc Dalton Trans 2731–2748
120. Tran NT, Powell DR, Dahl L (2000) Nanosized $Pd_{145}(CO)_x(PEt_3)_{30}$ containing a capped three-shell 145-atom metal-core geometry of pseudo icosahedral symmetry. Angew Chem Int Ed 39:4121–4125
121. Tran NT, Kawano M, Dahl L (2003) Nanosized $[Pd_{69}(CO)_{36}(PEt_3)_{18}]$: metal-core geometry containing a linear assembly of three face-sharing centered Pd_{33} icosahedra inside of a hexagonal-shaped Pd_{30} tube. Angew Chem Int Ed Engl 42:3533–3537
122. Bernardi A, Femoni C, Iapalucci MC, Longoni G, Ranuzzi F, Zacchini S, Zanello P, Fedi S (2008) *o*Synthesis, molecular structure and properties of the $[H_6[Ni_{30}C_4(CO)_{34}(CdCl)_2]^{n-}$ (n=3–6) bimetallic carbide carbonyl cluster: a model for the growth of noncompact interstitial metal carbides. Chemistry 14:1924–1934
123. Longoni G, Femoni C, Iapalucci MC, Zanello P (1999) In: Braunstein P, Oro LA, Raithby PR (eds) Metal clusters in chemistry. Wiley-VCH, Weinheim, pp 1137–1158
124. Longoni G, Iapalucci MC (1994) In: Schmid G (ed) Clusters and colloids. Wiley-VCH, Weinheim, pp 91–177
125. Herzberg G (1967) Molecular structure and molecular spectra, vol III. Van Nostrand, New York, pp 276–419
126. Nunokawa K, Ito M, Sunahara T, Onaka S, Ozeki T, Hirokazu Chiba H, Yasuhiro Funahashi Y, Hideki Masuda H, Tetsu Yonezawa T, Nishihara H, Nakamoto M, Yamamoto M (2005) A new 19-metal-atom cluster $[(Me_2PhP)_{10}Au_{12}Ag_7(NO_3)_9]$ with a nearly staggered–staggered M_5 ring configuration. J Chem Soc Dalton Trans 2726–2730
127. Washecheck DM, Wucherer DJ, Dahl LF, Ceriotti M, Sansoni M, Chini P (1979) Synthesis, structure and stereochemical implication of the $[Pt_{19}(CO)_{22}]^{4-}$ a bicapped triple decker all metal sandwich of idealized fivefold (D5h) geometry. J Am Chem Soc 101:6110–6112
128. Mingos DMP, Lin Z (1989) Site preference effects in heteronuclear clusters. Comments Inorg Chem 9:99–122

Struct Bond (2014) 162: 67–90
DOI: 10.1007/430_2013_139
© Springer International Publishing Switzerland 2014
Published online: 18 February 2014

Interfacial Structures and Bonding in Metal-Coated Gold Nanorods

Ruth L. Chantry, Ivailo Atanasov, Sarah L. Horswell, Z.Y. Li, and Roy L. Johnston

Abstract We present a topical review of research in the field of metal-coated gold nanorods. By combining synthetic design strategies with state-of-the art characterisation techniques (particularly aberration-corrected scanning transmission electron microscopy) and molecular dynamic simulations, we demonstrate the potential to gain a fundamental understanding of, and control over, the atomic detail of metal–metal bonding at the interfaces of core–shell nanorods and nanoparticles. This analysis is facilitated by making comparisons between the related bimetallic systems: AuRh, AuPd and AuPt.

Keywords Bimetallic core–shell structures · Electron microscopy · Gold · Molecular dynamic simulations · Nanoparticles · Nanorods · Synthesis · Scanning transmission electron microscopy

Contents

1 Introduction 68
2 Synthetic Strategies 69
3 Mechanisms for Metal-on-Metal Growth 70

R.L. Chantry and Z.Y. Li (✉)
Nanoscale Physics Research Laboratory, School of Physics and Astronomy, University of Birmingham, Edgbaston, Birmingham B15 2TT, UK
e-mail: z.li@bham.ac.uk

I. Atanasov
School of Chemistry, University of Birmingham, Edgbaston, Birmingham B15 2TT, UK

Institute of Electronics, Bulgarian Academy of Sciences, 72 Tzarigradsko Chaussee, 1784 Sofia, Bulgaria

S.L. Horswell and R.L. Johnston (✉)
School of Chemistry, University of Birmingham, Edgbaston, Birmingham B15 2TT, UK
e-mail: r.l.johnston@bham.ac.uk

4 Characterisation Techniques 72
    4.1 Imaging 72
    4.2 Spectroscopy 74
5 Computational Methods 75
6 Case Studies of AuM Nanorods (M=Rh, Pd or Pt) 76
    6.1 Characterisation of Interfacial Structures of AuRh and AuPd Nanorods 77
    6.2 Simulations of Deposition of Rh, Pd and Pt on Au Surfaces 79
    6.3 Synthetic Manipulation of Pt on Au Nanorod Growth 83
7 Conclusions 85
References 86

1 Introduction

Nanoalloys (NAs) are nanoscale clusters of two or more metallic elements. They can be generated in the vapour phase, in solution, on substrates or in matrices, with varying degrees of control over size, composition, structure and chemical ordering [1–4]. Nanoalloys are of great interest both for their unique chemical and physical properties and their tunability. For example, catalysis by NAs is a thriving area of research as the catalytic activity and/or selectivity of metal nanoparticles can often be dramatically improved as a result of synergistic interactions between the component metals in an alloy [5, 6].

Four main types of chemical ordering have been identified for NAs [7]. *Core–shell-segregated NAs* consist of a shell of one type of atom (B) surrounding a core of another (A), though there may be some mixing between the shells. *Layered-segregated NAs* consist of distinct A and B sub-clusters, such as so-called "Janus" particles [8]. *Mixed NAs* may be either ordered or random. *Multishell NAs* have onion-like alternating concentric -A-B-A- shells. Factors influencing chemical ordering include metal–metal bond strengths, surface energies, atomic size mismatch, charge transfer, substrate and surfactant binding and specific electronic or magnetic effects. The chemical ordering observed for a particular NA depends critically on the balance of these factors, as well as on the synthetic method and experimental conditions [7, 9].

The focus of this topical review is on a subset of core–shell nanoalloys, bimetallic nanorods (NRs). Rod-shaped nanoparticles, especially those of gold, have been extensively studied for their interesting opto-electronic properties, which can be varied by control of aspect ratio and metal distribution [10]. An additional advantage of the study of NRs is their more varied faceting compared with spherical nanoparticles, which is interesting for examining the growth pattern of one metal on another [10, 11].

The metal–metal interactions that drive the structure and properties of core–shell NAs are most concentrated at the buried core–shell interface. Thus understanding the nature and structures of these interfaces is of key importance for the effective exploitation of core–shell bimetallic NRs. In this review we summarise the current state of work in this topic. We compare three nanorod systems, employing examples drawn mainly from our recent investigations of the metal–metal interfacial structures of Rh, Pd and Pt deposited on gold nanorods. We show how the balance between thermodynamic properties (e.g. surface energy, bulk cohesive energy), lattice constant and kinetics influences their structure and, consequently, properties and stability. Before presenting a detailed review of this work, we introduce briefly the synthetic strategies, characterisation techniques and computational methodologies employed.

2   Synthetic Strategies

The synthesis of nanoalloys for particular applications requires control over the size and shape of the nanoparticles as well as the distribution of the metals within them (e.g. core–shell vs solid solution). Fine control of size can be achieved through the use of cluster beam sources, which are typically used to investigate the effects of the sizes of mono-elemental clusters on their properties [12–14]. In some cases of bimetallic nanoclusters, the control of chemical ordering has also been successful [15, 16]. The high yield of clusters required for many potential applications using this approach is a challenge and the synthesis of NAs via wet chemical methods has been very widely used.

Control of size can also be achieved with wet chemical methods. A metal salt is reduced with a chemical reducing agent in the presence of capping agents, surfactants or ions and the temperature and concentrations of the species are varied [17, 18]. For example, the kinetics of reduction can be controlled by altering either the concentration of the metal salt or the relative rates of nucleation and growth by varying the concentration and identity of a capping agent [17].

Control of the shape of nanoparticles has also attracted much interest [18, 19]. The shape of Au nanoparticles has a profound influence on their optical properties [19–21] and the shape of particles also determines the crystal facets present, which influences their catalytic properties [22, 23]. Shape control has been most extensively studied for Au nanoparticles, with spheres, cubes [24, 25], triangles [25], prisms [25], bipyramids [26], polyhedra [27] and rods [19, 28] all being reported. The control of particle shape is generally achieved via seed-mediated growth methods, where additives, such as specifically adsorbing ions or metals and shape-forming surfactants, are used to control the relative rates of growth of particular facets [25, 29]. For example, cetyltrimethylammonium bromide (CTAB) is commonly employed in the synthesis of cubes and rods, whereas the use of the corresponding chloride generally generates spherical particles [21] and the inclusion of varying amounts of Ag^+ ions in solution has been utilised to control the shape of Au nanoparticles as a result of the differences in affinity of Ag metal for different facets of Au [25]. These methods have been reviewed and discussed elsewhere [21, 25, 29].

The synthesis of bimetallic or multimetallic nanoparticles is more complex, with a range of parameters determining the final distribution of the metals involved. Co-reduction of two metal salts can lead to alloying of the two metals or core–shell distribution, depending on relative surface energies, cohesive enthalpies vs. mixing enthalpy, entropy and lattice constant [7]. Kinetics also influences the final structure as often one metal salt may be more readily reduced than another. To synthesise core–shell nanoparticles, where the desired structure is not the thermodynamically favoured structure, a species that favours binding to the desired "shell" metal can be added. For example, if Ag and Pd are co-reduced, clusters with Pd core and Ag shell normally result. If ammonia is added to the reaction mixture, its stronger binding to Pd than to Ag results in particles with Ag cores and Pd shells [30]. Similarly, PdRh structures can be controlled by varying the reaction conditions: the presence of oxygen favours an Rh-enriched shell, whereas the presence of carbon monoxide favours a Pd-enriched shell [31]. It should also be noted that the distribution of metals within nanoalloy catalysts can be affected during operation [31]. For example, PtNi nanoparticles have been shown with in situ X-ray absorption spectroscopy (XAS) and Density Functional Theory (DFT) to vary in surface composition, with enrichment of the surface in Ni during hydrogen adsorption [32]. An alternative means to achieve a core–shell structure that is the inverse of the thermodynamically favoured product is to synthesise seed nanoparticles of the "core" metal and to deposit the "shell" metal onto the seeds in a second step. This method is particularly useful when a particular shape is desired because shape control is simpler in monometallic systems. The wide variety of shapes and sizes available for gold nanoparticles makes gold core–M shell nanoparticles an ideal case study for the examination of the mechanism of the deposition of a metal shell onto pre-formed seed clusters.

3   Mechanisms for Metal-on-Metal Growth

To understand the complex structures that can form in bimetallic core–shell nanoparticles, one can start by studying the growth modes that may be followed by one metal deposited on the flat surface of another. Figure 1 shows a schematic example of three possible growth modes.

Considering only the key parameters of the relative lattice constants and bond strengths, Volmer–Weber island growth is expected in systems with a high lattice mismatch and low tendency to form inter-metallic bonds and layer-by-layer Frank van der Mei we growth in systems with a low lattice mismatch and a high preference for inter-metallic bonds [33]. However, for systems between these extremes, growth modes can become more complex, such as Stranski–Krastanov layer-plus-island growth, wherein a tendency to form inter-metallic bonds initially favours layer growth but accumulated strain caused by lattice mismatch then induces a second phase of 3D island growth. The point at which the growth changes from layer to island depends upon the balance between bond strength and lattice strain, so the exact growth mode followed by any given system may not be easy to predict, even in this simplified example. In real systems, the range of influencing factors is greater, for example, layered overgrowth can occur even when there is significant lattice mismatch by proceeding in a different orientation or with a different crystal

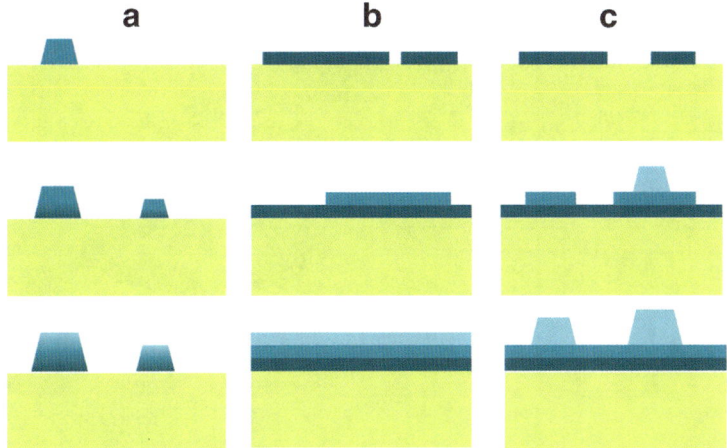

Fig. 1 Simple scheme for the categorisation of growth modes. (**a**) Volmer–Weber island growth, (**b**) Frank van der Merwe layer-by-layer growth and (**c**) Stranski–Krastanov layer-plus-island

structure to the substrate [34] or through the formation of systematic defects to relieve lattice strain [35].

Other parameters that may also be important for the formation of particular structures in bimetallic systems include the relative surface energies of the constituent metals, their respective rates of surface diffusion, the conditions prevailing during or after deposition and the existence of non-flat surfaces. These vary considerably between systems; for example, lattice mismatch can range from almost no mismatch in a system such as AuAg to a mismatch as great as 12% between Cu and Ag [36]. In addition, while lattice mismatch might be the principal driving factor for one size or structure of particle in one bimetallic system, other parameters such as surface or cohesive energy may override the influence of lattice mismatch and become more dominant for others. The wealth of influencing factors makes complete structural prediction a more challenging task in these systems; however, it gives more potential routes to manipulating structure and thus properties. For example, Au and Pd are miscible in bulk form [37], but AuPd nanoparticles have been synthesised with both segregated [38–41] and alloyed [39, 42] structures, by varying synthesis conditions. Transformations between these structures have been linked to variations in catalytic reactivity and can be induced by heating the samples [42].

For core–shell nanoparticles, the metal–metal interactions occur predominantly in sub-surface sites, at their buried interfaces. The interfacial structures of core–shell systems could influence their surface structure if the shell is thin, and thus their surface properties [43]. The connection between core–shell structure and resulting properties has been demonstrated by recent work conducted on AgPd [44] and PtNi [45] nanoparticles, where enhanced catalytic reactivity was attributed to the influence of lattice strain at the core–shell interface. Core–shell structure has also been

Fig. 2 Scanning transmission electron microscope (STEM) imaging detectors and spectrometers. Schematic diagram illustrating the location of the high angle annular dark field (HAADF) and bright-field (BF) imaging detectors and the energy dispersive X-ray (EDX) and electron energy loss spectroscopy (EELS) spectrometers. Reproduced from [49]

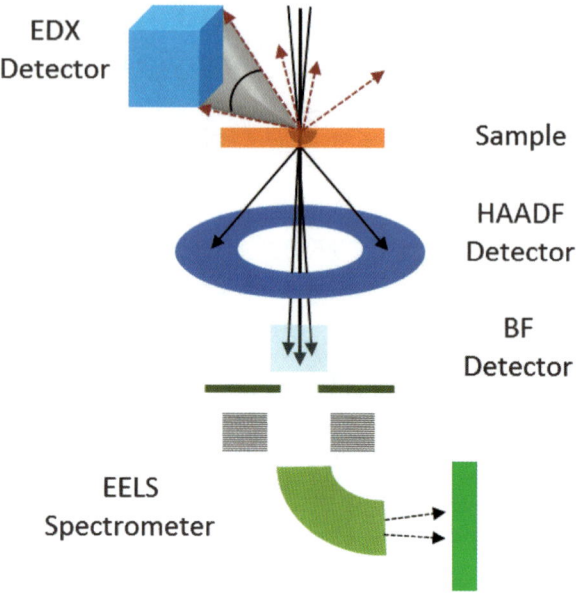

found to impart additional stability to nanoparticles, delivering greater durability compared with monometallic equivalents, which is an important factor in many industrial applications [46].

4 Characterisation Techniques

Gaining fundamental understanding of the driving forces behind the properties of bimetallic nanoparticles requires the use of techniques that are capable of obtaining atomically resolved elemental detail from sub-surface locations. *Aberration-corrected scanning transmission electron microscopy (ac-STEM)* is one of the few techniques with the potential to carry out this work [47, 48]. It has the capability to obtain a wide range of characterisation data, with atomic resolution and single-atom sensitivity, in a single instrument.

4.1 Imaging

STEM employs a focussed probe of electrons scanned across an area of a sample. Figure 2 shows a schematic representation of the principal detectors used in ac-STEM that can be employed to collect both elemental and structural data.

Fig. 3 STEM-HAADF
image showing Z-contrast
between the segregated Au
core and Pd shells.
Reproduced from [51].
Reproduced by permission
of the Royal Society of
Chemistry. [For the
electronic version: http://dx.
doi.org/10.1039/
C3NR02560H]

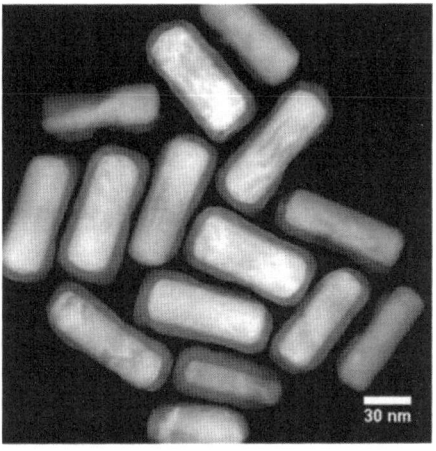

The high-angle annular dark-field (HAADF) detector collects electrons that are scattered to wider angles through interaction with the sample. These scattered electrons can be regarded as incoherent. The cross section of the incoherently scattered electrons depends on mainly sample thickness and the atomic numbers of the elements, which makes interpretation of HAADF image contrast relatively straightforward [50]. Where the sample is of uniform thickness, HAADF image contrast variation can be directly related to sample elemental variation. Figure 3 shows a typical HAADF image of segregated AuPd nanorods [51], demonstrating a clear difference in image intensity between the Au core ($Z = 79$) and the Pd shell ($Z = 46$).

In addition to the HAADF detector, ac-STEM also has the capability for bright-field (BF) imaging through the detection of electrons scattered to small forward angles by the sample, and, as such, assumed to retain their coherence. BF image contrast can be related to phase changes caused by interactions between the beam electrons and the sample. They are sensitive to extremely fine structural detail, such as the presence of lattice strain, and are also more affected by lighter elements, such as the amorphous carbon substrate of TEM grids, than DF images. Figure 4 shows simultaneously acquired STEM-HAADF and STEM-BF "as taken" images from the edge of an Au nanorod and illustrates the difference between DF amplitude contrast images and BF phase contrast images. Atomic resolution is apparent in both images; however, the greater sensitivity of BF imaging to lighter atoms means the grain of the underlying amorphous carbon support is apparent in the BF image but not in the HAADF image.

Imaging information can be maximised by using the two detectors simultaneously. This allows the simplicity of HAADF interpretation to aid interpretation of more complex BF images, while the finer structural detail available in BF images can complement the superior atomic resolution of HAADF images. Recently, Goris et al. further demonstrated the potential of ac-STEM in the characterisation of core–shell interfaces by employing tomography techniques to examine the core–shell AuAg nanorods from several different imaging angles [52]. Although tomography

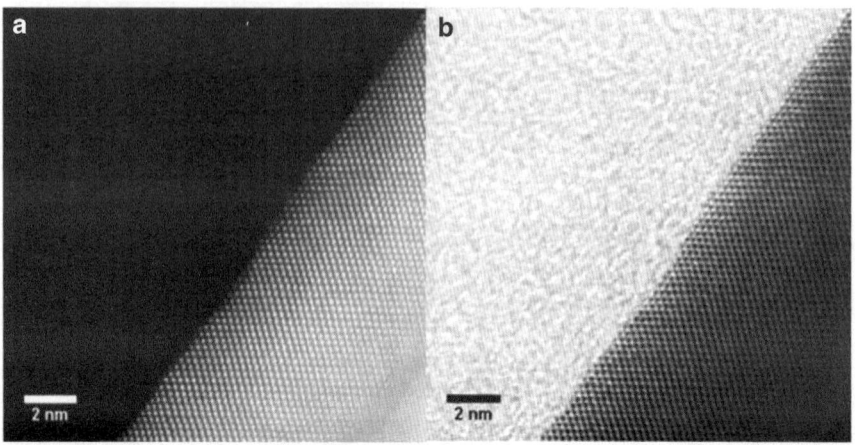

Fig. 4 Simultaneously acquired STEM-HAADF (**a**) and STEM-BF (**b**) images taken from the edge of an Au nanorod. Reproduced from [49]

is a time-consuming process, this study showed that it can be used to build a 3D representation of interfacial structure in individual bimetallic nanoparticles.

4.2 Spectroscopy

Energy-dispersive X-ray spectroscopy (EDX) is a form of spectroscopy that utilises the detection of X-ray emissions that occur when beam electrons interact with the sample. These X-ray emissions are entirely elementally specific. Given sufficient X-ray counts, statistical methods can be used to calculate accurately the relative quantities of elements present in the sample from EDX spectra [53]. The technique of elemental mapping is particularly useful for analysing bimetallic samples and has been used extensively in this regard.

In contrast to EDX, electron energy loss spectroscopy (EELS) tends to have a better signal to noise ratio, thus improving spatial resolution. EEL spectra show the energy lost by individual beam electrons as they are transmitted through the sample. As the electronic interactions are derived from the dielectric response, they are elementally specific [54]. Thus EELS can be an effective tool to measure the elemental composition of bimetallic samples, particularly when used in combination with elementally sensitive STEM-HAADF imaging. The better signal to noise ratio of EELS compared to EDX results in improved spatial resolution. However, in comparison to EDX, EELS spectra can be more complex to interpret, as the position and appearance of spectral features can be affected by the electronic structure of the sample. EELS is also limited to signals that are within the range of the spectrometer and so is of less use in measuring high energy loss signals (over 1,000 eV energy loss), which is typically the case for core loss signals of heavier elements such as Au. For these signals, EDX is the preferred technique.

5 Computational Methods

In the field of nanoalloys, computer simulation is becoming increasingly important, both to predict the structures that may form through metal–metal interactions in bimetallic nanoparticles and to support the interpretation of experimental data.

While high-level ab initio molecular orbital calculations, including electron correlation, have been carried out for small nanoalloys with up to around ten atoms [55], owing to their computational expense, they rapidly become unfeasible for larger clusters. DFT calculations [56, 57], which scale much better with cluster size than ab initio methods, while typically showing good agreement with experiment, have become increasingly popular for studies of NAs with tens or even hundreds of atoms [58, 59], though rigorous structural searches at the DFT level (typically combined with genetic algorithm and basin-hopping Monte Carlo methods) have only been performed for smaller particles with up to around 20–30 atoms [60].

For this reason, studies of large NAs have tended to use computationally cheaper empirical atomistic potential energy functions, which do not explicitly include electronic effects. There are a variety of empirical atomistic potentials which have been applied to study the structures, dynamics and thermodynamics of NAs [7, 61]. These empirical potentials typically have parameters which are fitted to experimental (or high-level theoretical) data for bulk metals and alloys and sometimes for small clusters. There are several empirical many-body potentials based on the second moment approximation to tight-binding theory [62], the most widely applied for studying NAs being the Gupta potential [63, 64]. This potential has been used to study static and dynamic properties of NAs with hundreds or thousands of atoms [58, 65–68].

In the Gupta potential [63, 69], the configuration energy of a system of N atoms is obtained as

$$E = \sum_{i=1}^{N} \left(E_i^R + E_i^B \right) \tag{1}$$

where the contribution of each atom, i, comprises two terms:

$$E_i^R = \sum_{j \neq i}^{N} A \cdot \exp\left[-p\left(\frac{r_{ij}}{r_0} - 1 \right) \right] \tag{2}$$

$$E_i^B = -\sqrt{ \sum_{j \neq i}^{N} \xi^2 \exp\left[-2q\left(\frac{r_{ij}}{r_0} - 1 \right) \right] } \tag{3}$$

The repulsive term, E_i^R, is the pairwise Born–Mayer-type interaction of atom i with its neighbours. The binding or cohesive term, E_i^B, is proportional to the width (square root of the second moment) of the d-band of the electron density of states,

Table 1 Key parameters for Rh, Pd, Pt and Au

	Rh	Pd	Pt	Au	Reference
Atomic number	45	46	78	79	[36]
Lattice constant (Å)	3.80	3.89	3.92	4.08	[36]
Average surface energy (J m^{-2})	2.7	2.0	2.5	1.5	[72]
Bulk cohesive energy (eV atom^{-1})	5.75	3.89	5.84	3.81	[36]
Au–M diatomic binding energy (eV)	2.03	1.90	1.81	1.55	This study[a]

[a]The Au–M diatomic binding energies have been calculated at the LDA–DFT (PAW) level using the VASP package [73, 74]

expressed in terms of the so-called hopping integrals, ξ, between atom i and its neighbours. This term incorporates the many-body nature of the interaction. All interactions are assumed to decay exponentially with the interatomic distance r_{ij}. The parameters $\{A, \xi, p, q\}$ depend on the nature of the two atoms. Three parameter sets are thus needed for a binary system: two sets for the two homonuclear interactions and one for the heteronuclear interaction.

Typically, the parameters of the homonuclear interactions are fitted to reproduce the experimental values of the cohesive energy of the metal and its elastic constants. As a first approximation, the parameters of the heteronuclear interaction can be taken as the mean values of the corresponding parameters of the two elements. Although, in many cases, this has proved to be a successful strategy [70], a better representation of the mixing properties of the two elements can be achieved by fitting the heteronuclear parameters to the dissolution energies of each element into a bulk matrix of the other [71].

6   Case Studies of AuM Nanorods (M=Rh, Pd or Pt)

Gaining a fundamental understanding of the origin of the nanoscale properties of bimetallic nanoparticles is a challenging issue. We describe our recent work on three AuM systems, where M is Rh, Pd or Pt, to illustrate how the application of experimental structural characterisation, computer simulation and chemical synthesis techniques is possible to gain an understanding of, and control over, the factors that drive the formation of the interfacial structure in core–shell nanorods.

Key structural and energetic parameters for the elements Rh, Pd, Pt and Au are summarised in Table 1. These parameters, and the relative differences between them for the constituent metals of each system, underpin the metal–metal interactions that are instrumental in the formation and stability of structure through metal-on-metal growth.

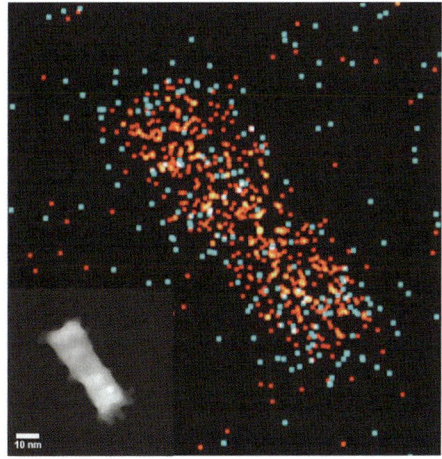

Fig. 5 EDX maps of Au (red dots) and Rh (blue dots) taken from a AuRh nanorod. The insert is the STEM-HAADF image of the nanorod. Reproduced from [80]. Copyright 2012 American Chemical Society

6.1 Characterisation of Interfacial Structures of AuRh and AuPd Nanorods

The deposition of Rh on Au-seeded nanorods is of interest, partly because of the contrasting physical and chemical properties of these two elements. While Rh is catalytically active in the bulk form, Au has only been found to demonstrate catalytic activity in nanoscale systems. From Table 1, it can be seen that the large 7% lattice mismatch and the higher bulk cohesive energy and surface energy of Rh compared with Au suggest a preference for the $Rh_{core}Au_{shell}$ configuration. This has indeed been observed for nanoparticles formed on $TiO_2(110)$ surfaces by physical vapour deposition of either Rh followed by Au or vice versa [75, 76]. In these studies, the morphology of the bimetallic nanoparticles was examined using scanning tunnelling microscopy, while the chemical composition was characterised with low-energy ion scattering. Both techniques are surface sensitive; hence, information about the metal–metal interaction at the sub-surface region is rather limited. Despite the complete immiscibility of Au and Rh in the bulk [77], chemical synthesis of both segregated [78] and alloyed AuRh NAs [79] has also been reported.

In order to gain mechanistic understanding of metal–metal interactions at the atomic level, we have applied aberration-corrected STEM, as described in Sect. 4, to AuRh nanorods synthesised using a seed-mediated sequential growth method via a wet chemical route. The main results have been reported in two recent publications [51, 80]. Figure 5 shows EDX maps of Au and Rh overlaid on the same image. Despite the limited EDX counts, a clear correlation in the pattern of Au and Rh signals supports the formation of the $Au_{core}Rh_{shell}$ structure. As discussed in Sect. 3, a system of Rh sequentially deposited onto Au nanorods may not be in thermodynamic equilibrium. The atomic details at the interface can be revealed by simultaneously acquired DF and BF STEM images, shown in Fig. 6a, b, respectively, from

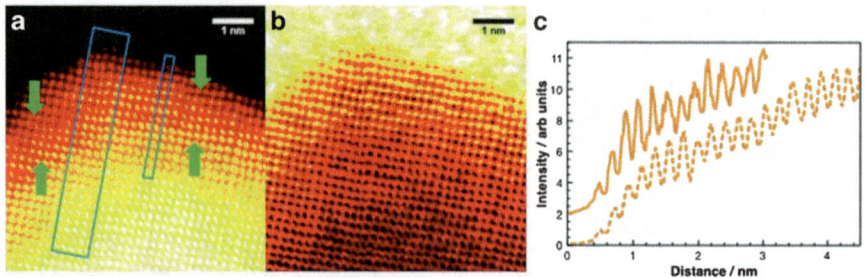

Fig. 6 Atomically resolved STEM-HAADF and BF images of the Au-Rh interface. Simultaneously acquired (**a**) STEM-HAADF and (**b**) STEM-BF images of an $Au_{core}Rh_{shell}$ nanorod, with (**c**) line intensity profiles taken as indicated in (**a**) over 4 atomic columns (70 pixels) and a single atomic column (16 pixels). Reproduced from [49]

the end of an AuRh rod. Although atomic columns are clearly resolved, there is no clear Z-contrast intensity variation as expected for Au ($Z = 79$) and Rh ($Z = 46$). Instead, random intensity variation from column to column is seen, as indicated by arrows in Fig. 6a, or more clearly by the intensity profiles shown in Fig. 6c. This result can be interpreted as a sign of intermixing or alloying at the interface. Further evidence can be seen below from the comparative study of Pd on Au nanorods. In addition, the atomic column spacing from the centre to the edge of the nanorod remains consistent with the Au lattice spacing, despite the 7% lattice mismatch between Rh and Au (see Table 1). Furthermore, there is no phase contrast feature in the AuRh BF image, suggesting that any mismatch strain may have been relieved by intermixing at the Au-Rh interface. The sequential method followed in synthesising these nanorods means that any mixing should have happened during or immediately after Rh deposition.

In contrast to AuRh, Z-contrast intensity variation is seen for AuPd nanorods, as shown earlier in Fig. 4. Here, comparable atomically resolved DF and BF images taken from a corner of an AuPd nanorod are displayed in Fig. 7a, b, where an abrupt contrast change is apparent that is consistent with the presence of a strain contrast feature in the BF image. This suggests that Au and Pd are well segregated, although Au and Pd are readily miscible in the bulk [37]. Core–shell segregation has been observed in other studies of AuPd nanoparticles synthesised using similar seed-mediated sequential methods [35, 38, 81].

Given that Rh and Pd have comparable atomic numbers, 46 vs 47, the clear difference in Z-contrast imaging obtained for these two systems highlights the difference in metal–metal interaction at the interfaces. The comparative approach adopted in this study of two systems with similar relative differences in key structural and elemental parameters (Table 1) highlights the important interplay between energetics, kinetics and thermodynamics in bimetallic nanostructure formation and shows the considerable potential that exists for structural manipulation through controlling kinetic parameters during synthesis. However, the driving force responsible for the interfacial structure of any given system cannot easily be determined by simply considering these parameters in isolation, so we have used

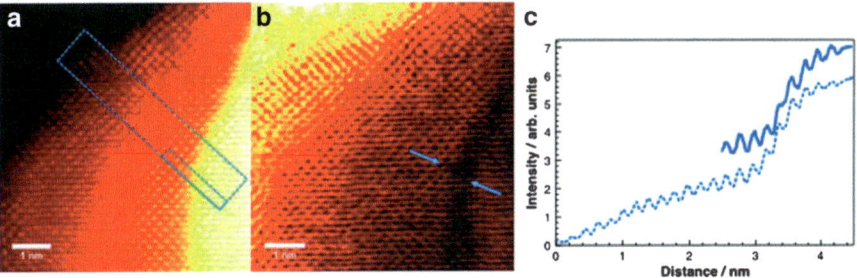

Fig. 7 Atomically resolved STEM-HAADF and BF images of the Au-Pd interface. Simultaneously acquired (**a**) STEM-HAADF and (**b**) STEM-BF images of an $Au_{core}Pd_{shell}$ nanorod, with (**c**) line intensity profiles taken as indicated in (**a**). Reproduced from [51]. Reproduced by permission of the Royal Society of Chemistry. [For the electronic version: http://dx.doi.org/10.1039/C3NR02560H]

computer simulations to assist with the analysis and interpretation of the experimental results. In addition to the AuPd and AuRh systems mentioned above, we have also considered the related AuPt system, for which structural and chemical studies have recently been carried out [82, 83].

6.2 Simulations of Deposition of Rh, Pd and Pt on Au Surfaces

The characteristic length of metal-coated nanorods is of the order of tens of nanometres. Modelling of deposition on flat surfaces should therefore be representative of the key processes occurring at the interface between the nanorod and the coating material during the process of deposition.

The computational studies reported below are based on molecular dynamic (MD) simulations, using the Gupta potential, of the deposition of metal atoms, Rh, Pd or Pt, onto an Au surface (henceforth denoted M/Au), in order to explore the initial stages of interfacial structure formation in these systems. The most instructive crystallographic orientations for simulation are the two low-index {111} and {100} orientations, because they typically have the lowest surface energies.

The Gupta potential parameter sets adopted for our simulations were taken from the literature. The Rh and Au parameters were taken from [63, 84] for Rh and Au, respectively, and the mixed interaction parameters were fitted to experimental dissolution energies of each element in a matrix of the other [77] (Ferrando R (2013) Personal communication). The AuPd parameters were taken from [85, 86]. Finally, the Au and Pt parameters were taken from [63], while the heteronuclear AuPt parameters were derived as the mean of the corresponding Au and Pt parameters [70].

The Nosé–Hoover thermostat [87] was applied in all MD simulations in order to maintain a constant temperature throughout the deposition process. Au {100} and

{111} slabs were modelled using six atomic layers, the bottom one being fixed at mean bulk positions, while the atoms of the other five layers are free to experience thermal motions. Each crystallographic plane contains 128 and 90 atoms for the {100} and {111} orientations, respectively. The thermal expansion of Au predicted by the Gupta potential was used to set the lattice constant for a given temperature. A deposition rate of 1 atom/ns was applied. Figure 8 shows snapshots from the MD simulations of the deposition of 1 monolayer (ML) of metal M (M=Rh, Pd and Pt) on both the {100} and {111} Au surfaces, for temperatures of 300 and 500 K.

6.2.1   Rh/Au

From the first column in Fig. 8, it is evident that Rh forms surface clusters on the Au substrate. Their average size depends on the surface mobility of the adatoms (and on the deposition rate). For instance, two sub-clusters can be distinguished for {100}/300 K, where the surface mobility is relatively low because of the low temperature and the higher coordination of an adatom on the {100} surface compared with the {111} surface. The surface cluster is more compact at $T = 500$ K. A single Rh cluster is observed on the {111} substrate for both temperatures. The formation of 3D Rh clusters on Au surfaces is consistent with the strong immiscibility of the two elements in the bulk. If low-coordinated Au atoms exist on the surface, they display a tendency to attach themselves to the Rh clusters and to climb onto their surface. The illustration at the bottom of column 1 of Fig. 8 shows a snapshot of a Rh cluster on the Au{111} substrate annealed at 500 K for 20 ns together with 8 Au adatoms. This effect, which is consistent with the Au–Rh exchange previously observed [75, 88, 89], results from the greater Au–Rh bond strength compared with that of the Au–Au bond and the much lower surface energy of Au than Rh (Table 1). If low-coordinated Au atoms are present on the surface, they would aggregate together at step or kink sites of the growing Rh island and could be buried by newly deposited Rh atoms. This is a possible scenario to explain the observed mixing at the interface of Rh-coated Au nanorods [51].

6.2.2   Pd/Au

The second column in Fig. 8 shows MD snapshots for Pd deposition on Au. In contrast to Rh, Pd effectively wets the Au surfaces at this deposition rate, even at low temperatures. This makes it possible for an ideal layer-by-layer growth to take place until the accumulated strain resulting from the lattice mismatch causes some surface reconstructions. If, however, at this point the Pd layer is sufficiently thick the system may be kinetically trapped in a state which is far from equilibrium, where a mixed bulk phase and a surface enriched in Au are expected. This could be a kinetically driven mechanism for the formation of the very sharp Au–Pd interface observed in Fig. 7. It is evident from Fig. 8 that the energy barrier for the interchange of Pd adatoms and surface Au atoms on Au{100} is low enough to

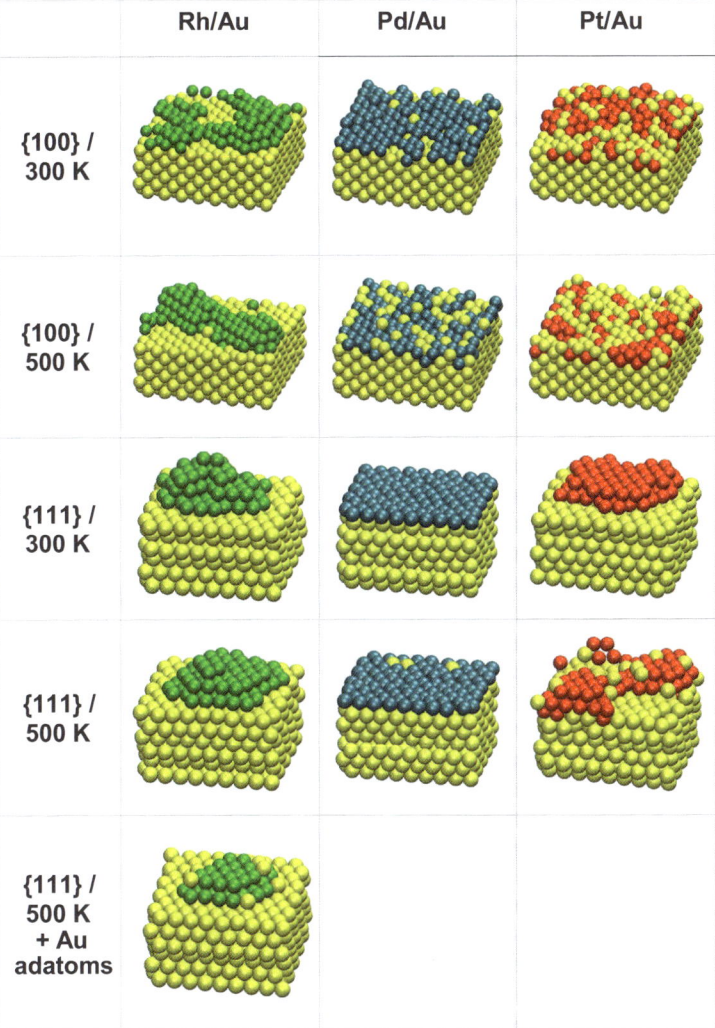

Fig. 8 Snapshots of molecular dynamic computer simulations of Rh, Pd and Pt vapour deposition on Au substrates at temperatures of $T = 300$ K and $T = 500$ K. The deposition rate is at a rate of 1 atom/ns. The *bottom panel* of column 1 is a MD snapshot of a Rh cluster on the Au{111} substrate annealed at 500 K for 20 ns with 8 Au adatoms

occur occasionally even at room temperature. The closer packed Au{111} surface exhibits much greater kinetic stability. In order for mixing at the interface to be prevented, the rate of Pd deposition should be sufficient to form a stabilising wetting layer before any substitutions between the metals can occur. This is consistent with other theoretical calculations conducted for small AuPd clusters, which indicate that increasing shell thickness can improve the stability of segregated structures [68].

6.2.3 Pt/Au

The third column in Fig. 8 shows MD snapshots for Pt deposition on Au. A significantly higher percentage of displacement of {100} surface Au atoms by deposited
Pt atoms is observed compared with Pd/Au. Pt and Au, which are mostly immiscible in the bulk, display a greater degree of surface mixing than Pd and Au, which
are completely miscible. This situation, which is analogous to that for Rh/Au, can
be attributed to the greater difference of the surface energies between the substrate
and the deposited element in the case of Pt/Au than in the case of Pd/Au and is
consistent with previous simulations by Haftel et al. who modelled the early stages
of growth of Pt/Au{100} and Au/Pt{100} surfaces [90]. The apparent mixing in the
Pt/Au{100} system is characterised by a certain degree of clustering, although not
as pronounced as for Rh/Au. Raising the temperature from 300 to 500 K results in a
smoother Pt/Au{100} interface, which is not observed for Rh/Au{100}. Pt deposited onto the Au{111} substrate forms 3D clusters, similarly to Rh/Au{111}, but
displays greater wetting of the substrate. Two main differences are evident when
comparing the effect of temperature on the Pt/Au{111} and Rh/Au{111} systems.
First, there are a greater number of displaced Au surface atoms at 500 K for Pt/Au,
which shows that this temperature is sufficient to overcome the energy barrier for
this elementary process even on the close-packed {111} surface, which supports the
hypothesis put forward for the morphological changes in AuPt nanorods discussed
below (Sect. 6.3). Second, the growth at 500 K is characterised by smaller surface
clusters than at 300 K. This tendency is opposite to that observed for Rh/Au{100}
and results from the very different kinetic picture of the active processes taking
place on the surface in both cases. For instance, several Pt atoms implanted in the
Au{111} surface at 500 K become nucleation centres, around which 3D surface
clusters form, while the geometries of the Rh/Au{100} surface clusters are mainly
controlled by the rate of surface diffusion. This also illustrates that the characteristics of surfaces formed by kinetic effects may change in a non-uniform way with
temperature, especially in heteroatomic systems such as those considered here.

It should be emphasised that these MD simulations correspond to the physical
progress of vapour deposition. They reveal only certain internal properties of the
system and tendencies of the deposition process and do not give a comprehensive
picture of the growth kinetics, which may be different in the case of chemical
deposition. Also, modelling deposition on flat surfaces does not account for the
effect of curvature of the nanorod surface, which may result in different strain
conditions when the coating film becomes sufficiently thick. Nevertheless, the
results from these simulations underline the importance of taking special care in
kinetic modelling of such systems to determine the key kinetic processes and to
verify whether or not they change within the temperature range under study.

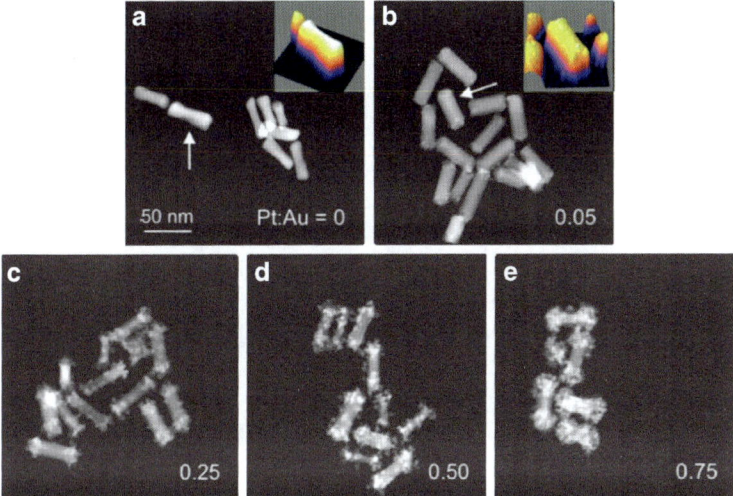

Fig. 9 STEM-HAADF images of AuPt nanorods. (**a**) Bare Au NR seeds and (**b**) to (**e**) Pt-coated Au NRs with Pt:Au ratios, respectively, of 0.05, 0.25, 0.50 and 0.75. The 3D intensity profiles from an as-grown Au NR (*arrowed*) and a Pt-coated NR (*arrowed*) are inset in (**a**) and (**b**), respectively. Reproduced from [83]. Copyright 2012 American Institute of Physics

6.3   Synthetic Manipulation of Pt on Au Nanorod Growth

As shown in Table 1, the lattice constant of Pt is very similar to that of Pd and its average surface energy lies between those of Rh and Pd. These factors determine similar surface segregation properties at equilibrium as in the AuPd and AuRh systems, namely, that surfaces are expected to be enriched in Au. Pt, like Rh, has a much higher cohesive energy than Au, whereas Pd has a similar cohesive energy. With respect to its miscibility properties in the bulk, the AuPt system is interesting because although Au and Pt are immiscible for most compositions, they form solid solutions in the very Au-rich part of the phase diagram [77]. As a result of this, AuPt can be said to bridge the miscible AuPd system and the more immiscible AuRh system.

Our own study and the work of others conducted on AuPt nanorods indicates a tendency for clustered growth of Pt, with a marked preference for Pt growth on the nanorod end facets [83, 91]; examples from our work are shown in Fig. 9. Growth of Pt (or indeed Rh) on the outside of an Au NR represents a kinetically controlled product rather than the thermodynamically most stable product. Our study has shown that marked changes in the morphology of Pt-coated Au NRs were observed over the course of 18 months when stored under ambient conditions; see Fig. 10 [83]. It seems that, unlike the AuRh and AuPd systems discussed in Sect. 6.1 above, the Au–Pt interface is not stable. Similar morphology changes were observed for AuPt NRs with different amounts of Pt coating. This process can be accelerated by annealing AuPt nanorods at 200°C [83]. It is possible that the mechanism of the

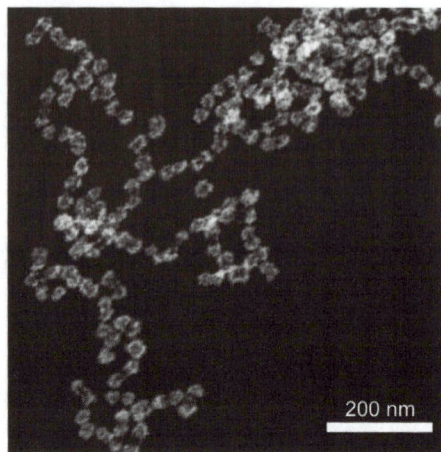

Fig. 10 An overview of the AuPt nanorods showing evolution of morphology. Reproduced from [83]. Copyright 2012 American Institute of Physics

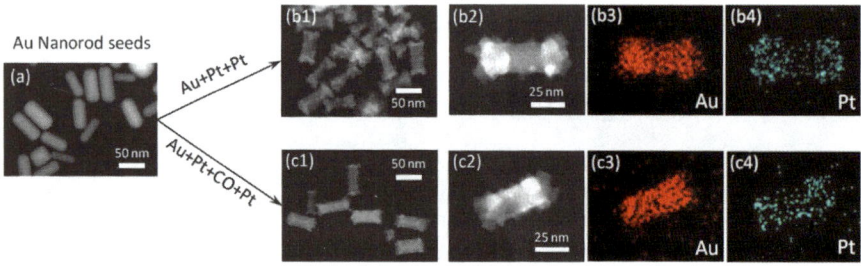

Fig. 11 STEM-HAADF images and EDX elemental maps of AuPt nanorods showing the influence of CO blocking on Pt overgrowth. (**a**) Bare Au NRs seed, (**b1-4**) Au–Pt NRs prepared without the CO blocking step, and (**c1-4**) Au–Pt NRs of same composition prepared with the CO blocking step. Reproduced from [82]. Copyright 2013 American Chemical Society

change in morphology is a result of initial Pt clustering growth that leaves part of the Au surface uncovered especially in the middle of the Au nanorod. The thermo-dynamic preference for Au-coated Pt probably drives Au to diffuse onto the outside of the Pt, thinning the centre of the NR until it breaks into two nanoparticles. While the AuPt nanorods were kept in solution, the samples were stable, suggesting that the surfactant remaining on the surface may inhibit this diffusion process.

Working on the hypothesis that the incomplete Pt coverage might enable Au diffusion to the outside of $Au_{core}Pt_{shell}$ NRs, a selective blocking method was developed to even out Pt coverage across the surface of Au NRs. The results from this study are summarised in Fig. 11 [82]. The method involved sequential Pt deposition steps and exploited the difference in binding energies of carbon monoxide (CO) on Au and Pt surfaces. STEM-HAADF, EDX and electrochemical measurements all indicated that using CO blocking after the first deposition step resulted in the coverage of remaining exposed Au sites with Pt deposited in the subsequent step rather than continued growth at the ends of the NRs. The effect of smoothing Pt deposition and increasing its coverage of Au also resulted in greater stability of the NRs. STEM-HAADF carried out on the same samples after more than 18 months showed that the NRs prepared with CO blocking retained their structure [82].

Our study also showed that the different surface compositions of the Au–Pt NRs prepared with CO blocking and without CO blocking had an influence on catalytic activity and selectivity towards the electrochemical reduction of oxygen. This reaction was more selective towards water as a product and had higher rate constant and specific activity (per area Pt) when carried out with NRs prepared with CO blocking, as a result of the higher Pt:Au surface ratio [82].

7 Conclusions

In this brief review, we have shown the potential to gain a fundamental understanding of, and control over, the atomic detail of metal–metal bonding at the interfaces of core–shell nanorods and nanoparticles. Using a combination of experimental and computer simulation techniques and by making comparisons between related bimetallic systems has proved to be very effective.

Using examples from our recent work, we have shown that experimental structural characterisation can be conducted to great effect by exploiting the atomically resolved capabilities of ac-STEM imaging. For bimetallic nanosystems comprising elements from different rows in the period table (such as AuRh and AuPd), STEM-HAADF imaging can provide elemental information about the interface at the atomic scale. For systems such as AuPt with similar atomic numbers, ac-STEM in combination with EDX elemental mapping provides a very effective method for investigating interfacial structures. The direct visualisation of the interface at the atomic scale highlights the complex range of factors that drive the metal–metal interactions in bimetallic systems and the need for targeted computer simulation of metal-on-metal growth to provide insight into the mechanisms of the observed interfacial structures.

Using molecular dynamic simulations, we have shown the key role played by kinetics in forming the interfacial structures in these AuM systems. Although the complexity of real systems cannot be fully reproduced in computer simulation, this work shows how improved understanding can be achieved through the effective use of computer simulation. [It should be noted that an alternative kinetic Monte Carlo approach could be used to study growth over longer timescales, though this approach is better for fixed-lattice systems and where there is good epitaxy.] Our simulation results have also revealed the potential for kinetic control of synthesis structure during chemical synthesis. Our study using CO blocking during the chemical synthesis of AuPt nanorods has demonstrated experimentally how particle structure can be controlled at the atomic scale by manipulating kinetics to realise significant reactivity gains.

Research into metal-on-metal growth and the resulting interfacial structures in bimetallic nanoparticles is still rather limited. The nature of interfacial structures impacts directly on reactivity. This review has shown that by taking a combined approach of experimental structural characterisation of interfacial structures, computer simulation of metal-on-metal growth and controlled sample synthesis to manipulate interfacial structure, alongside catalytic reactivity measurements, it is possible to make substantial steps towards the ultimate goal of the rational design of bimetallic nanocatalysts.

Acknowledgements We acknowledge financial support from EPSRC grant numbers (EP/D056241/1, EP/G070326/1), COST Action MP0903, and the FP7 "ELCAT" grant no. 214936–2. R.L.C. thanks the EPSRC for a Ph.D. studentship and I.A. thanks the Marie Curie Actions (FP7/2007–2012). The authors thank R. Ferrando (University of Genoa, Italy) for the AuRh interatomic potentials and for helpful discussions, J.B.A. Davis (University of Birmingham, UK) for the calculated Au–M binding energies, and other co-workers mentioned in the references.

The STEM used in this research was obtained through Birmingham Science City with support from Advantage West Midlands and partially funded by the European Regional Development Fund. Calculations were performed on the University of Birmingham's BlueBEAR high-performance computer (http://www.bear.bham.ac.uk).

References

1. Alloyeau D, Mottet C, Ricolleau C (eds) (2012) Nanoalloys – synthesis, structure and properties. Springer, London
2. Calvo F (ed) (2013) Nanoalloys – from fundamentals to emergent applications. Elsevier, Amsterdam
3. Johnston RL, Wilcoxon J (eds) (2012) Metal nanoparticles and nanoalloys. Elsevier, Amsterdam
4. Mariscal M, Oviedo O, Leiva E (2013) Metal clusters and nanoalloys. Springer, New York
5. Cui C, Gan L, Heggen M, Rudi S, Strasser P (2013) Compositional segregation in shaped Pt alloy nanoparticles and their structural behaviour during electrocatalysis. Nat Mater 12:765–771
6. Wang D, Xin HL, Hovden R, Wang H, Yu Y, Muller DA, DiSalvo FJ, Abruña HD (2012) Structurally ordered intermetallic platinum-cobalt core-shell nanoparticles with enhanced activity and stability as oxygen reduction electrocatalysts. Nat Mater 12:81–87
7. Ferrando R, Jellinek J, Johnston RL (2008) From theory to applications of alloy clusters and nanoparticles. Chem Rev 108:845–910
8. Parsina I, Baletto F (2010) Tailoring the structural motif of AgCo nanoalloys: core/shell versus Janus-like. Phys Chem C 114:1504–1511
9. Langlois C, Li ZY, Yuan J, Alloyeau D, Nelayah J, Bochicchio D, Ferrando R, Ricolleau C (2012) Transition from core-shell to Janus chemical configuration for bimetallic nanoparticles. Nanoscale 4:3381–3388
10. Chen H, Shao L, Li Q, Wang J (2013) Gold nanorods and their plasmonic properties. Chem Soc Rev 42:2679
11. Hou S, Hu X, Wen T, Liu X, Wu W (2013) Core-shell noble metal nanostructures template by gold nanorods. Adv Mater 25:3857

12. Kwon G, Ferguson GA, Heard CG, Tyo EC, Yin C, DeBartolo J, Seifert S, Winans RE, Kropf AJ, Greeley J, Johnston RL, Curtiss LA, Pellin MJ, Vajda S (2013) Size-dependent subnanometer Pd cluster (Pd_4, Pd_6 and Pd_{17}) water oxidation electrocatalysis. ACS Nano 7:5808–5817
13. Li ZY, Young NP, Di Vece M, Palmer RE, Bleloch AL, Curley BC, Johnston RL, Jiang J, Yuan J (2008) Three-dimensional atomic-scale structure of size-selected gold nanoclusters. Nature 451:46–48
14. Nesselberger M, Roefzaad M, Fayçal Hamou R, Biedermann PU, Schweinberger FF, Kunz S, Schloegl K, Wiberg GKH, Ashton S, Heiz U, Mayrhofer KJH, Arenz M (2013) The effect of particle proximity on the oxygen reduction rate of size-selected platinum clusters. Nat Mater 12:919–924
15. Belic D, Chantry RL, Li ZY, Brown SA (2011) Ag-Au nanoclusters: structure and phase segregation. Appl Phys Lett 99:17194
16. Yin F, Wang ZW, Palmer RE (2011) Controlled formation of mass-selected Cu-Au core-shell cluster beams. J Am Chem Soc 133:10325–10327
17. Daniel MC, Astruc D (2004) Gold nanoparticles: assembly, supramolecular chemistry, quantum-size related properties, and applications toward biology, catalysis and nanotechnology. Chem Rev 104:293–346
18. Lisieck I (2005) Size, shape, and structural control of metallic nanocrystals. J Phys Chem B 109:12231–12244
19. Pérez-Juste J, Pastoriza-Santos I, Liz-Marzán LM, Mulvaney P (2005) Gold nanorods: synthesis, characterisation and applications. Coord Chem Rev 249:1870–1901
20. Myroshnychenko V, Rodríguez-Fernández J, Pastoriza-Santos I, Funston AM, Nova C, Mulvaney P, Liz-Marzán LM, García de Abajo FJ (2008) Modelling the optical response of gold nanoparticles. Chem Soc Rev 37:1792–1805
21. Sharma V, Park K, Srinivasarao M (2009) Colloidal dispersion of gold nanorods: historical background, optical properties, seed-mediated synthesis, shape separation and self-assembly. Mat Sci Eng R 65:1–38
22. Erikson H, Sarapuu A, Alexeyeva N, Tammeveski K, Solla-Gullón J, Feliu JM (2012) Electrochemical reduction of oxygen on palladium nanocubes in acid and alkaline solutions. Electrochim Acta 59:329–335
23. Vidal-Iglesias FJ, Aran-Ais RM, Solla-Gullon J, Herrero E, Feliu JM (2012) Electrochemical characterization of shape-controlled Pt nanoparticles in different supporting electrolytes. ACS Catal 2:901–910
24. Hernandez J, Solla-Gullón J, Herrero E, Aldaz A, Feliu JM (2007) Electrochemistry of shape-controlled catalysts: oxygen reduction reaction on cubic gold nanoparticles. J Phys Chem C 38:14078–14083
25. Langille MR, Personick ML, Zhang J, Mirkin CA (2012) Defining rules for the shape evolution of gold nanoparticles. J Am Chem Soc 134:14542–14544
26. Personick ML, Langille MR, Wu JS, Mirkin CA (2013) Synthesis of gold hexagonal bipyramids directed by planar-twinned silver triangular nanoprisms. J Am Chem Soc 135:3800–3803
27. Burt JL, Elechiguerra JL, Reyes-Gasga J, Montejano-Carrizales JM, Yacaman MJ (2005) Beyond Archimedean solids: star polyhedral gold nanocrystals. J Cryst Growth 285:681–691
28. Nikoobakht B, El-Sayed MA (2003) Preparation and growth mechanism of gold nanorods (NRs) using seed-mediated growth method. Chem Mater 15:1957–1962
29. Lohse SE, Murphy CJ (2013) The quest for shape control: a history of gold nanorod synthesis. Chem Mater 25:1250–1261
30. Goia DV, Matijevic E (1998) Preparation of monodispersed metal particles. New J Chem 22:1203–1215
31. Tao F, Grass ME, Butcher DR, Zhang Y, Renzas JR, Liu Z, Chung JY, Mun BS, Salmeron M, Somorjai GA (2008) Reaction-driven restructuring of Rh-Pd and Pt-Pd core-shell nanoparticles. Science 322:932–934

32. Hoffmannova H, Okube M, Petrykin V, Krtil P, Mueller JE, Jacob T (2013) Surface stability of Pt₃Ni nanoparticulate alloy catalysts in hydrogen adsorption. Langmuir 29:9046–9050
33. Seifert W, Carlsson N, Miller M, Pistol ME, Samuelson L, Wallenberg LR (1996) In-situ growth of quantum dot structures by the Stranski-Krastanov growth mode. Prog Cryst Growth Ch 33:423–471
34. Gong J (2012) Structure and surface chemistry of gold based metal catalysts. Chem Rev 112:2987–3054
35. Ding Y, Fan F, Tian Z, Wang ZL (2010) Atomic structure of Au-Pd bimetallic alloyed nanoparticles. J Am Chem Soc 132:12480–12486
36. Kittel C (2008) Introduction to solid state physics. Wiley, New Jersey
37. Okamoto H, Massalski TB (1985) The Au-Pd (gold-palladium) system. J Phase Equlib 6:229–235
38. Akita T, Hiroki T, Tanaka S, Kojima T, Kohyama M, Iwase A, Hori F (2008) Analytical TEM observation of Au-Pd nanoparticles prepared by sonochemical method. Catal Today 131:90–97
39. Dash P, Bond T, Fowler C, Hou W, Coombs N, Scott RWJ (2009) Rational design of supported PdAu nanoparticle catalysts from structured nanoparticle precursors. J Phys Chem C 113:12719–12730
40. Ferrer D, Torres-Castro A, Gao X, Sepulveda-Guzman S, Ortiz-Mendes U, Jose-Yacaman M (2007) Three-layer core/shell structure in Au-Pd bimetallic nanoparticles. Nano Lett 7:1701–1705
41. Hu JW, Li JF, Ren B, Wu DY, Sun SG, Tian ZQ (2007) Palladium-coated gold nanoparticles with a controlled shell thickness used as surface-enhanced Raman scattering substrate. J Phys Chem C 111:1105–1112
42. Lee AF, Baddeley CJ, Hardacre C, Ormerod RM, Lambert RM (1995) Structural and catalytic properties of novel Au/Pd bimetallic colloid particles: EXAFS, XRD, and acetylene coupling. J Phys Chem 99:6096–6102
43. Campbell CT (1990) Bimetallic surface chemistry. Annu Rev Phys Chem 41:775–837
44. Yang J, Yang J, Ying JY (2012) Morphology and lateral strain control of Pt nanoparticles via core-shell construction using alloy AgPd core toward oxygen reduction reaction. ACS Nano 6:9373–9382
45. Gan L, Heggen M, Rudi S, Strasser P (2012) Core-shell compositional fine structures of dealloyed Pt$_x$Ni$_{1-x}$ nanoparticles and their impact on oxygen reduction catalysis. Nano Lett 12:5423–5430
46. Wang C, van der Vliet D, More KL, Zaluzec NJ, Peng S, Sun S, Daimon H, Wang G, Greeley J, Pearson J, Paulikas AP, Karapetrov G, Strmcnik D, Markovic NM, Stamenkovic VR (2011) Multimetallic Au/FePt₃ nanoparticles as highly durable electrocatalyst. Nano Lett 11:919–926
47. Brydson R (ed) (2011) Aberration-corrected analytical transmission electron microscopy. Wiley, Chichester
48. Li ZY (2012) Scanning transmission electron microscopy studies on mono- and bimetallic nanoclusters. In: Johnston RL, Wilcoxon J (eds) Metal nanoparticles and nanoalloys. Elsevier, Amsterdam
49. Chantry RL (2013) Characterising the structure and properties of bimetallic nanoparticles. Ph. D. Thesis, University of Birmingham, UK
50. Williams DB, Carter CB (2009) Transmission electron microscopy. Springer, New York
51. Chantry RL, Siriwatcharapiboon W, Horswell SL, Khanal BP, Zubarev ER, Atanasov I, Johnston RL, Li ZY (2013) An atomistic view of the interfacial structures of Au-Core Rh and Pd-Shell nanorods. Nanoscale 5:7452–7457
52. Goris B, De Backer A, Van Aert S, Gómez-Graña S, Liz-Marzán LM, Van Tendeloo G, Bals S (2013) Three-dimensional elemental mapping at the atomic scale in bimetallic nanocrystals. Nano Lett 13:4236–4241

53. Herzing AA, Watanabe M, Edwards JK, Conte M, Tang ZR, Hutchings GJ, Kiely CJ (2008) Energy dispersive X-ray spectroscopy of bimetallic nanoparticles in an aberration corrected scanning transmission electron microscopy. Faraday Discuss 138:337–351
54. Egerton RF (2011) Electron energy-loss spectroscopy in the electron microscope, 3rd edn. Springer, New York
55. Lin YC, Sundholm D, Juselius J, Cui LF, Li X, Zhai HJ, Wang LS (2006) Experimental and computational studies of alkali-metal coinage-metal clusters. J Phys Chem A 110:4244–4250
56. Hohenberg P, Kohn W (1964) Inhomogeneous electron gas. Phys Rev 136:B864–B871
57. Kohn W, Sham LJ (1965) Self-consistent equation including exchange and correlation effects. Phys Rev 140:A1133–A1138
58. Bochicchio D, Ferrando R (2010) Size-dependent transition to high-symmetry chiral structures in AgCu, AgCo, AgNi, and AuNi nanoalloys. Nano Lett 10:4211–4216
59. Ferrando R, Fortunelli A, Johnston RL (2008) Searching for the optimum structures of alloy nanoclusters. Phys Chem Chem Phys 10:640–649
60. Heiles S, Johnston RL (2013) Global optimization of clusters using electronic structure methods. Int J Quantum Chem 113:2091–2109
61. Baletto F, Ferrando R (2005) Structural properties of nanoclusters: energetic thermodynamics and kinetic effects. Rev Mod Phys 77:371–423
62. Ducastelle F, Cyrot-Lackmann F (1970) Moments developments and their application to the electronic charge distribution of d bands. J Phys Chem Solids 31:1295–1306
63. Cleri F, Rosato V (1993) Tight-binding potentials for transition metals and alloys. Phys Rev B 48:22–33
64. Rosato V, Guillope M, Legrand B (1989) Thermodynamical and structural properties of f.c.c. transition metals using a simple tight-binding model. Philos Mag A 59:321–336
65. Baletto F, Mottet C, Ferrando R (2003) Growth of three-shell onion-like bimetallic nanoparticles. Phys Rev Lett 90:135504
66. Bochicchio D, Ferrando R (2013) Morphological instability of core-shell metallic nanoparticles. Phys Rev B 87:165435
67. Kuntová Z, Rossi G, Ferrando R (2008) Melting of core-shell Ag-Ni and Ag-Co nanoclusters studied via molecular dynamics simulations. Phys Rev B 77:205431–205438
68. Logsdail AJ, Johnston RL (2012) Interdependence of structure and chemical order in high symmetry (PdAu)N nanoclusters. RSC Adv 2:5863–5869
69. Gupta RP (1981) Lattice relaxation at a metal surface. Phys Rev B 23:6265–6270
70. Logsdail AJ, Paz-Borbón LO, Johnston RL (2009) Structures and stabilities of platinum-gold nanoclusters. J Comput Theor Nanosci 6:857–866
71. Baletto F, Mottet C, Ferrando R (2002) Growth simulations of silver shells on copper and palladium nanoclusters. Phys Rev B 66:155420
72. Tyson WR, Miller WA (1977) Surface free energies of solid metals-estimation from liquid surface-tension measurements. Surf Sci 62:267
73. Kresse G, Furthmüller J (1996) Efficient iterative schemes for ab initio total-energy calculations using a plane-wave basis set. Phys Rev B 54:11169
74. Kresse G, Joubert D (1999) From ultrasoft pseudopotentials to the projector augmented-wave method. Phys Rev B 59:1758
75. Ovari L, Bugyi L, Majzik Z, Berko A, Kiss J (2008) Surface structure and composition of Au-Rh bimetallic nanoclusters on $TiO_2(110)$: a LEIS and STM study. J Phys Chem C 112:18011–18016
76. Óvári L, Berkó A, Baláza N, Majzik Z, Kiss J (2010) Formation of Rh-Au core-shell nanoparticles on $TiO_2(110)$ surface studies by STM and LEIS. Langmuir 26:2167–2175
77. Okamoto H (1987) Phase diagrams of binary gold alloys. ASM International, Materials Park
78. Sneed BT, Kuo CH, Brodsky CN, Tsung CK (2012) Iodide-mediated control of rhodium epitaxial growth on well-defined noble metal nanocrystals: synthesis, characterization, and structure-dependent catalytic properties. J Am Chem Soc 134:18417–18426

79. Essinger-Hileman ER, DeCicco D, Bondi JF, Schaak RE (2011) Aqueous room temperature synthesis of Au-Rh, Au-Pt, Pt-Rh, and Pd-Rh alloy nanoparticles: fully tunable composition within the miscibility gaps. J Mater Chem 21:11599–11604
80. Chantry RL, Siriwatcharapiboon W, Horswell SL, Logsdail AJ, Johnston RL, Li ZY (2012) Overgrowth of rhodium on gold nanorods. J Phys Chem C 116:10312–10317
81. Ferrer D, Blom DA, Allard LF, Mejia S, Perez-Tijerina E, Jose-Yacaman M (2008) Atomic structure of three-layer Au/Pd nanoparticles revealed by aberration corrected scanning transmission microscopy. J Mater Chem 18:2442–2446
82. Fennell J, He DS, Tanyi AM, Logsdail AJ, Johnston RL, Li ZY, Horswell SL (2013) A selective blocking method to control the overgrowth of Pt on Au nanorods. J Am Chem Soc 135:6554–6561
83. He DS, Han Y, Fennell J, Horswell SL, Li ZY (2012) Growth and stability of Pt on Au nanorods. Appl Phys Lett 101:113102
84. Ismail R, Ferrando R, Johnston RL (2013) Theoretical study of the structures and chemical ordering of palladium-gold nanoalloys supported on MgO(100). J Phys Chem C 117:293–301
85. Baletto F, Ferrando R, Fortunelli A, Montalenti F, Mottet C (2002) Crossover among structural motifs in transition and noble-metal clusters. J Chem Phys 116:3856–3863
86. Pittaway F, Paz-Borbón LO, Johnston RL, Arslan H, Ferrando R, Mottet C, Barcaro G, Fortunelli A (2009) Theoretical studies of palladium—gold nanoclusters: Pd—Au clusters with up to 50 atoms. J Phys Chem C 113:9141–9152
87. Hünenberger PH (2005) Thermostat algorithms for molecular dynamics simulations. In: Holm C, Kremer K (eds) Advanced computer simulation, Advances in polymer science. Springer, Berlin, pp 105–149
88. Altman EI, Colton RJ (1994) Growth of Rh on Au(111): surface intermixing of immiscible metals. Surf Sci Lett 304:L400–L406
89. Chado I, Scheurer F, Bucher JP (2001) Absence of ferromagnetic order in ultrathin Rh deposits grown under various conditions on Gold. Phys Rev B 64:094410
90. Haftel MI, Rosen M, Franklin T, Hettermann M (1996) Molecular-dynamics investigation of early film growth of Pt/Au(100) and Au/Pt(100) and an interdiffusive growth mode. Phys Rev B 53:8007–8014
91. Grzelczak M, Pérez-Juste J, Rodriguez-Gonzalez B, Liz-Marzán LM (2006) Influence of silver ions on the growth mode of platinum on gold nanorods. J Mater Chem 16:3946–3951
92. Hernández-Fernández P, Rojas S, Ocón P, Gómez de la Fuente JL, San Fabión J, Sanza J, Peña MA, García-García FJ, Terreros P, Fierro JLG (2007) Influence of the preparation route of bimetallic Pt-Au nanoparticle electrocatalysts for the oxygen reduction reaction. J Phys Chem C 111:2193–2923

Struct Bond (2014) 162: 91–138
DOI: 10.1007/430_2013_135
© Springer International Publishing Switzerland 2013
Published online: 10 November 2013

Model Catalysts Based on Au Clusters and Nanoparticles

Niklas Nilius, Thomas Risse, Shamil Shaikhutdinov, Martin Sterrer, and Hans-Joachim Freund

Abstract Small Au particles have been shown to exhibit interesting catalytic properties. In an attempt to parallel catalytic studies on powder supports we have undertaken a series of model studies using oxide films as support. We address the formation of Au aggregates as a function of size starting from Au atoms to clusters and islands of larger size and as a function of the support. In addition we have studied different support materials such as alumina and iron oxide and we compare ultrathin and thicker oxide films of the same material (MgO). From a comparison of charge transfer through ultrathin films with the situation encountered in thicker films, we propose the use of dopants in bulk materials to control particle shape. We include the study of carbon monoxide adsorption on Au clusters of varying size. It is demonstrated how chemical modification (hydroxylation) of oxide supports influence particle growth and properties. Finally, we report on effects to study the processes involved in particle growth by wet impregnation in order to bridge the gap to catalyst preparation under realistic conditions. On that basis one may now compare properties of supported particles prepared in ultrahigh vacuum using physical vapor deposition with those prepared by wet impregnation.

Keywords Adsorption · Au clusters · Chemical modification · Dopants · Electronic properties · Oxide films · Wet impregnation

Contents

1 Introduction 92
2 Binding of Single Metal Atoms and Small Clusters to Oxide Surfaces 93
   2.1 General Considerations 93
   2.2 Role of Charge Transfer for Binding Single Adatoms to Oxide Thin Films 95

N. Nilius, T. Risse, S. Shaikhutdinov, M. Sterrer, and H.-J. Freund (✉)
Department of Chemical Physics, Fritz Haber Institute of the Max Planck Society, Faradayweg 4-6, 14195, Berlin, Germany
e-mail: freund@fhi-berlin.mpg.de

    2.3   Charge-Mediated Growth of Metallic Chains on Oxide Thin Films   99
    2.4   Development of Two-Dimensional Metal Islands on Oxide Thin Films 105
    2.5   Metal Growth on Doped Oxide Materials 110
3 CO Adsorption on Supported Au Particles 116
    3.1   CO Adsorption on Au on Thin MgO/Ag Films 117
    3.2   CO Adsorption on Au on Ultrathin Alumina and Iron Oxide Films 122
4 Nucleation and Bonding of Au on Chemically Modified Oxide Surfaces 124
    4.1   Gold Nucleation at Hydroxyl Groups 125
    4.2   Surface Science Approach to Supported Au Catalyst Preparation 129
5 Synopsis 132
References 132

1   Introduction

In 1984 Michael Mingos published two papers important for the present review [1, 2]. One paper was published in Account of Chemical Research in which he summarized the state of affairs concerning the prediction of structures of cluster compounds based on their total number of valence electrons. He and Ken Wade pioneered in the early 1970s [3] what has been called the Polyhedral Skeletal Electron Pair Theory, an extension of the VSEPR (valence shell electron pair repulsion theory) [4] of molecular compounds. This led then to setting up of electron counting rules, which connect the number of available valence electrons with stability criteria for particular cluster geometries. The second paper was published in Gold Bulletin entitled "Gold cluster compounds: are they metals in miniature?" The results of his thoughts about surface-clusters analogies may be summarized with two quotes from his paper: "The writer's personal view is that such analogies bear the same deficiencies as attempting to relate benzene to graphite and strained hydrocarbons such as cubane, C_8H_8, to diamond" and "These results suggest that those scientists studying the catalytic properties of crystallites of gold on inert supports should recognize that the clusters may change their geometries as a result of introducing substrates and changes in environment." The latter, in particular, is an important observation that many researchers in catalysis, even today, have not taken into consideration, seriously. It is one of the messages that this paper wants to convey: the flexibility of nanoparticles to respond to the environment, for example, the support, is the source for their reactivity. By controlling the environment and using the flexibility of nanoparticles one may change not only morphologies and structures but also charge states of active components. For the latter, the metal-oxide interface plays an important role and thus cannot be taken into consideration by the single crystal approach, so successful in the past, culminating in the 2007 Nobel Prize for Gerhard Ertl [5]. While solid-state theory has provided fundamental insights into processes and intermediates on single crystal surfaces using density functional calculations with periodic boundaries these approaches have to be substantially modified to allow for successful treatment of nanoparticles of relevant size on an oxide support. In this respect, also, calculations on free clusters may only be used for reference purpose. Our group has pointed this out in a number of reviews on modeling heterogeneous catalysts [6–9]. The present review goes somewhat beyond previous ones, as we attempt to make contact between model systems of various degrees of sophistication and the preparation of the same model systems via wet impregnation techniques. We have chosen Au supported on simple, non-reducible oxides as the example, as Au shows a pronounced catalytic activity when it exhibits nanoparticulate morphology and is in interaction with a support so beautifully pointed out in the late 1980s by Masatake Haruta [10, 11].

The chapter is organized as follows.

After a short introduction into our philosophy for preparing model systems of supported dispersed metal catalysts, we proceed by showing how one may build up Au particles from the individual atom to intermediate-sized Au clusters towards nanoparticles of a few nanometers in size. We will then proceed to larger islands and address the question how one may be able to learn about the convergence of cluster size towards representation of a fully developed metal. As we are working with thin oxide films as supports in order to be able to use charged (entities) particles as information carriers we also address the question of film thickness to represent the bulk situation. In fact, out of this approach, the interesting observation of the influence of dopants in the support arose and this problem will be addressed as well. A brief intermezzo on CO adsorption will also be included. As a first step towards the study of particle preparation via wet chemical techniques we have studied substrate modification, basically by hydroxylation, and its influence on the electronic and structural properties of metal deposits. Finally, we compare properties of deposits prepared via wet impregnation on oxides with systems prepared by physical vapor deposition on oxides using ultrahigh vacuum (UHV)-based techniques, and at the end conclusions within a synopsis will be presented.

2 Binding of Single Metal Atoms and Small Clusters to Oxide Surfaces

2.1 General Considerations

Adsorption on defect-free oxides is generally weak given the high degree of bond saturation at their surface and the large gap that governs their electronic structure [12–18]. Metal atoms deposited onto pristine oxides have essentially two means to interact with the surface. The first one, being accessible to all atoms, arises from van der Waals or dispersive forces, i.e., the adatom gets polarized in the Madelung field of the oxide and experiences dipolar coupling to the surface. Depending on the atom polarizability, the resulting adsorption energies are of the order of 0.5 eV or below for a single atom. The second interaction channel that is relevant for open-shell d and f-elements is direct hybridization between orbitals of the ad-species and the

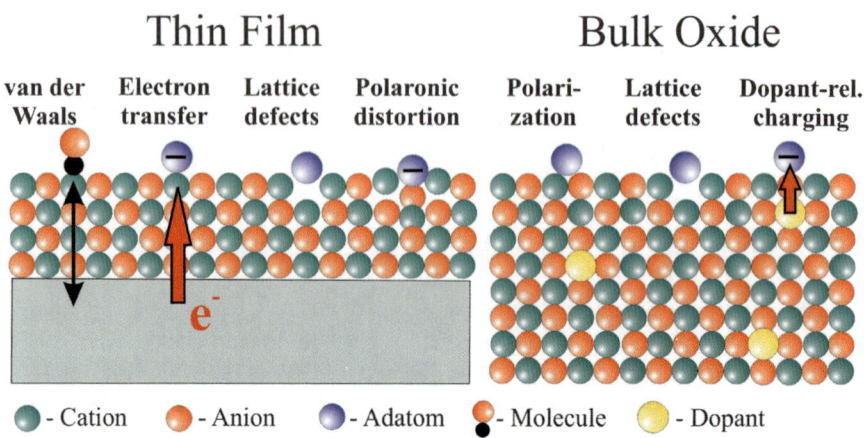

Fig. 1 Different binding mechanisms of metal adatoms on thin films and bulk oxides

oxide surface. Especially the overlap between the d-states of transition metal atoms and the 2p orbitals of the surface oxygen plays an important role and enables an increase in the metal-oxide bond strength to more than 1.0 eV [19–21]. Naturally, this channel is dominant for metals with partly filled d-shells (Cr, Mn, Fe) and loses influence for semi-noble and noble metals, e.g., Pt, Cu, Ag, and Au.

Oxide materials are never perfect and therefore surface defects need to be considered as potential binding sites for metal atoms [22]. Oxide defects often given rise to considerable variations in the electrostatic potential, which originate from unbalanced charges and cannot be screened due to the low density of free carries in the insulating material. In covalently bound oxides, dangling bond states may emerge at the defect site, reflecting the rigid lattice structure of the system that does not support bond reorganization. Whereas dangling bond states are highly susceptible to form covalent bonds to metal adatoms, electrostatic forces and charge transfer processes become relevant in the presence of charged defects in ionic oxides. Oxygen vacancies in MgO, for example, are able to exchange electrons with metallic adsorbates, which enable strong Coulomb attraction between both partners. A model case for this scenario is the interaction between a doubly occupied O-defect (F^0 center) and an Au atom, in which the gold turns anionic and binds with more than 3.0 eV to the surface [23]. In the opposite scenario, electron flow into an electron trap in the oxide lattice (e.g., an F^{2+} center) is observed for electropositive ad-species and governs for instance the adsorption of alkali metals to different surfaces. Defect-mediated interaction schemes exceed the binding potential of the regular surface by up to a factor of three, underlining the significance of such lattice irregularities for the nucleation and growth of metals on oxide materials.

A particularly interesting approach to modify the metal-oxide adhesion without generating surface defects is the insertion of charge sources directly into the oxide lattice (Fig. 1). Two approaches have been proposed in the literature and were successfully realized in experiments. In a first scenario, the bulk oxide support is transformed into a thin film grown on a metal single crystal [24–26]. The metal substrate acts as infinite charge reservoir and on readily exchange electrons with adatoms bound to the surface of the thin film. A different electron potential in both subsystems is a requirement for the charge transfer to occur; however, this situation is often fulfilled as thin oxide films usually alter the work function of the pristine metal beneath [27]. In the second scenario, doping with aliovalent impurity ions may be exploited to introduce charges into the interior of thick films and even bulk oxides. Also in this case, charge transfer into the ad-species has been revealed, the direction of which is given by the nature of the dopant in conjunction with the electronegativity of the adsorbate. Again, charge exchange is connected with a considerable increase of the metal-oxide adhesion. To give an estimate for the efficiency of charge-mediated interaction schemes, DFT calculations have revealed an increase in binding energy from 0.9 eV for neutral Au on defect-free MgO to 2.3 eV for Au^- species on thin MgO/Mo(001) films [28]. In the following, we will substantiate these general considerations with a number of experiments performed in our group. Whereas in the first part, the relevance of charge transfer for binding of single atoms to oxide surfaces is discussed, consequences on the growth and electronic properties of metal aggregates are in the focus of the later sections.

2.2 Role of Charge Transfer for Binding Single Adatoms to Oxide Thin Films

A first hint for the formation of charged adsorbates on thin oxide films came from low-temperature scanning tunneling microscopy (STM) experiments performed on 3 ML MgO/Ag(001) exposed to small amounts of Au [15]. Whereas on bulk oxides, gold shows a strong tendency for aggregation, mainly isolated atoms are detected on the thin film even at high gold exposure. Moreover, the ad-species self-assemble into a hexagonal superlattice on the film surface, with the Au–Au distance depending exclusively on the coverage and not on the MgO crystal structure (Fig. 2a). This unexpected ordering effect provides first evidence for the charged nature of the Au atoms on MgO films, as equally charged species tend to maximize their interatomic distance in order to reduce the mutual Coulomb repulsion. A similar phenomenon was found for alkali atoms on metal and semiconductor surfaces before and was assigned to a positive charging upon adsorption [29–31]. Conversely, the Au atoms on the MgO films charge up negatively, as their 6s orbital gets filled with electrons from the Ag(001) support below the oxide spacer layer. The charge exchange is enabled by the high electronegativity of gold in combination with a small work function of the Ag-MgO system, both effects promoting an electron release towards ad-gold [27, 28]. The experimentally deduced charging scenario was corroborated by DFT calculations that computed a Bader charge of $-1|e|$ at the Au atoms in addition to the expected increase in binding energy [32].

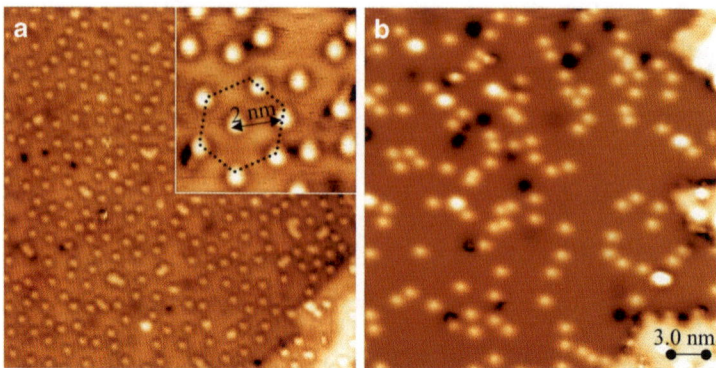

Fig. 2 STM image of (**a**) single Au and (**b**) single Pd adatoms on a 3 ML thick MgO film grown on Ag(001) (30 × 30 nm^2). While the negatively charged Au atoms self-assemble into a hexagonal pattern (see *inset*), the neutral Pd atoms are randomly distributed on the oxide surface and show a large tendency to aggregate into clusters. A negative charge on the Au species is also compatible with a distinct sombrero-like shape observed in low-bias STM images. *Black dots* in both images are attributed to defects in the oxide film [32]

However, charge-mediated adsorption on oxide thin films is rather the exception than the rule and requires a favorable electronic structure of the adsorbate. Counterexamples are Pd atoms that, in contrast to Au atoms, do not possess a low-lying affinity level that can be filled with extra electrons from the MgO/Ag support. In fact, the Pd 5s, as the lowest-unoccupied atomic orbital, is located well above the Fermi level of the thin-film system and therefore not accessible to electron transfer. As a result, the Pd atoms remain neutral upon adsorption and do not experience any self-ordering on the MgO surface (Fig. 2b) [33]. Apart from their random distribution, the Pd atoms exhibit a high tendency to assemble into small aggregates even at low coverage, mimicking the anticipated binding behavior of metals on bulk oxides in the absence of charge transfer.

Charge-driven adsorption schemes were observed for many other thin-film systems, using not only metal adatoms but also molecular species [34, 35]. Particularly interesting in this context are experiments on a 5 Å thick alumina film grown on NiAl(110) [36], as the results provide direct insight into the direction of the charge transfer and the number of exchanged electrons (Fig. 3) [37]. Although charge transfer into the Au atoms prevails also in this case, the interaction involves a certain bond reorganization in the oxide and therefore deviates from the more simple MgO/Ag(001) case [32]. The Au atoms bind exclusively to Al^{3+} ions in the alumina surface, and no adsorption to anionic lattice sites is revealed (Fig. 3a). Upon bond formation, the Al ion below the gold is lifted above the surface plane. This upward motion of the Al^{3+} leads to a homolytic rupture of the bond to the oxygen in the layer beneath. The Au atom takes up the electron donated by the Al ion, while the electron deficiency at the oxygen is balanced by back-bonding it to an Al atom in the NiAl(110) metal surface. The involvement of the metal substrate thus reinforces the interaction between adatom and surface, whereby not only

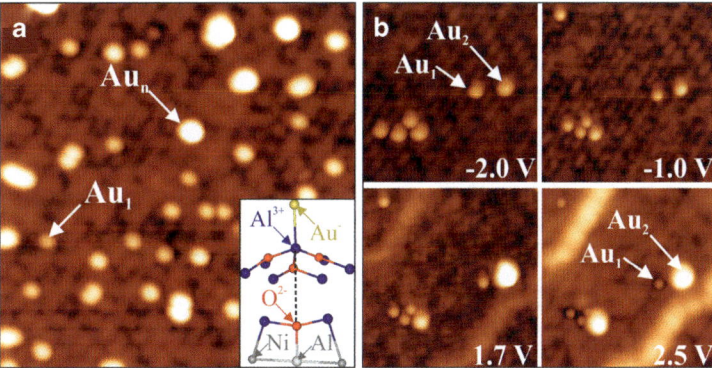

Fig. 3 (**a**) STM image of Au monomers and small aggregates on alumina/NiAl(110) (-1.5 V, 18×18 nm^2). The inset shows a ball-stick model of a monomer bound to an Al surface ion. (**b**) Image series, showing a small alumina region as a function of bias voltages (15×15 nm^2). The various Au species undergo large contrast changes as a function of the imaging conditions [39]

charges are shuttled through the oxide, but the film reorganizes its internal bonding network in order to accommodate the gold. In numbers, a participation of the NiAl support leads to an increase of the Au/alumina adsorption strength from 1.0 to 2.1 eV, emphasizing once more the importance of charge transfer [37]. It should be added that the Au-induced bond cleavage in the alumina film can be considered as an extreme form of a polaronic lattice distortion, being a typical response of an ionic oxide to stabilize charged adsorbates on its surface [38].

The extra charges localized in alumina-bound Au atoms are responsible for a distinct contrast evolution in STM images taken as a function of sample bias (Fig. 3b) [39]. At negative voltage (probing the occupied sample states), Au monomers are imaged as circular protrusions of $\sim$6 Å diameter and $\sim$1 Å apparent height. This height value decreases at positive polarity and finally reverses at 3.0 V, when the Au atoms show up as shallow depressions in the surface (Fig. 4a). At intermediate voltages (between +2.0 and +3.0 V), the ad-species are surrounded by a characteristic sombrero ring, again being the typical fingerprint of charged adsorbates (Figs. 2a, 3b) [40, 41]. According to the Tersoff–Hamann theory for vacuum tunneling, a negative contrast indicates a lower state density of the adatom as compared to the surrounding oxide, which forces the tip to approach the surface in order to maintain the preset tunnel current [42]. This decrease in state density is easily explained with a local upward bending of the oxide bands, in response to the negatively charged Au atom. Whereas tunneling is efficient in regions away from the Au, as the oxide conduction band provides suitable final states, it remains blocked at the adatom site due to upshift of the band onset at this location (Fig. 4b, inset). Such upward bending is compatible with the accumulation of negative surface charges, exerting a repulsive interaction on the alumina electronic states [43], while positive charges would result in a downward bending of the bands

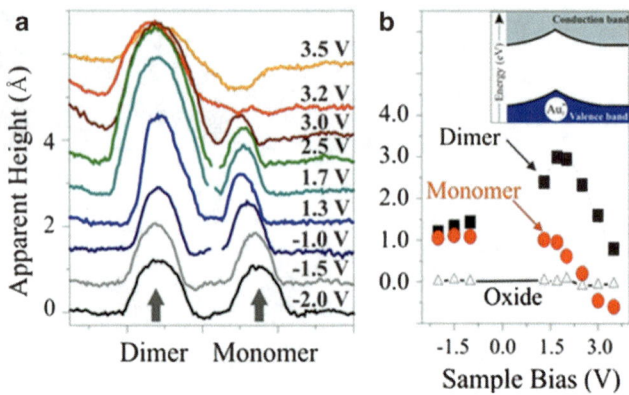

Fig. 4 (**a**) Line profiles and (**b**) apparent heights of an Au monomer and dimer on alumina/NiAl (110) taken as a function of the bias voltages. The monomer turns into a depression above 3.0 V, as its negative charge triggers an upward bending of the alumina bands (see *inset*). The dimer appears bright due to tunneling through its LUMO located at 2.5 eV [37]

and does not explain the experimental observations [44]. Assuming a screened Coulomb interaction:

$$V = -\frac{q}{4\pi\varepsilon_0\varepsilon_r r} \tag{1}$$

with an alumina dielectric constant of $\varepsilon_r = 10$ and an Au excess charge of $q = -1|e|$, the upward bending calculates to +1.8 eV in distance $r = 1$ Å from the anion [43]. This value matches the bias window of 3.0–4.5 V, in which the Au atom appears with negative contrast in the STM images. The contrast reversal observed for Au atoms on alumina thin films therefore provides additional evidence for the charge transfer that accompanies the interaction of gold with oxide thin films [28]. We note in passing that no contrast reversal is observed for Au dimers, which appear bright over the entire bias range (Fig. 4b). The reason is the Au_2 electronic structure, which exhibits an intrinsic electronic state at 2.5 eV that compensates for the LDOS reduction due to band bending [39].

Charge transfer across oxide thin films does not necessarily lead to the formation of anionic gold species. In fact, the direction of charge transfer depends on the position of the Au affinity level (the Au 6s orbital) with respect to the Fermi level of the metal-oxide system. Because the level position of individual Au adatoms is largely governed by the vacuum energy, which is, in turn, given by the work function ϕ of the metal-oxide system, it is the latter quantity that determines the direction of the charge flow [38]. In general, low work function systems promote an electron flow into the gold, while the opposite trend is revealed for systems with high ϕ values. The two oxide films discussed so far, MgO/Ag and Alumina/NiAl, are characterized by a low work function, because the oxide layer prevents electron spill out from the metal surface and therewith removes a main reason for the high work function of non-alkali metals. A counterexample is the FeO thin film that can be grown on Pt(111). Already the bare Pt(111) surface features an exceptionally high work function and this situation does not change upon FeO deposition. The reason is that the FeO film is of polar nature and features an intrinsic surface dipole with the negative side pointing towards the surface (oxygen termination) [45]. In this particular case, the Au atoms lose electrons to the film and become positively charged upon adsorption [46]. Although the direction of the charge flow is opposite to the one on magnesia and alumina, the resulting binding principles are similar and arise from a combination of electrostatic interactions and polaronic lattice distortion of the ionic oxide in presence of the charged ad-species.

2.3 Charge-Mediated Growth of Metallic Chains on Oxide Thin Films

The charge transfer that governs the binding of Au monomers to oxide thin films affects also the aggregation behavior of the gold at higher exposure. At cryogenic temperature, aggregation of equally charged adatoms is inhibited due to the Coulomb repulsion, giving rise to the formation of ordered adatom patterns as shown in Fig. 2a. The repulsion is overcome, however, when dosing the gold at elevated temperature, e.g., 100 K. In this case, the Au atoms assemble into 1D atom chains at low coverage, as observed on both magnesia [47, 48] and alumina films [40]. The smallest aggregate is a flat-lying dimer with 9 Å apparent lengths, while longer chains contain between three and seven atoms and are 12–22 Å long (Fig. 5). The development of 1D chains on thin films is in contrast to the common behavior on bulk oxides, where 3D clusters are energetically favorable at any coverage due to the weak metal-oxide adhesion [26, 49].

The development of Au atom chains seems unexpected at first glance as the number of stabilizing Au–Au bonds is small with respect to 2D and 3D aggregates. However, the linear atom arrangement is in agreement with the charged nature of the aggregates. Similar to the monomer case, electron transfer through the oxide film into the Au cluster is active and increases the electrostatic coupling to the oxide lattice. Conversely, the charge transfer leads to a Coulomb repulsion in the aggregate that may be minimized when the extra electrons are separated over large distances. Minimization of Coulomb repulsion in electron-rich aggregates is therefore the fundamental reason for the development of 1D cluster shapes, and no 1D chains are to be expected in the absence of charge transfer through the oxide films [47, 48].

Also for Au aggregates bound to MgO/Ag(001) [47] and MgO/Mo(001) films [50], the concept of charge transfer has been verified by DFT calculations. Two configurations have been identified for the Au dimer, a neutral and upright one that binds to an O^{2-} ion in the MgO surface and a flat-lying, negatively charged species (Bader charge $-0.8|e|$) that bridges two Mg^{2+} or two hollow sites. The flat-lying

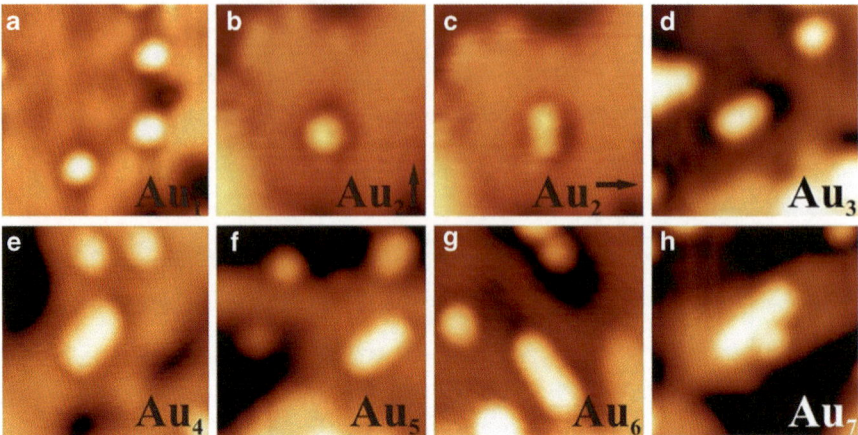

Fig. 5 STM images of Au monomers, dimers, and different chains on 2 ML MgO/Ag(001) (-0.4 V, 19×19 nm^2). For the dimer, an upright (Au$_2\uparrow$) and a flat-lying isomer is observed (Au$_2\rightarrow$) [48]

configuration is thermodynamically preferred by 0.34 eV, a finding that explains its abundance in STM images of the MgO film (Fig. 5) [47]. The preference of charged over neutral Au clusters becomes even more pronounced for larger aggregates [48]. A Au$_3^-$ chain for example (Bader charge $-1|e|$) has a higher binding energy than various neutral isomers (Fig. 6). This finding holds for Au tetramers, where a doubly charged Au$_4$ chain (Bader value: $-1.6|e|$) is energetically preferred over different rhomboidal structures that can be found on bulk MgO(001). Note that the two extra electrons in the Au$_4$ chains are localized at the terminal atoms, emphasizing the shape-determining role of the Coulomb repulsion. With increasing atom count, 2D Au aggregates gain stability with respect to 1D configurations, as the formation of additional Au–Au bonds outweighs the energy surplus due to charge delocalization [48]. Whereas, a Au$_5^{2-}$ chain - having two negative charges - is still iso-energetic with a flat-lying Au$_5^-$ sheet on MgO/Ag(001), a 2D Au$_6$ island has already a lower energy than the corresponding linear structure (Fig. 6). The critical atom number at which the 1D $\rightarrow$ 2D dimensionality crossover occurs has been determined to five/six, both in experiment and theory.

So far, the amount of charge transfer into oxide-supported Au clusters has proven mainly by DFT calculations. However, this quantity can be accessed also by experiment, if the electron filling of specific quantum-well states (QWS), in particular, of the highest-occupied (HOMO) and the lowest-unoccupied (LUMO) state, can be measured with STM conductance spectroscopy. How this technique is exploited to quantify the charge transfer into Au chains on alumina thin films is demonstrated in the following (Fig. 7) [37, 40].

Also on alumina films, gold spontaneously forms linear clusters that are 8–22 Å long and contain between two and seven atoms [37, 40]. These Au chains exhibit a particularly simple electronic structure that arises from the overlap of Au 6s and 6p

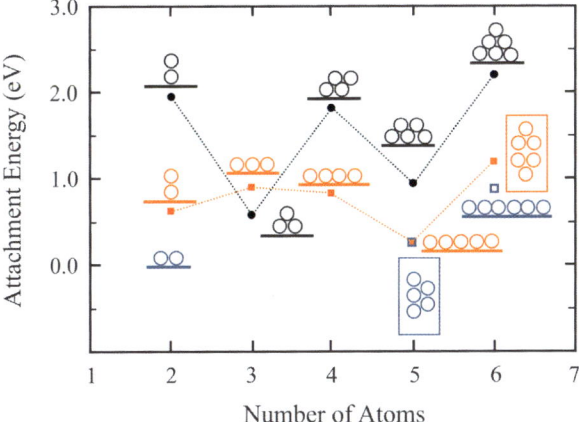

Fig. 6 Attachment energies to form the displayed Au clusters, calculated for bulk MgO(001) (*black circles*) and 2 ML MgO/Ag(001) films (*orange squares*). The *blue squares* depict the second most stable isomers on the thin film. On the oxide film, linear configurations are energetically preferred up to Au₅, when planar isomers become more stable. The odd-even oscillations in the attachment energy are related to the high energy of open-shell systems containing an unpaired electron. They are less pronounced on thin MgO films due to the screening influence of the Ag support [48]

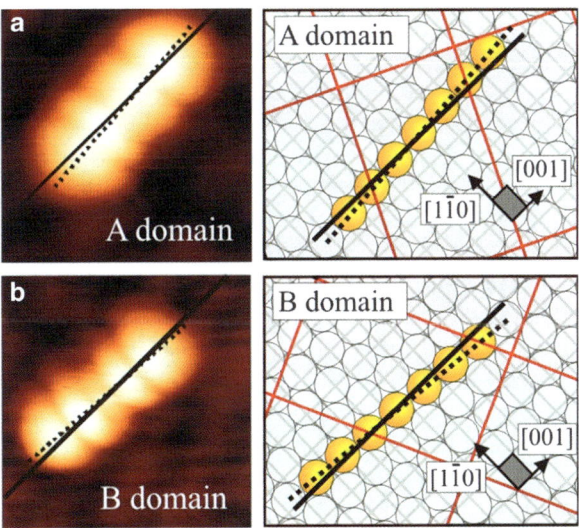

Fig. 7 STM image and structure model for Au₇ chains bound to the two reflection domains in the alumina/NiAl(110) film (5 × 5 nm²). The chain orientation and the NiAl [001] direction are indicated by *dashed* and *solid lines*, respectively. Although the two domains are tilted against each other by 48°, the chains closely follow the NiAl [001] direction in both cases. The sketches on the *right side* show the Al_2^{3+} top layer (*open spheres*), the Au chain atoms (*yellow, filled spheres*), as well as the NiAl and alumina unit cells, being depicted with *grey* and *red lines*, respectively [40]

Fig. 8 Spin-averaged
LDOS and orbital shapes for
(**a**) an Au monomer, (**b**) a
dimer, and (**c**) a trimer on
alumina/NiAl(110), as
calculated with DFT. The
black line marks the total Au
LDOS; the *blue* and *red
lines* denote the s- and
p_z-contributions,
respectively. The *dashed
line* depicts the alumina
states. (**d**, **e**) Experimental
dI/dV spectra and
topographic images of an Au
monomer and a dimer.
(**f**) Trimer spectrum with
dI/dV images taken at the
two peak positions. All
images are 4.5×4.5 nm^2 in
size; the set point for
spectroscopy was set to
3.0 V

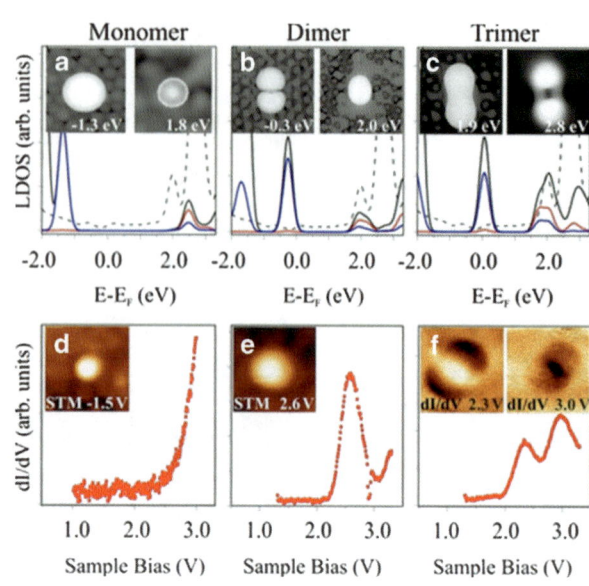

orbitals to QWS formed in the 1D potential well [43, 51]. There, the HOMO of the
Au monomer derives from the Au 6s orbital and is doubly occupied ($E_{bind} = -1.3$ eV, Fig. 8a). This double occupancy reflects the negative charge on the
adatom, as gas-phase Au features an Au 6s^1 ground state. The LUMO is a p_z-like
state that locates at +2.5 eV above E_F, hence inside the alumina conduction band. In
Au dimers, the two 6s orbitals hybridize and form two new states at -0.3 and
-1.5 eV. Both states are filled, which brings the total number of s-electrons in the
dimer to four. As two of these electrons are transfer electrons from the NiAl
substrate, the Au$_2$ is twofold negatively charged. The Au$_2$ LUMO shifts to
+1.9 eV due to the superposition of the two p_z states of each monomer (Fig. 8b).
In linear Au$_3$, a third s-like state appears directly at the Fermi level, while the two
other QWS shift to -1.8 and -2.5 eV. Due to the half-filled nature of the HOMO,
five s-electrons occupy the Au$_3$ valence orbitals, three intrinsic Au 6s electrons and
two transfer electrons from the metal beneath. The lowest-unoccupied Au$_3$ states
are the symmetric and antisymmetric combination of the 6p$_z$ orbitals, giving rise to
two QWS at +1.9 and +2.8 eV, respectively.

In extension of this concept, the electronic structure of an m-atom Au chain
arises from the consecutive splitting of the 6s- and 6p-like states and results in the
development of m mainly occupied s-derived QWS and m empty states with
p_z-character (Fig. 9). The s-like states display the characteristic properties of a
particle-in-a-box system with infinite walls [43]. The eigenfunctions are sinusoidals
($\psi_n = \sin k_n x$), defined by a wave-number k_n that is proportional to the inverse
box-length L and a quantum number n: ($k_n = n\pi/L$). The resulting electron

Fig. 9 (a) Conductance spectra of an Au_7 chain and the bare oxide film. The *insets* display dI/dV maps taken at the peak positions of the respective spectrum (5×5 nm²). (b) Calculated LDOS for an Au_7 chain. The *blue line* denotes the seven Au 6s-like QWS (termed S1–S7), while the *red line* depicts the 6p state density. The quantized nature of the 6p states is not resolved due to the small energy separation between adjacent levels. The energy position is however marked by arrows (P1–P7). The calculated orbital shapes for the fourth to sixth s-like and the first to third 6p-like QWS are shown as *insets* [39]

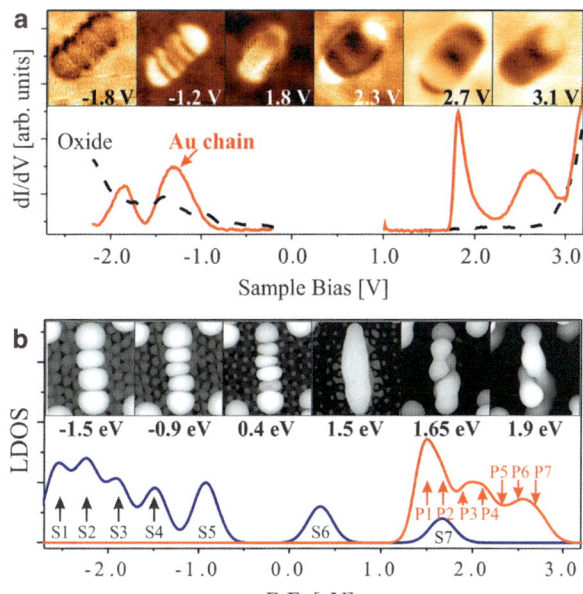

distribution in the nth QWS (ψ_n^2) is given by n maxima separated by n-1 nodal planes along the chain axis, while the eigen-energies follow a parabolic dispersion relation:

$$E_n = E_0 + \frac{\hbar^2}{2m_{\text{eff}}} \left(\frac{\pi n}{L}\right)^2 \tag{2}$$

The electronic signature of ultra-small Au aggregates discussed above has been reproduced experimentally by STM conductance spectroscopy (Fig. 8d–f) [39]. The Au dimer displays one dI/dV maximum at +2.6 V, while two maxima at +2.3/+3.0 V are detected for the trimer. These peaks are compatible with the lowest p_z-like QWS in the Au aggregates. This assignment is supported by dI/dV maps of the trimer taken at the respective peak positions. While a homogeneous dI/dV intensity distribution is revealed for the lower state, reflecting the constant density probability of the ground-state p_z orbital, a region of suppressed conductance marks the nodal plane in the first excited p_z-derived state (Fig. 8f). Note that the LUMO of the monomer was not detected, most likely because of the overlap of this state with the alumina conduction band.

Also conductance spectra of longer Au chains are in agreement with the electronic structure sketched above [37, 39]. In positive-bias spectra, two dI/dV peaks are observed for Au_4 and Au_5 chains, while three maxima are resolved for the gold heptamer (Fig. 9) [40]. Based on their energy position, the maxima are assigned to

HOMO   Topography   HOMO   Topography   Model

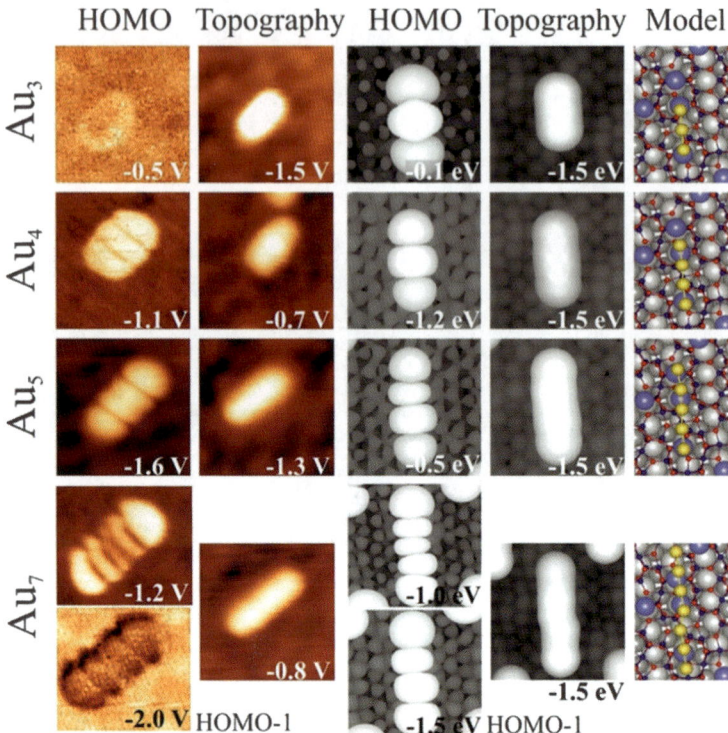

Fig. 10 Experimental and calculated topography, HOMO shape, and model structure for Au$_3$ to Au$_7$ chains on alumina/NiAl(110). All images are 5.0 × 5.0 nm^2. The HOMO-1 for the Au$_7$ is shown in addition. Measured chain lengths are 9, 12, 15, and 22 Å; calculated distances between first and last chain atom amount to 5.3, 7.8, 10.5, and 15.5 Å. To compare theoretical and exp. lengths, 2–3 Å should be added to both sides of the chain to account for the diffusivity of the 1D orbitals [37]

the unoccupied 6p$_z$-like QWS, the onset of which has been determined to 1.8 eV. As the energy spacing between adjacent 6p$_z$-levels is relatively small, the corresponding dI/dV maps do not show the unperturbed symmetry of isolated QWS but always result from a superposition of adjacent levels. In contrast to ultra-small aggregates, filled QWS are detected in the longer Au chains as well as the tetramer displays two dI/dV peaks below E_F, whereby the upper one at -1.1 V (the HOMO) exhibits three electron-density maxima and two nodal planes along the chain axis (Fig. 10). The HOMO is therefore assigned to the third s-like QWS, being characterized by three density maxima. Assuming double occupancy of all QWS below the Fermi level, the total number of s-electrons in this ad-chain would be six. While four electrons are provided by the incoming Au atoms, the other two originate from the NiAl support, rendering the Au$_4$ chain twofold negatively charged. With the same arguments, the Au$_5$ chain was found to contain three extra electrons, as its HOMO at -1.6 V exhibits four intensity maxima and is thus assigned to the fourth QWS. Only five of the eight s-electrons are intrinsic to the Au atoms in this case, while the other three originate from the substrate again. For the Au_7, being the longest chain observed experimentally, the fourth and fifth QWS are detected at -1.8 and -1.3 V, respectively. In agreement with the particle-in-the-box model, the lower state displays four density maxima and the higher five, along the chain axis in the dI/dV maps (Fig. 10). The Au_7 chain has consequently five occupied QWS carrying ten s-electrons in total, which brings the number of transfer electrons from the NiAl to three again. It should be mentioned that no six-atom chains have been identified in the experiment, which might be explained with an unfavorable magnetic ground state of the linear Au_6 arising from an unpaired electron at E_F.

Both the experimental and theoretical behaviors of s-like QWS in different Au chains are in good agreement with the particle-in-the-box model sketched above. Fitting the computed level energies to a parabolic dispersion relation yields a potential depth of $E_0 = -2.65$ eV and an effective electron mass $m_{eff} = 0.85$ m_e. It is interesting to note that Au chains assembled directly on the metallic NiAl(110) support have a smaller electron mass of 0.5 m_e [51, 52]. This finding indicates higher electron mobility in the metal-supported chains, despite a somewhat larger Au–Au distance in that case (2.89 Å on NiAl versus 2.6 Å on alumina). The difference reflects the role of indirect coupling between the Au atoms, mediated by electronic states in the NiAl substrate. Naturally, this contribution is missing on the insulating alumina film [53].

2.4 Development of Two-Dimensional Metal Islands on Oxide Thin Films

The charge transfer through a thin oxide spacer prevails also for larger Au aggregates and keeps controlling their geometry and electronic structure. On 2 ML MgO/Ag(001) films, for example, gold first forms flat, single-layer islands and develops a nearly complete wetting layer with increasing exposure (Fig. 11) [15]. The formation of 2D islands is in sharp contrast to the 3D growth that is typically observed on bulk oxides [16, 26, 54]. It reflects the tendency of gold to increase the contact area with the oxide film, as this maximizes the charge transfer into the gold affinity levels and therewith the strength of the metal-oxide adhesion. Similar to Au monomers and chains, the reinforcement of the Au bonding results from increased electrostatic and polaronic interactions upon charge transfer [55]. As a rough number, the average charge transfer per adatom has been calculated to be $-0.2|e|$ for a close-packed Au layer on 2 ML MgO/Ag(001) [56].

Especially in larger 2D gold islands, the excess electrons are not homogenously distributed but show a preference to accumulate at the island perimeter [57]. The reason for this particular charge distribution is similar to the one that leads to the development of 1D chains at small atom numbers. The excess electrons repel each other, which increases the internal Coulomb repulsion in the island and hence the total energy. To minimize this Coulomb term, the extra charges maximize their

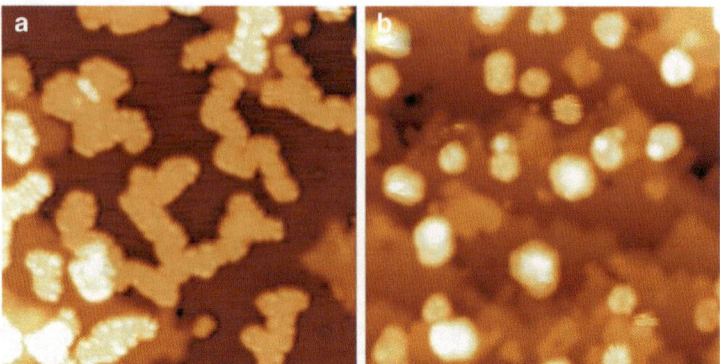

Fig. 11 STM images of Au deposits on (**a**) 3 ML and (**b**) 8 ML MgO/Ag (001) (30×30 nm^2). Due to the charge-mediated adhesion, Au wets the ultrathin MgO film. In the case of thicker films, the electron transfer is inhibited and 3D gold particles form on the oxide surface [15]

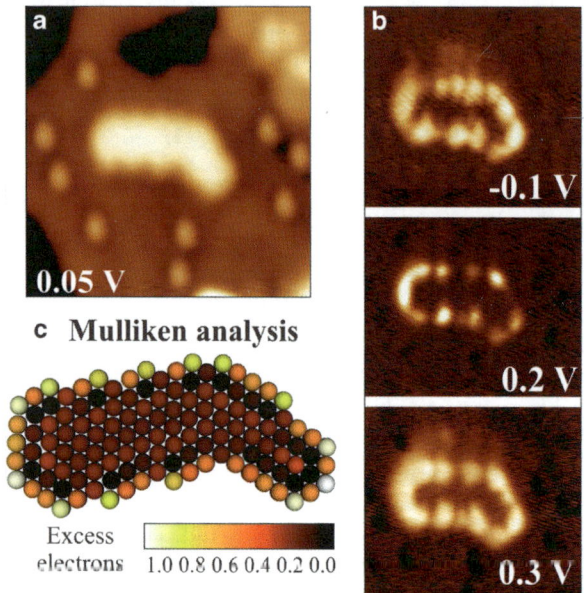

Fig. 12 (**a**) STM topographic image and (**b**) dI/dV maps of a planar Au island on 2 ML MgO/Ag (001) taken at the indicated bias voltage (US = 0.2 V, 25×25 nm^2). (**c**) Corresponding charge distribution as calculated with a Density Functional Tight Binding approach [57]

mutual distance, which promotes their localization along the island edges. The presence of a negatively charged perimeter can be deduced directly from STM conductance maps of the 2D Au islands, as shown in Fig. 12. Especially around the Fermi level, the island edges appear with enhanced dI/dV contrast, indicating a higher density of states as compared to the island interior. The contrast enhancement at the edge is compatible with specific edge states that enable accommodation of the extra electrons that have been transferred through the oxide film into the gold sheets. This mechanism has been corroborated by tight-binding DFT calculations, which allow for an explicit treatment of the atomic structure and the edge configuration even of extended metal nanostructures [57]. Also, the computed charge-density plots unambiguously demonstrate the localization of electronic states along the island perimeter (Fig. 12). These states are able to store one extra electron per low-coordinated edge atom and get filled up with transfer electrons although the island interior remains neutral. This charge localization in the low-coordinated edge atoms renders the 2D Au islands supported on thin oxide films particularly interesting for adsorption and chemical reactions involving electron-accepting molecules (Lewis acids) [58].

Also for flat Au islands on MgO/Ag(001), STM conductance spectroscopy can be exploited to determine their charge state [56]. As discussed before, the development of well-defined QWS is the necessary precondition for this analysis, whereby symmetry and electron filling of the QWS have to be compared to the findings in neutral aggregates. As electron confinement in a 2D potential is more sensitive to structural irregularities, Au islands with suitable QWS need to be highly symmetric and free of defects.

A particularly instructive example shows an ultrasmall Au cluster with ~10 Å diameter and 0.8–0.9 Å apparent height grown on a 2 ML MgO/Ag(001) film (Fig. 13) [56]. In low-bias STM images, mainly the cluster morphology is revealed, as no eigenstates of the aggregate are available in the probed energy window. At slightly higher bias, the apparent cluster height doubles and flowerlike protrusions emerge in the image. This bias-dependent contrast change provides evidence that tunneling is now governed by the electronic and not the topographic properties of the nanostructures. More precisely, a distinct eigenstate, the LUMO of the Au aggregate, becomes accessible to the tunneling electrons and dominates the image contrast at positive bias. A similar observation is made at negative bias, when the Au HOMO moves into the bias window and a comparable "nano-flower" becomes visible in the STM. The two observed QWS closely resemble the eigenstates of a free-electron gas confined in a 2D parabolic potential (Fig. 14). They are derived from the Au 6s states of the participating atoms and preserve their characteristic symmetry, as they neither mix with the states of the wide-gap oxide material nor with the Au 5d and 6p states positioned at much lower or higher energy, respectively [43, 59]. The symmetry of the states is defined by the angular momentum quantum number m, being a measure of the number of nodes in the 2D electron-density probability [60]. The two lowest QWS in Fig. 13 feature four nodal planes, which corresponds to an m of 4 or, equivalently, to a state with G-symmetry.

Again, the STM reveals not only the symmetry of the eigenstates, but via the spectroscopic channel also their energy position. For the Au cluster shown in Fig. 13, the HOMO and LUMO are clearly identified as dI/dV peaks at -0.4 and $+0.8$ V, respectively, separated by a region of zero conductance of 1.0 V width.

Fig. 13 (**a**) Topographic
and (**b**) dI/dV image of a
symmetric Au cluster on
2 ML MgO/Ag(001) taken
at the given sample bias
$(3.9 \times 3.9 \text{ nm}^2)$.
The corresponding dI/dV
spectra are shown in
addition (*blue* and *cyan*
curves: *top* and *left part* of
the cluster). (**c**) Calculated
HOMO and LUMO shape
as well as structure model of
an Au_{18} cluster on MgO/Ag
(001). Perfect match
between experimental and
theoretical cluster
properties indicate
the identity of both
aggregates [56]

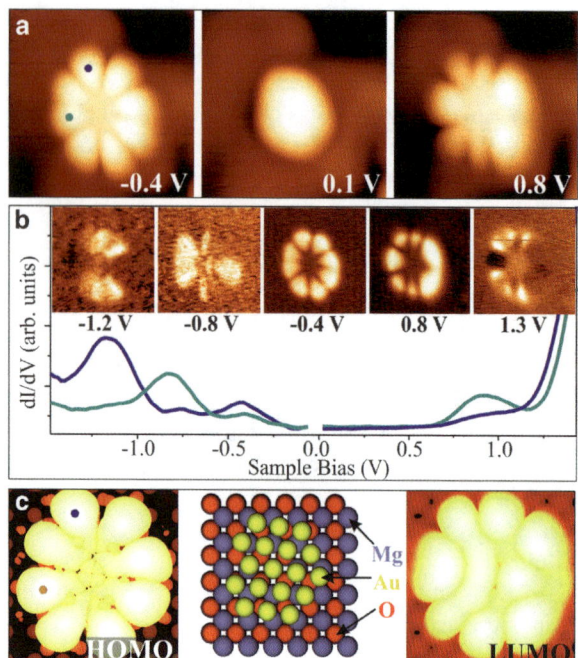

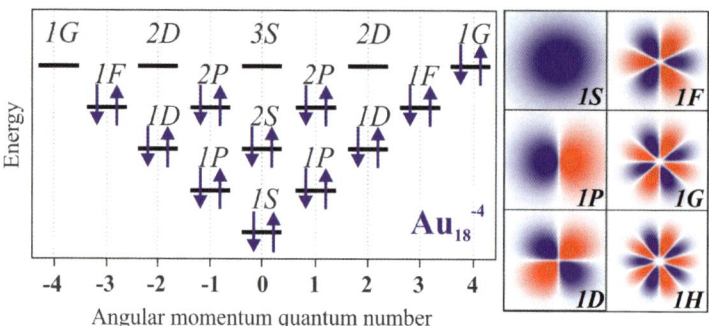

Fig. 14 Energy levels (*left*) and orbital shapes (*right*) of the free electron eigenstates in a 2D parabolic potential. The electron occupancy determined for a particular Au cluster on 2 ML MgO/Ag(001) (here the Au18 cluster) is depicted by *arrows*

Also higher QWS show up in the dI/dV spectra of the cluster. The HOMO-1 at
−0.8 V and the HOMO-2 at −1.2 V are both of P-symmetry with the nodal plane
pointing in two orthogonal directions. Note that higher and lower states cannot be
resolved as an increasing overlap with the MgO electronic states renders their
identification difficult.

With this experimental input, the electronic structure of the gold nanoisland and
more importantly its charge state can be determined by comparing the measured

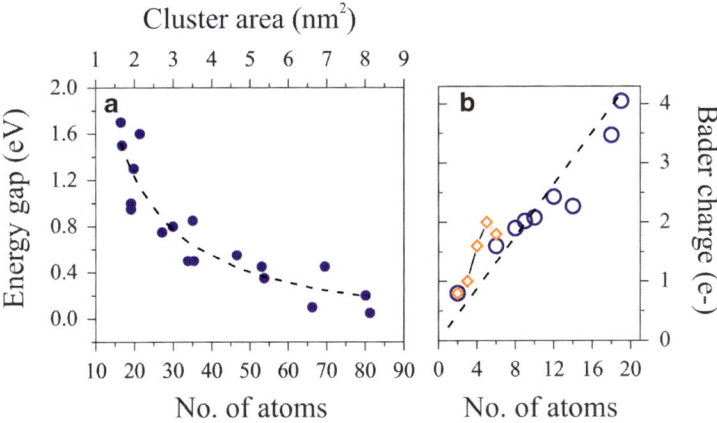

Fig. 15 (**a**) Experimental HOMO–LUMO gap for differently sized Au clusters on MgO/Ag(001). The atom number is deduced from the measured particle area Ω, using DFT results as a reference for the de-convolution procedure. The *dashed line* is a fit of the data to the inverse cluster size. (**b**) Calculated number of excess electrons in linear (*diamonds*) [48] and 2D Au clusters on 2 ML MgO/Ag(001) [56]. The *line* depicts the charge accumulation of 0.2|e| per atom, being computed for a compact Au layer on 2 ML MgO/Ag(001)

orbital shapes with DFT calculations for possible sample clusters [56]. Following an extensive theoretical search, the experimental signature shown in Fig. 13 could be matched on the properties of a planar Au_{18} cluster. Its structure is derived from a magic-size Au_{19} cluster with one missing corner atom. In agreement with experiment, the HOMO and LUMO are two 1G orbitals located in the fifth shell of the harmonic potential, whereas the HOMO-1 and HOMO-2 are the 2P-like QWS in the fourth shell (see level scheme in Fig. 14). The missing atom with respect to a symmetric Au_{19} gives rise to a slight asymmetry in the orbital shapes that becomes particularly evident for the LUMO. Counting the number of orbitals from the 1S-like ground state to the 1G HOMO, as plotted in Fig. 14, we find eleven filled valence states for the Au_{18} cluster being occupied by a total number of 22 s-electrons. As each of the 18 Au atoms adds only a single 6s electron to the delocalized QWS, there is a difference of four electrons to the total electron count determined by experiment. As discussed in detail for the Au chains, the missing four electrons are introduced via charge transfer from the MgO/Ag interface into the Au island. This charging effect is corroborated by a DFT Bader analysis, yielding a value of $-3.54|e|$ for the Au_{18} cluster, but matches also the average transfer of 0.2|e| per atom as calculated for dense-packed Au layers on thin MgO films [56].

Similar procedures have been carried out for many other Au aggregates on the MgO/Ag(001) system [56]. In all cases, a negative charging has been revealed, verifying the charge-mediated binding concept for Au islands on thin oxide films [7, 38, 49]. The number of transfer electrons was found to be proportional to the atom number in the 2D cluster or synonymously to the interfacial contact area (Fig. 15). Interestingly, the accumulated charge per Au atom is generally be lower in 2D aggregates as compared to 1D chains, which might be rationalized by the higher efficiency of the linear systems to distribute the excess electrons and lower the internal Coulomb repulsion [48]. The energetic preference for chains diminishes, however, for clusters containing more than 6–7 atoms, inducing a crossover to 2D island shapes and later to 3D structure. It should be mentioned that also the HOMO–LUMO gap shows a strong dependence on the atom number per cluster (Fig. 15). Whereas Au aggregates containing 10–15 atoms have experimental gap sizes of around 1.5 eV, the HOMO–LUMO gap closes for more than 100 atoms in the cluster, reproducing earlier results for Au clusters on TiO_2 [61]. The gap size hereby follows the inverse cluster area Ω according to $E_g \propto \Omega^{-1}$, as expected for the energy separation of eigenstates in a 2D harmonic potential [43, 60].

In summary, the charge-mediated binding scheme of gold to oxide films is closely related to the possibility to transfer electrons through the insulating spacer. The dominant transport mechanism is electron tunneling from the Fermi level of the metal substrate through the oxide layer. The tunneling probability depends exponentially on the oxide thickness, and to a smaller extent on the gap size, as already discussed by Cabrera and Mott [55, 62]. As an interesting consequence the growth regime of Au on thin oxide films might be altered simply by increasing the film thickness until tunneling transport becomes impossible (Fig. 11) [15]. Whereas only 2D islands are found on 3 ML MgO/Ag(001) films, a dimensionality crossover to 3D particles occurs on 8 ML films that do not support charge transfer anymore [16]. These experiments have been reproduced by DFT calculations, addressing the shape of Au_8 clusters on 2 and 5 ML thick MgO/Ag(001) films [55]. With increasing oxide thickness, the amount of charge transfer was found to decrease by 50%, causing the interfacial adhesion to drop from 0.7 to 0.35 eV per interfacial atom. As a result, the initially planar Au_8 cluster turned 3D on the thick film in order to increase the number of Au–Au bonds and to counterbalance the loss in interfacial adhesion. Both experiments and theory therefore confirm the importance of charge transfer in determining the growth shape of metals on thin-film systems.

2.5   Metal Growth on Doped Oxide Materials

The former paragraphs have demonstrated how the equilibrium shape of metal particles on oxide supports depends on the charge state of the deposits. It turned out that charged species, either single atoms or small aggregates, tend to bind stronger to ionic oxides, as they exhibit additional electrostatic and polaronic interaction channels that are not available for neutral entities [28, 55]. However, the concept of charge-mediated control of the metal-oxide adhesion seems to be restricted to ultrathin films, as the extra electrons are provided by a metal substrate and need to tunnel through the insulating spacer layer. In contrast, the request to tailor the equilibrium shape of metal particles becomes particularly large on bulk oxides as used in heterogeneous catalysis, as the chemical properties of metal-oxide systems were shown to depend sensitively on the particle shape [63–66]. The relation between chemical reactivity and geometry of the active metal species has been explored in detail for gold [8]. Raft-shaped Au islands on iron oxide, for example, have been identified as the active entities in the low-temperature oxidation of CO [67, 68]. Also in the Au/TiO_2 system, bilayer deposits turned out to be the most active [61, 69]. Both results suggest a special role of the perimeter sites of metal deposits, which enable molecules to interact simultaneously with the metal and the oxide support. As those sites are most abundant on flat metal islands, a close interrelation between structure and reactivity is not surprising. We note in passing that the shape affects also other fundamental properties of metal deposits, e.g., their electronic structure and optical response [70, 71], which renders a careful shape control relevant for applications in microelectronics, nano-optics, and photocatalysis as well.

One possibility to extend the concept of charge-mediated particle growth to bulk oxides is the insertion of suitable charge sources directly into the oxide material, preferentially into a near-surface region to allow for charge exchange with adsorbates. By this means, all advantageous effects of charge control could be maintained for oxide slabs of arbitrary thickness. The fundamental approach to insert charge centers into a material is doping, and the underlying concepts have been introduced and brought to perfection already in the mature field of semiconductor technology. Surprisingly, the art of doping is less advanced to what oxide concerns, which relates to a number of peculiarities in these materials. Oxides are subject to self-doping either by native defects or unwanted impurities, the concentration of which is difficult to control experimentally [72]. Both lattice defects and impurity ions may adopt different charge states in the oxide lattice [73, 74], a variability that leads to pronounced compensation effects and is less common in semiconductors. And finally, the dopants may be electrically inactive in a wide-gap insulator, as thermal excitation is insufficient to rise the electrons from defect states into the bulk bands. As a result, the excess charges remain trapped at the host ions and are unavailable for charge transfer. The following examples demonstrate, however, that doping is a versatile approach to control the growth of metals even on bulk-like oxide materials [17, 18, 75–79]. The underlying concepts are thereby similar to the charge-transfer picture developed for thin films before.

In general, doping is carried out with impurity ions that adopt either a higher or lower valence state than the substituted ions in the oxide lattice. In rare case, also charge-preserving doping is realized, and geometric and strain effects and not charge transfer become relevant in these cases. Whereas high-valence dopants may serve as charge donors and provide extra electrons, undervalent dopants have acceptor character and may accommodate electrons from suitable adsorbates. Based on the above considerations we now expect that charge donors in an oxide lattice have a similar influence on the particle shape as the metal support below a thin oxide films.

The impact of doping on the growth morphology of gold has first been realized for crystalline CaO(001) doped with Mo in the sub-percent range [80]. On the doped oxide, the gold was found to spread out into extended monolayer islands, while the conventional 3D growth regime prevailed on the pristine, non-doped

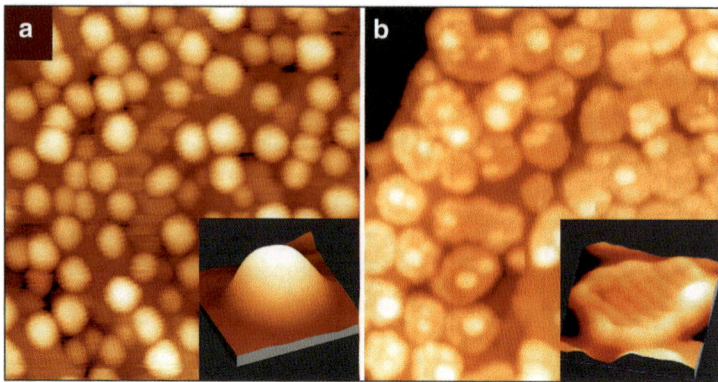

Fig. 16 STM images of 0.7 ML Au dosed onto (**a**) pristine and (**b**) doped CaO films (4.5 V, $50 \times 50 \, nm^2$). The *insets* display close-ups of two characteristic particles (-5.0 V, $10 \times 10 \, nm^2$)

material (Fig. 16). Evidently, the donor character of the Mo dopants is responsible for the 2D growth morphology, as the bare CaO(001) surface interacts with gold only weakly. The Mo impurity ions mainly occupy Ca substitutional sites and, in the absence of gold, adopt the typical 2+ charge state of the rock salt lattice in order to maintain charge neutrality. In the 2+ configuration, four Mo 4d electrons are localized in the dopant, three of them occupying (t_{2g}-α) crystal field states and one sitting in a (t_{2g}-β) level close to the upper end of the CaO band gap (Fig. 17) [80]. Especially the latter one is in an energetically unfavorable position and therefore susceptible to be transferred into an acceptor state with lower energy. Such states are indeed provided by the Au atoms that exhibit half-filled Au 6s levels at lower energy. DFT calculations therefore reveal a spontaneous transfer of the topmost Mo 4d-electron into the Au 6s affinity level, resulting in the formation of an Au$^-$ anion (Fig. 17). As discussed for the thin films, the charged gold experiences reinforced bonding to the CaO surface, reflected by the increase of the binding energy from ~1.5 eV without to ~3.5 eV with Mo dopants in the film. We emphasize that the charge transfer does not require the presence of a Mo ion in the surface and remains active over relatively large Mo-Au distance of up to ten atomic planes.

The increase of the metal-oxide adhesion due to dopant-induced charge transfer fully explains the 2D growth regime of gold observed in the experiment. Gold tends to wet the CaO surface in an attempt to maximize the number of exchanged electrons, hence the interfacial interaction. Further DFT calculations suggested that also a Mo^{3+} species that has already lost one electron remains a potential donor, as two of the residual d-electrons are still higher in energy than the Au 6s affinity level [81]. Consequently, even a second and a third electron may be transferred into the ad-metal, leaving behind thermodynamically stable Mo^{4+} and Mo^{5+} ions in the CaO lattice. It is this behavior of the Mo ions that is responsible for the robust donor behavior of Mo-doped CaO [80].

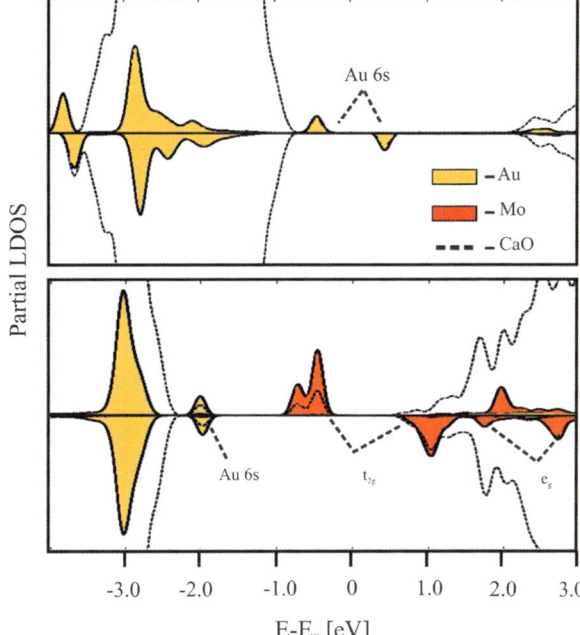

Fig. 17 PBE projected state density calculated for non-doped (*top*) and doped (*bottom*) CaO films in presence of an Au adatom [80]

The presence of suitable dopants is, however, not the only requirement for a stable donor characteristic, the interplay between dopants and host oxide determines the redox activity. This shall be demonstrated for two similar systems, Cr-doped MgO and Mo-doped CaO, which still exhibit an entirely different behavior [81–83]. Chromium has a similar electronic structure as Mo, i.e., the same number of d-electrons, but is a 3d and not a 4d metal. Surprisingly, it is unable to influence the Au growth behavior on the MgO support (Fig. 18). Even at a high Cr concentration, the 3D growth of gold prevails and hardly any 2D islands are found on the surface. The reason for this poor behavior is the low energy position of the Cr t_{2g} levels in the MgO band gap, which originates from a substantial stabilization of the Cr 3d-electrons in the MgO crystal field. Note that the crystal field in MgO is substantially stronger than in CaO, due to the reduced lattice parameter [84]. In addition, the ionization energies of Cr atoms are higher than for Mo, which makes formation of Cr^{4+} and Cr^{5+} ions energetically more expensive [85]. As a result, Cr is able to donate at maximum one single electron to gold, which compares to three for the Mo dopants in CaO [81].

Moreover, this electron may not even reach the ad-metal, because it is likely to be captured by parasitic electron traps that are present in real oxides. Typical electron traps are cationic vacancies, e.g., Mg defects, domain boundaries, or dislocation lines [86]. The cation defects (*V*-centers) preferentially develop in oxides with a large number of high-valance dopants, as they are able to compensate the augmental charge state of the impurities with respect to the intrinsic ions [84]. According to DFT calculations, the formation energy of electron trapping

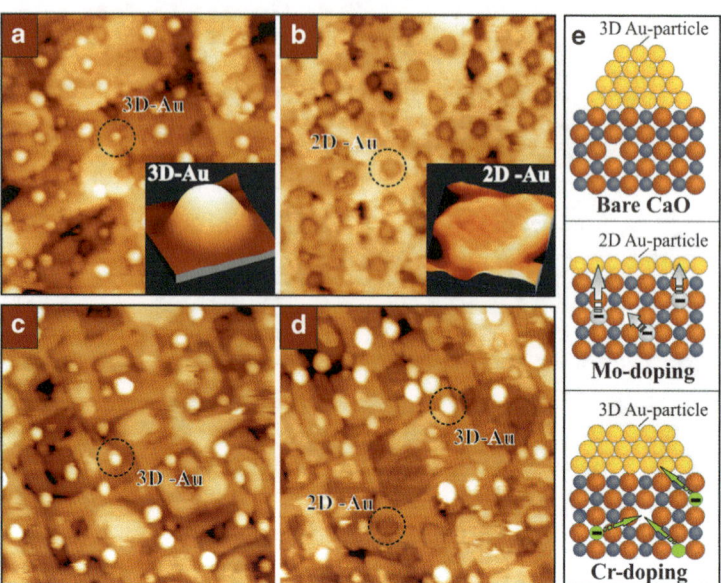

Fig. 18 STM images of (**a**) bare and (**b**) Mo-doped CaO(001) films of 60 ML thickness after dosing 0.5 ML of Au (60×50 nm^2, +6.0 V). The 2D Au islands appear as depressions on the insulating oxide, because electron transport into the gold is inhibited at the high positive bias needed for scanning. The *insets* show typical 3D and 2D islands taken at lower bias (3.5 V, 10×10 nm^2). Similar measurements on (**c**) bare and (**d**) Cr-doped MgO(001) films of 20 ML thickness. The particle shape is not affected by the dopants in this case. (**e**) Ball models explaining possible charge-transfer processes between the doped oxide and the ad-metal and their influence on the particle shape

center (*V*-centers) decreases from more than 8 eV in bare CaO and MgO to 1.0–1.5 eV in the presence of high-valence dopants [81]. As a result, internal compensation effects become favorable, in which every dopant in the lattice produces a balancing vacancy. While this internal compensation remains insufficient to cancel the donor potential of Mo ions, it annihilates the effect in Cr-doped MgO. Consequently, no charge-mediated changes in the Au growth morphology are observed for Cr-MgO, although the effect is huge for the Mo-doped CaO (Fig. 18). One conclusion of these results is that the donor behavior of a transition-metal-doped oxide cannot be predicted by simple valence arguments but needs to be checked in every single case.

The influence of overvalent dopants on the equilibrium morphology of metal particles can, to a certain extent, be annihilated by undervalent dopants in the oxide lattice (Fig. 19) [87, 88]. Dopants with lower charge state generate holes in the oxide electronic structure that are able to trap the extra electrons provided by the charge donors. This mutual compensation between donors and acceptors has been explored for CaO films co-doped with Li and Mo ions [89]. At low Li concentration with respect to Mo, most of the Au particles still adopt 2D shapes, indicating a prevailing charge transfer into the gold. With increasing Li doping level, however,

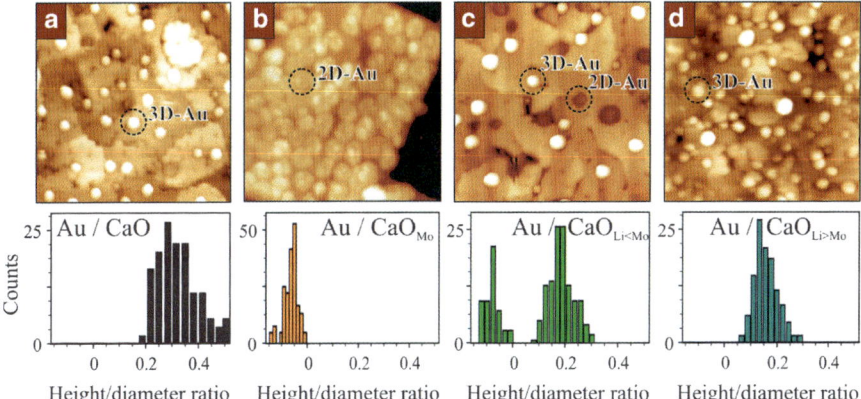

Fig. 19 STM images of 0.5 ML Au deposited onto (**a**) pristine CaO, (**b**) doped with 4% Mo, (**c**) doped with 4% Mo + 2% Li, and (**d**) doped with 4% Mo + 8 % Li ($50 \times 50\,\mathrm{nm}^2$, 6.0 V). Note that monolayer Au islands in (**b**) and (**c**) appear as depressions at high sample bias. Histograms of the particle shapes (height-to-diameter ratio) are plotted below. Note the transitions from 2D to 3D particles when the overvalent doping with Mo in (**b**) is balanced with increasing amounts of monovalent Li impurities

more deposits turn 3D, as the ad-particles are unable to accumulate enough excess charges from the Mo donors (Fig. 19c,d). The extra electrons are trapped by lattice defects that arise from the presence of Li in the lattice. Every monovalent Li^+ ion sitting in Ca^{2+} substitutional sites produces a deep hole in the O2p states of an adjacent oxygen forming an effective trap for the Mo 4d electrons. As a result, charge transfer to the surface ceases at a critical Li doping level and the Au deposits adopt the typical 3D geometry of pristine CaO films.

We note that pure hole doping could not be realized in the experiments so far. Neither MgO nor CaO films, doped exclusively with Li, were found to alter the morphology and electronic structure of gold, and no sign for the generation of positively charge ad-species was obtained. This finding is in agreement with the occurrence of intrinsic compensation mechanisms in the oxides that remove the energetically unfavorable holes produced by the Li^+ species. Even in well-prepared oxide films and at perfect vacuum conditions, competing electron sources are present, such as electron-rich oxygen vacancies (F^o centers) and donor-type adsorbates from the rest gas (water, hydrogen) [87, 88, 90, 91]. Hole doping of oxides as a means to tailor the properties of metal ad-particles is therefore more difficult to realize than electron doping with donor-type impurities.

Interestingly, the competition for excess electrons occurs not only in Mo and Li co-doped films but is observed also in the presence of electron-accepting species on the oxide surface. Gold forms monolayer islands if deposited onto a Mo-doped CaO at vacuum conditions, but 3D particles if oxygen is present during growth. The reason is that O_2 molecules bound to the surface act as electron acceptors as well and may trap charges in their O $2\pi^*$ antibonding orbitals. These electrons are lost for the Au islands, resulting in a gradual transition from a 2D to a 3D growth regime with increasing O_2 partial pressure. Whereas in 5×10^{-7} mbar of oxygen, 50% of the Au islands still adapt monolayer shapes, all Au deposits turn 3D grown in a 5×10^{-5} mbar O_2 background. The interplay between the observed growth mode and the composition of the gas environment emphasizes the pivotal importance of excess electrons from donor species for the reactivity of oxides towards adsorption of metallic and gaseous species.

In summary, doped bulk oxides display in many aspects similar adsorption properties as ultrathin oxide films. In both systems, excess electrons are transferred into the ad-species and open up specific charge-mediated adsorption schemes. Whereas for ultrathin oxide films, the extra electrons are provided by the metal substrate below, doped oxides contain intrinsic charge sources in the form of aliovalent impurity ions. Thin oxide films grown on metal single crystals are mainly of academic interest, as they provide easy access to the properties of oxide materials via conventional surface science techniques. Doped oxides, on the other hand, are of practical relevance for many processes in heterogeneous catalysis. As shown in this section, overvalent dopants are able to change the particles equilibrium shape from 3D to 2D, which is expected to change the reactivity pattern of the metal-oxide system. Moreover, charge transfer from electron-rich dopants might be a suitable pathway for the activation of small molecules, such as oxygen or methane. A possible activation mechanism would be the occupation of the antibonding orbitals, leading to a destabilization or even dissociation of the molecules. In a recent experiment, electron donation from Mo impurities was shown to produce super-oxo (O_2^-) species on a CaO surface, which might be a highly reactive precursor for subsequent oxidation reactions. Although the chemical reactivity of doped oxides is still subject of active research, it appears already that oxide doping forms a promising route to fabricate highly reactive and selective catalysts for the future.

3 CO Adsorption on Supported Au Particles

The binding of molecules to specific sites of a metal nanoparticle is thought to determine the reactivity of metal/oxide systems used in heterogeneous catalysis. The model systems, described above, represent ideal objects to study the influence of size, shape, and charge onto adsorbed molecules. Given the large body of information on CO as an adsorbate, this molecule represents the probe molecule of choice. Figure 20 shows the range of frequencies CO typically shows when bound to Au in different oxidation states [92].

This knowledge, together with the possibility of imaging, allows us to investigate, in detail, molecule-Au interactions. Before we discuss CO adsorption on nanoparticles we briefly discuss the frequency of species not fitting the scheme of Fig. 20, i.e., the IR spectrum of CO interacting with a single Au atom [93]. When small amounts of Au are deposited on a 10 monolayer thick MgO(100) film, the Au atoms remain neutral, as conclusively shown via EPR measurements on the system. CO adsorption leads to the appearance of a band at 1852 cm^{-1} in the IR

Fig 20 Connection between Au charge scale and the stretching frequency of adsorbed CO [92]

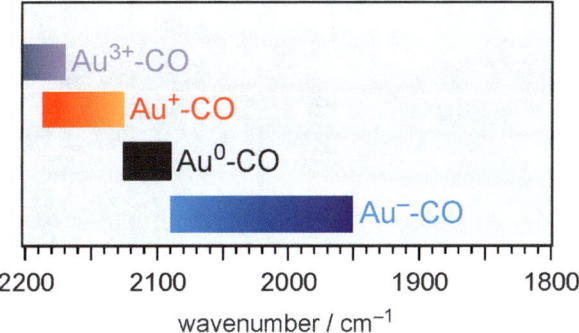

spectrum, which has been identified with the help of a DFT calculation. The large shift of 285 cm^{-1} with respect to the frequency of the gas-phase CO arises because the Au 6s valence electron is strongly polarized away from the MgO surface due to Pauli repulsion with the oxygen ions to which the Au preferentially binds. The concomitant transfer of electrons into the C–O antibonding $2\pi^*$ orbital leads to this dramatic weakening of the intermolecular bond, hence a red shift of the stretching frequency. This shift is even larger than for an Au atom residing on a color center holding one electron, i.e., there the shift is 220 cm^{-1} [93].

3.1 CO Adsorption on Au on Thin MgO/Ag Films

Molecular species can also be identified directly on oxide-supported gold nanostructures, using low-temperature STM. For example, single CO molecules are imaged as circular depressions on the MgO surface (Fig. 21a).

A similar imaging contrast is observed on most metal surfaces, where it originates from the absence of CO orbitals for tunneling transport close to the Fermi level (E_F) [94]. CO attachment to the tip gives rise to a distinct contrast change of the adsorbates, which now appear as dark rings of ~8 Å diameter (Fig. 21b). This imaging mode enables CO identification even on corrugated surfaces, e.g., next to an Au deposit, where the tiny CO-induced depressions obtained with metallic tips are no longer detectable. Figure 21c shows a corresponding STM image of an Au island saturated with CO. Apparently, the CO-induced features are exclusively localized along the island perimeter, while the center remains adsorbate-free. An alternative method to identify CO molecules on the oxide surface is inelastic-electron-tunneling spectroscopy (IETS) with the STM [95–97]. The peak/dip positions observed in the second-derivative d^2I/dV^2 spectra correspond to the vibrational energy modes of the molecules. The CO vibrational mode with the highest excitation cross section in STM-IETS is the frustrated rotation [95]. Consequently, bias voltages of 45 mV have been chosen to identify CO molecules next to the 2D Au islands via their vibrational signature (Fig. 22a–d).

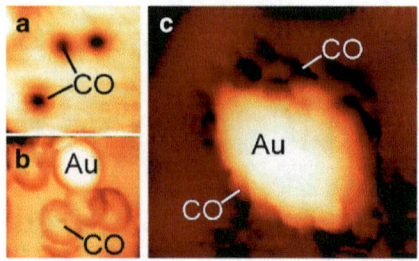

Fig. 21 STM topographic images of CO molecules on 2 ML MgO/Ag(001) imaged with (**a**) a metallic and (**b**) a CO-covered tip (3 × 3 nm², 150 mV, 3 pA). The CO(tip)−CO(sample) interaction leads to a ringlike appearance of the adsorbates in (**b**). (**c**) STM image of a CO-saturated Au island taken with a CO-covered tip (4 × 4 nm², 100 mV, 3 pA). CO-induced contrast is only revealed at the perimeter of the Au island [66]

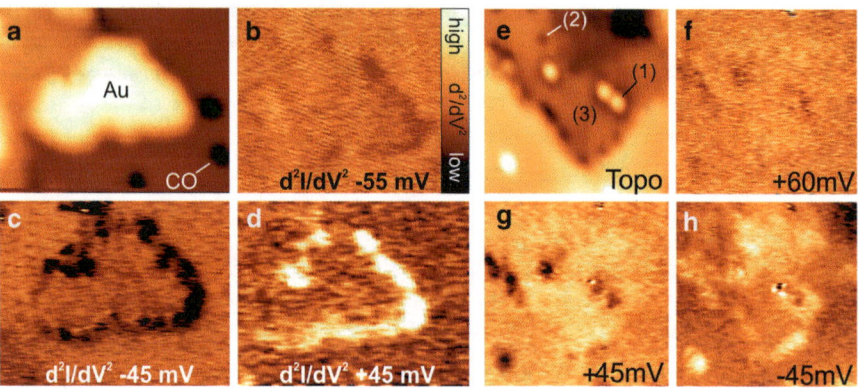

Fig. 22 (**a**) STM topographic image and (**b–d**) corresponding second-derivative maps of a planar Au island saturated with CO (7.0 × 5.5 nm², 10 pA). Due to the metallic tip state, the CO molecules are not resolved on the island but give rise to distinct energy-loss features at the island perimeter (**c, d**). The second-derivative contrast vanishes at bias voltages away from the CO vibrational modes (**b**), (**e–h**) corresponding information for single Au atoms [66, 94]

As mentioned before, no CO is visible on the aggregate in topographic images taken with a metallic tip. However, CO-related loss signals emerge in the d^2I/dV^2 maps and produce a bright/dark brim around the island at positive/negative sample bias (Fig. 22c, d). Control measurements taken at bias values above or below the CO frustrated rotation did not produce any d^2I/dV^2 contrast (Fig. 22b). The localization of the inelastic signals along the perimeter therefore corroborates the CO attachment to edge and corner atoms of the Au islands. For comparison, we show similar information for a single Au atom with adsorbed CO. Here we see a very similar spectral response indicating the a-top position of CO on the Au monomer (Fig. 22e–h).

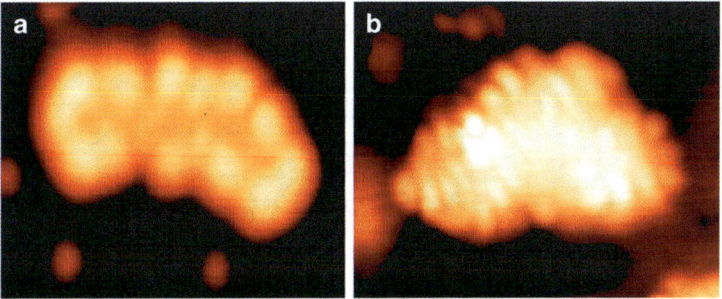

Fig. 23 STM topographic images of (**a**) a bare (see also Fig. 12) and (**b**) a CO-saturated Au island on 2 ML MgO/Ag(001) (7.0×5.5 nm^2, $V = 150$ mV). Whereas the bare island is surrounded by a bright rim, indicating charge accumulation at the perimeter, the CO-covered island exhibits charge-density waves in the interior. The standing electron waves are due to electron displacement from the island boundary upon CO adsorption [66]

The preferential CO binding to the island perimeter might be ascribed to two effects. On the one hand, the boundary atoms have a lower coordination than the ones enclosed in the Au plane. The same mechanism controls CO adsorption on stepped metal surfaces [98, 99]. On the other hand, electrons transferred from the MgO/Ag support are localized at the island perimeter, as discussed above, and may alter the CO binding characteristic. According to DFT calculations, the coordination effect clearly dominates the CO binding position. Whereas CO adsorbs with 0.72 eV to 3-fold coordinated edge atoms of a negatively charged Au island on MgO/Ag(001), it is nearly unbound to the 5-fold coordinated atoms in the interior of island. A similar binding enhancement is revealed for edge versus inner atoms of a neutral Au$_{12}$ island on bulk MgO [100]. The extra electrons localized on the perimeter of charged islands actually counteract this trend, as they impede charge donation from the CO 5σ orbital into the metal and hence increase the CO-Au repulsion [101]. Clearly, CO attaches to the island perimeter not because but in spite of the extra charge, and the binding preference with respect to the top facet is governed by the reduced coordination of the edge atoms.

This conclusion is supported by the experimental observation that CO adsorption modifies the charge distribution within the Au islands. The specific contrast at the rim of the islands (Fig. 23a) disappears after CO dosage, when standing electron waves emerge in the island center (Fig. 23b) [102]. Apparently, the CO removes the localized charges from the perimeter sites by pushing them into the interior of the island. Also this finding is in line with DFT results for charged Au$_{12}$ clusters on MgO/Ag(001) [100].

While on ultrathin films we resorted to IETS to probe the vibrations on adsorbed CO, on thick insulating films, the CO binding properties have been studied with IRAS, whereas STM still provides information on the nucleation density and size of the Au aggregates. Figure 24a displays an STM topographic image of a 14 ML thick MgO/Mo film, being exposed to electrons to produce F^0 and F$^+$ centers. Those point defects that mainly form along step edges and misfit induced dislocation lines are

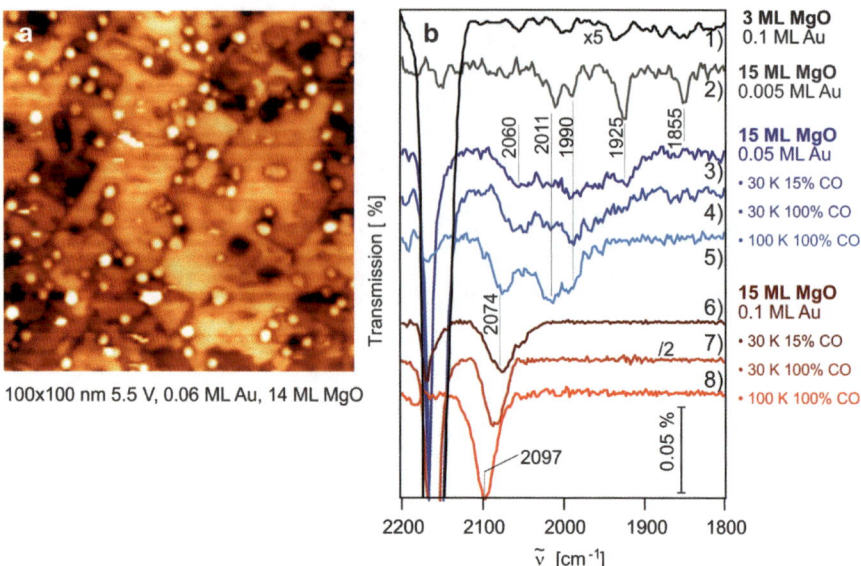

Fig. 24 (**a**) STM image of a 14 ML MgO film. 0.06 ML Au were deposited at room temperature after electron bombardment of the film. (**b**) Infrared absorption spectrum of CO adsorbed on a 3 ML MgO/Ag(001) film covered with 0.1 ML Au. (*2–8*) IR spectra taken on an electron-bombarded 15 ML thick MgO/Mo film covered with (*2*) 0.005 ML Au (CO saturation coverage, 100 K), (*3*) 0.05 ML Au (15% CO exposure, 30 K), (*4*) 0.05 ML Au (saturation coverage, 30 K), and (*5*) 0.05 ML Au (saturation coverage, 100 K). The spectra (*6–8*) were taken for 0.1 ML Au coverage and (*6*) 15 s CO exposure (30 K), (*7*) CO saturation coverage (30 K), and (*8*) after annealing the sample to 100 K [66]

the preferential Au nucleation sites on the oxide [103]. After dosing 0.06 ML Au at room temperature, Au particles develop on the surface that are one to two layers high and contain 30–50 atoms (Fig. 24a).

For the IRAS experiments shown in Fig. 24b, Au deposition was carried out at 30 K, which increases the nucleation density and decreases the mean particle size. For the given coverage range of 0.1–0.005 ML Au, the size of Au deposits is expected to vary between a few tens of atoms down to monomers and dimers. The IRAS data (spectra 2–8) in Fig. 24b show bands ranging from 2090 cm^{-1} all the way down to 1990 cm^{-1} as well as two isolated lines at 1925 and 1855 cm^{-1}. All bands are red-shifted compared to CO adsorbed on Au clusters on pristine MgO films, which are found between 2125 and 2100 cm^{-1} depending on the preparation conditions. The low-frequency lines that only occur at the smallest metal exposure (Fig. 24b spectrum 2, gray line) are assigned to Au atoms adsorbed to regular terrace sites (1855 cm^{-1}) and F-centers (1925 cm^{-1}) on the basis of results discussed above. All other bands can be explained by CO bound to negatively charged Au aggregates located on top of electron-rich surface defects [92]. In particular, the noisy features at ~1990 and ~2011 cm^{-1} are suggested to originate from CO adsorption to ultrasmall Au aggregates, e.g., dimers and trimers.

This assignment is corroborated by IR spectra taken at higher Au exposure (Fig. 24b, spectra 3–5, blue lines). In addition to the low-frequency bands discussed before, new blue-shifted lines emerge in the frequency range around 2060 cm^{-1} that shifts to 2070 cm^{-1} upon annealing to 100 K. For the highest coverage of 0.1 ML (Fig. 24b, spectra 6–8, red lines), the spectrum is dominated by a band at 2074 cm^{-1}, while the strongly red-shifted bands have disappeared. Annealing the system to 100 K induces again a blue shift of the line to 2097 cm^{-1}. Inspection of the two data sets shown in Fig. 24b, spectra 3–5 and spectra 6–8 reveals that neither the intensity nor the line shape of the Au-related bands changes when the CO coverage is increased from 15 to 100% of the saturation value. Annealing the systems to 100 K, on the other hand, leads to a blue shift of the lines that reaches 15 cm^{-1} for the bands at 2060 (Fig. 24b, spectrum 5) and 2080 cm^{-1} (Fig. 24b, spectrum 8). At 100 K, CO desorbs from the MgO film as evident from the reduction of the MgO-related CO-stretch bands between 2150 and 2180 cm^{-1}. In general, the CO stretching frequencies experience a decreasing red shift with higher Au exposure and increasing cluster size, a phenomenon that can be rationalized in the following way. Assuming that each Au aggregate nucleates on a single color center and takes up one electron, the density of excess charges decreases in the larger particles, which reduces the amount of π-back-donation into the CO and hence the red shift of the stretch mode. A similar conclusion was drawn from CO adsorption experiments on charged gas-phase clusters, where the CO-stretch [104] mode was found to increase when going from Au^- (2050 cm^{-1}) to Au^+ clusters (2150 cm^{-1}). This simple picture needs, however, to be modified in view of the earlier discussion. CO binds to the low-coordinated sites of the charged Au aggregates, which accommodate the extra electrons as well. The amount of π-back-donation now depends on the ability of the cluster to distribute the charge away from the CO adsorption site to lower the Pauli repulsion with the CO. Naturally, this ability diminishes with decreasing particle size, causing the charge transfer into the CO $2\pi^*$ to increase.

Based on these considerations, the experimental IR bands may be assigned. The bands above 2060 cm^{-1} that shift in an almost continuous fashion with Au exposure are produced by particles containing a few tens of atoms, whose charge density changes only gradually with size. The quasi-discrete bands at 1990 and 2010 cm^{-1}, on the other hand, are characteristic for ultrasmall clusters such as dimers. A direct assignment of these bands is difficult, as not only the size of the aggregate but also the nature of the oxide defect underneath determines the vibrational response. According to DFT calculations, Au dimers bound to F^0-centers interact only weakly with CO and produce a red shift of the stretching frequency of 194 cm^{-1}, whereas on F^+-centers the binding is sizable and the red shift is smaller (158 cm^{-1}). However, also negatively charged Au_3^- and Au_4^- clusters [100] produce a red shift of ~175 cm^{-1}. The experimental CO bands at 1990 and 2010 cm^{-1} are therefore compatible with several ultrasmall Au clusters.

In summary, whereas, on ultrathin films, the ad-particles charge up due to an electron transfer from the support, the charging on thicker films is realized by an electron donation from defects. In both cases, the excess electrons give rise to unusual

CO adsorption properties, as deduced from STM and IRAS measurements. On planar Au islands on thin MgO films, the CO exclusively binds to low-coordinated Au atoms along the island perimeter. The adsorption is accompanied by a redistribution of the cluster electrons away from the boundary towards the interior. On thicker films, Au charging gives rise to an increasing red shift of the CO-stretch mode with decreasing particle size, reflecting the gradual filling of the antibonding CO $2\pi^*$ orbital. The occurrence of sharp separated lines in the spectra, hence, indicates the presence of Au aggregates with well-defined atom counts.

3.2 CO Adsorption on Au on Ultrathin Alumina and Iron Oxide Films

In the following, we will briefly discuss CO adsorption on Au particles supported on other supports, such as alumina and iron oxide thin films, as we have also discussed their structure in Sect. 2 of this chapter.

TPD experiments of CO adsorbed on Au, deposited onto a thin alumina film at room temperature, revealed two clearly resolved peaks, with the first peak at around 120 K and another located at a higher temperature, depending on the Au coverage (see Fig. 25) [105]. Based on CO desorption results for stepped Au single crystal surfaces [106, 107], both peaks were attributed to desorption from gold particles. The morphology of the similarly prepared samples was studied by STM in another UHV chamber as a function of the Au coverage. At low coverages, gold forms very small clusters and even single atoms (see e.g., Figs. 3 and 7) if the sample has not been exposed to elevated temperatures. Apparently, gold species are found randomly distributed across the surface without preferential decoration of the line defects clearly seen in the bottom STM image in Fig. 25. At high coverage, gold forms hemispherical 3D nanoparticles about 3–5 nm in diameter. With increasing Au coverage and hence Au cluster size, the position of the high-temperature peak gradually shifts from 210 to 160 K, thus indicating a size effect such that small Au particles adsorb CO more strongly. In addition, the intensity of this CO desorption signal reached saturation at a fairly low coverage of gold and low exposure of CO, suggesting that CO only adsorbs on low-coordinated Au atoms, but not on the regular terraces. Increasing Au coverage basically leads only to the gradual development of flat (111) terraces on the gold particles, which bind CO less strongly.

On FeO(111) grown on Pt(111), the TPD data of CO adsorption again showed a strong size effect for Au particles deposited at ~90 K, with desorption states observed at temperatures as high as 300 K for the lowest gold coverage [108]. After annealing to 500 K, however, a reduction in signal intensity was observed, and the size effect is completely suppressed, as TPD spectra show two peaks similar to those observed in Fig. 25 for alumina-supported particles, independent of the amount of gold deposited. This finding indicates that after annealing particles have reached some critical size beyond which their adsorption behavior is size independent.

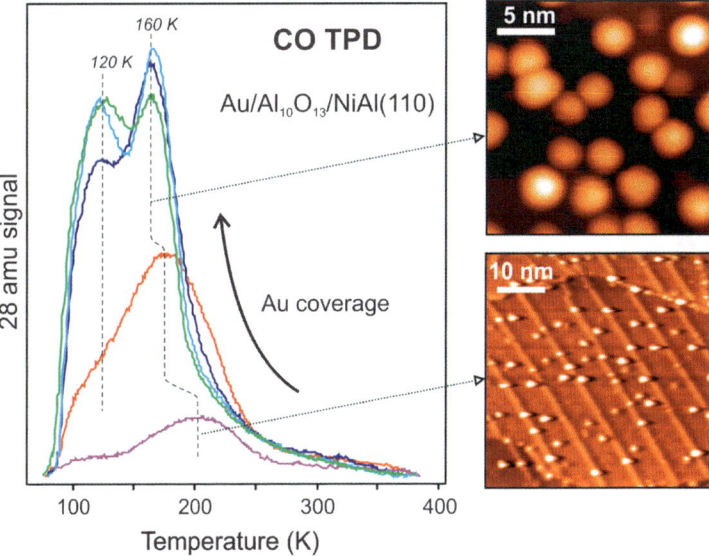

Fig. 25 TPD spectra of CO exposed to Au/alumina film/NiAl(110) surface at 90 K as a function of Au coverage. Note that CO does not adsorb on the alumina film above 60 K. Typical morphology of similarly prepared surfaces is shown in the respected STM images taken in another experimental setup (see text)

Au coverage-dependent IRA- spectra taken after CO saturation (adsorption and deposition at 90 K) are shown in Fig. 26 (left panel). At the lowest coverage studied here, a feature at 2165 cm^{-1} and a second at 2131 cm^{-1} are observed with approximately equal intensity. With increasing coverage, the low-frequency peak gains intensity and shifts from 2131 to 2108 cm^{-1}, while the high-frequency peak gradually attenuates. As shown in the right panel of Fig. 25, annealing to 500 K in UHV results in a single state at 2108 cm^{-1} independent of the Au coverage. Bearing in mind the sintering effects of annealing, the peak at 2108 cm^{-1} can be straight-forwardly assigned to CO adsorbing on metallic gold. The species giving rise to the peak at ~2165 cm^{-1} is CO adsorbed on very small clusters, which – as shown above – undergo charge transfer. In this case a specific interaction with the support leads to positively charged Au species [46, 109].

Interestingly, even monolayer islands formed on a FeO(111) film surface upon annealing display the same CO adsorption behavior as large, three-dimensional particles (see the STM images on the right side of Fig. 26) or gold single crystals [108]. We thus concluded that the observed dependences are due to a higher percentage of highly uncoordinated gold atoms found for smaller particles, formed at low temperatures, which favor CO adsorption. Indeed, the single IRAS peak observed for the annealed samples did not shift and simply grew monotonically at increasing Au coverage. This supports the TPD results indicating no apparent change in the possible adsorption states for CO on monolayer islands as compared

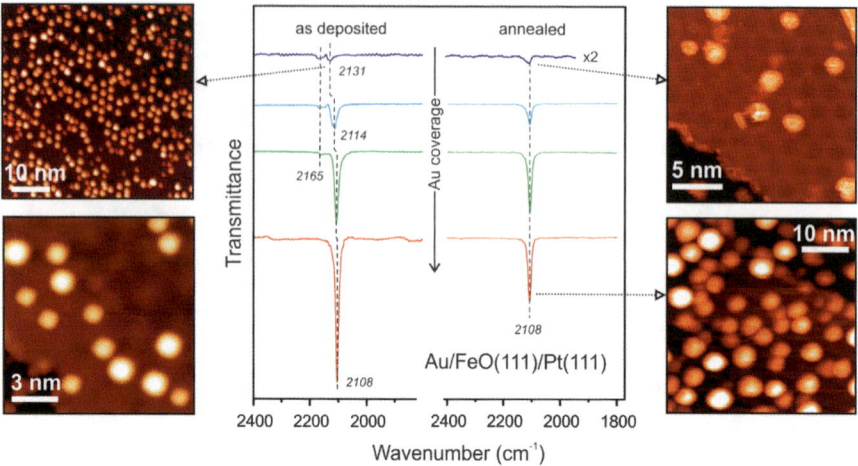

Fig. 26 IRA spectra of CO exposed to Au/FeO(111)/Pt(111) surface at 90 K as a function of Au coverage. *Left panel*: as deposited particles at ~90 K. *Right panel*: after annealing to 500 K in UHV. The STM images of the similarly prepared samples are shown. At very low temperatures (~10 K) and low coverages, single Au atoms dominate the surface. Upon annealing, gold forms single-layer islands with about 2.5 nm diameter. At high Au coverage, only hemispherical 3D nanoparticles are observed

to large particles. Therefore, the results fit well the picture of exclusive CO adsorption on low-coordinated surface atoms.

CO adsorption studies on Au deposited on alumina and iron oxide films performed in our group clearly show stronger CO binding on small particles. For the smallest Au coverage, CO may desorb at temperatures close to 300 K, which is in the temperature range of working Au catalysts for low-temperature CO oxidation and has never been observed for Au single crystals. The presence of such a state is likely due to the presence of highly uncoordinated gold atoms.

Inspection of CO-TPD spectra collected for gold particles deposited on various oxide [110] and carbon [111] films show that, for a given particle size (~3 nm as measured by STM), the interaction of CO with gold particles on different supports is virtually identical. However, the finding that annealing strongly attenuates size effects indicates that the nature of the support and its defect structure may be critical for nucleation, growth, and stabilization of very small Au clusters.

4 Nucleation and Bonding of Au on Chemically Modified Oxide Surfaces

In Sect. 2 we demonstrated the enormous potential of combining surface science experiments with DFT calculations in providing details about the interplay between electronic and structural properties of oxide-supported Au atoms, clusters, and particles. Catalysis of supported gold strongly depends on these parameters, in particular on the particle size and the structure of the Au-support interface, and it is at the heart of catalyst preparation/synthesis to develop recipes that allow those units that are responsible for catalytic activity to be stabilized on the catalysts' surface. Indeed, much of the recent interest in gold surface chemistry arose from Masatake Haruta's observation of the unusual catalytic activity of highly dispersed Au supported by oxides, after finding a preparation method that resulted in stable, nano-sized Au particles [112]. In most cases, catalyst synthesis involves at least one wet-chemical preparation step, which adds some degree of chemical complexity to the systems that is typically not covered by surface science experiments carried out under ultrahigh vacuum conditions.

In the following we present two examples of surface science investigations into gold nucleation and bonding on oxide surfaces, which attempt to include some of the complexity of a real-world catalyst. The first one investigates the specific role of hydroxyl groups present on an oxide support on the morphological and chemical properties of Au nanoparticles. The second example presents an approach to study the preparation of an oxide-supported gold model catalyst using a single-crystalline oxide support in combination with a wet-chemical preparation procedure.

4.1 Gold Nucleation at Hydroxyl Groups

We have chosen to use MgO(001) as a model surface to investigate the effect of hydroxyl groups on the properties of supported Au particles. Along the lines described in the previous sections, MgO(001) films were grown on a Ag(001) or Mo(001) substrate, but thicker films on the order of 10–15 ML thickness were used to exclude any influence of the metallic substrate. As shown previously, hydroxylation of MgO(001) requires the surface to be exposed to elevated pressures of water vapor, typically in the range of 10^{-4} to 1 mbar [113]. The formation of a stable hydroxyl layer on MgO is suggested to proceed via hydrolysis of Mg-O surface bonds and leads to microscopic roughening of the MgO surface [114, 115]. The results presented in this section were obtained from MgO thin-film samples that were hydroxylated by exposure to 10^{-3} mbar D_2O at room temperature in a dedicated elevated pressure cell, which leads to a surface hydroxyl coverage of ~0.4 ML as estimated from quantitative XPS measurements [115].

A most obvious influence of hydroxyl groups on the properties of gold on the MgO surface is seen in Fig. 27, which compares STM images of Au-MgO(001) (Fig. 27a) and Au-MgO$_{hydr}$ (Fig. 27b) [116]. The first set of images was taken directly after Au deposition at room temperature (top), while the other set shows the surface state after subsequent heating to 600 K (bottom).

The Au particles are moderately dispersed over the MgO(001) surface at room temperature (Fig. 27, top), but Ostwald ripening and particle coalescence lead to particle growth and a strong reduction of particle density after the elevated temperature treatment (Fig. 27a, bottom), which is expected for this weakly interacting system. By contrast, the Au particle size and the particle density remain unaffected when the same thermal treatment is applied to Au-MgO$_{hydr}$ (Fig. 27b).

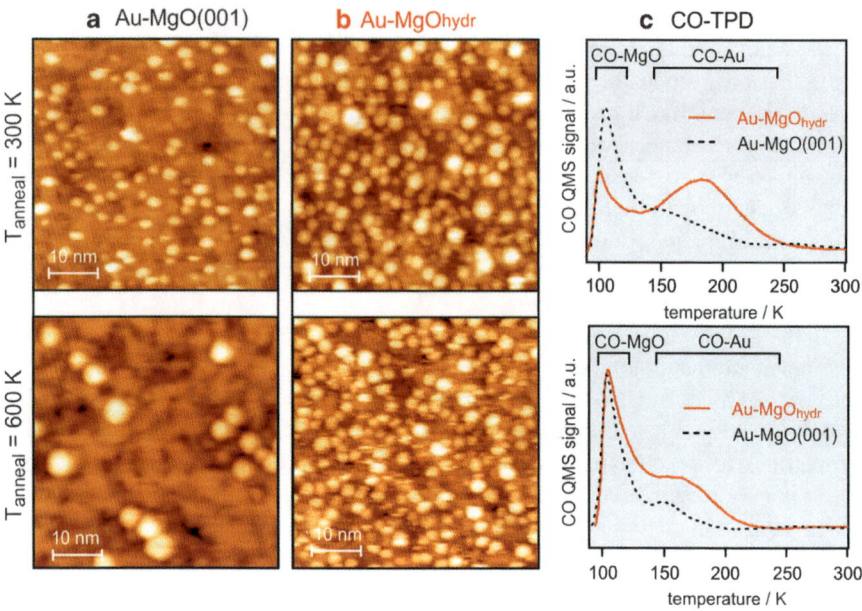

Fig. 27 (**a**) STM micrographs of 0.2 ML Au deposited on MgO(001) at room temperature taken immediately after Au deposition (*top*) and after subsequent heating to 600 K (*bottom*). (**b**) Same as (**a**) but for 0.02 ML Au deposited onto hydroxylated MgO (MgO$_{hydr}$). (**c**) Comparison of CO-TPD spectra from 0.02 ML Au deposited on MgO(001) (*black traces*) and MgO$_{hydr}$ (*red traces*) for Au particles grown at room temperature (*top*) and the same samples after heating to 600 K (*bottom*). CO was dosed at 100 K

The enhanced stability of Au-MgO$_{hydr}$ towards sintering is also reflected in the CO adsorption capacity of the Au particles. The comparison of CO-TPD spectra from Au-MgO(001) and Au-MgO$_{hydr}$ taken for particles grown at room temperature (Fig. 27c, top) and after subsequent heating to 600 K (Fig. 27c, bottom) clearly shows the enhancement of CO adsorption on Au-MgO$_{hydr}$, in line with the larger Au surface area of the more dispersed Au particles on MgO$_{hydr}$ [117]. This is a significant finding in light of the importance of small Au particle size for the catalytic activity of oxide-supported Au catalysts. Indeed, chemical functionalization of a TiO$_2$ support with hydroxyls prior to the deposition of gold particles was found to have a strong enhancing effect on the CO oxidation activity [118].

What is the origin of the enhanced sinter resistance of Au particles on the hydroxylated MgO surface? As mentioned above, hydroxylation of MgO(001) is accompanied by microscopic roughening of the surface because of hydrolysis of Mg-O surface bonds. The morphological changes of the MgO surface may lead to increased barriers for Au atom diffusion and thus limited mobility, even at high temperature. On the other hand, additional or new nucleation centers may be created upon hydroxylation, which could enhance the adhesion of the Au particles because of a distinct chemical interaction with the substrate. While the first

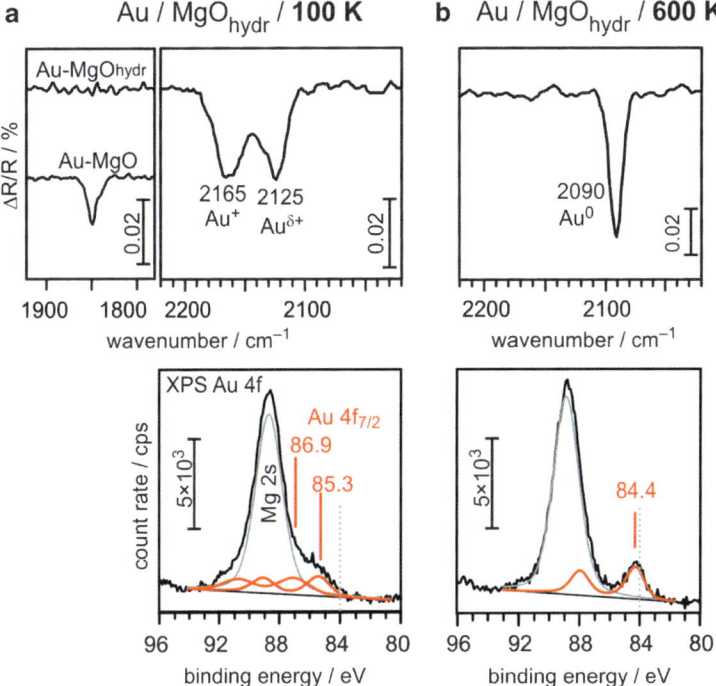

Fig. 28 (**a**) CO-IRAS spectra (*top*) and Au 4f XPS (*bottom*) spectrum from 0.05 ML Au deposited at 100 K onto MgO$_{hydr}$. (**b**) Same as (**a**) taken after annealing of the samples to 600 K

argument cannot be conclusively substantiated, direct evidence for the second one can be obtained with the help of spectroscopic methods.

Infrared spectroscopy using CO as a probe molecule has in the past been shown to provide valuable information about the charge state of gold particles [92]. While the stretching frequency of CO adsorbed on neutral Au particles is typically found in a narrow spectral region around 2100 cm^{-1}, the presence of positively (negatively) charged Au particles results in a shift to higher (lower) wave number as a result of reduced (increased) back-donation into the antibonding $2\pi^*$ orbital of CO as has been discussed in the previous section. We noted that for CO adsorbed on neutral, single gold atoms on the MgO(001) surface, a ν(CO) of 1852 cm^{-1} was observed as a result of internal charge reorganization in the MgO-Au-CO adsorption complex (see Sect. 3 above, Fig. 28a) [93]. This very particular characteristic represents an easily accessible spectroscopic indicator for the presence of single Au atoms on MgO and its presence or absence provides information about the nucleation of Au on modified MgO surfaces, such as the hydroxylated one investigated here. Indeed, this signal cannot be recovered for gold deposited at 100 K onto MgO$_{hydr}$, which provides an initial evidence for the different nucleation of gold on this surface relative to clean MgO(001). More specific information about the state of gold nucleated on MgO$_{hydr}$ is obtained by looking into the spectral region around

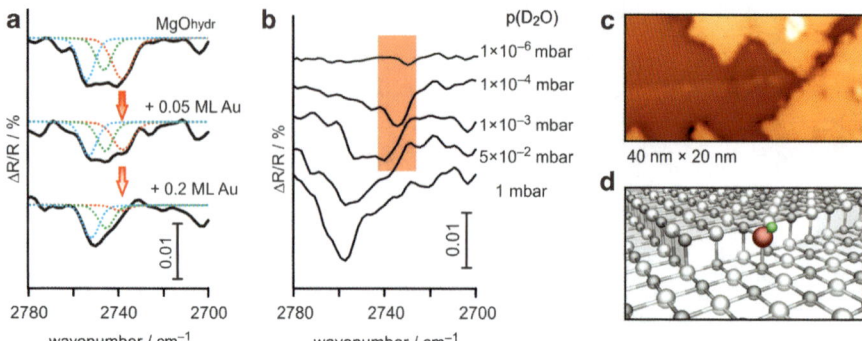

Fig. 29 (**a**) IRAS spectra showing the OD spectral detail of MgO$_{hydr}$ (*top*) and after subsequent deposition 0.02 and 0.2 ML Au. (**b**) IRAS spectra showing the OD spectral detail of MgO$_{hydr}$ after hydroxylation in water vapor at various pressures. (**c**) STM micrograph of the surface of a MgO (001) thin film grown on Ag(001). (**d**) Model of a di-coordinated OH group located on a step site, which is proposed to be the initial nucleation site of Au on MgO$_{hydr}$

2100 cm^{-1}, where CO vibrational signals from Au particles are expected. Two bands at 2165 and 2125 cm^{-1} are observed, which are significantly blue-shifted relative to CO on neutral gold and are indicative for the presence of cationic or oxidized Au species on MgO$_{hydr}$ [116]. The cationic nature of Au is also inferred from the corresponding XPS spectrum taken from this sample [116], which exhibits two Au 4f$_{7/2}$ signals at 86.9 and 85.3 eV binding energy (Fig. 28a) – significantly shifted to higher binding energy relative to bulk gold (84.0 eV).

Since hydroxyl groups are obviously involved in the nucleation of Au on MgO$_{hydr}$, an impact of Au nucleation is also expected in the hydroxyl IR spectra. The topmost spectrum in Fig. 29a displays the OD spectral detail of the infrared spectrum obtained from MgO$_{hydr}$ for the specific hydroxylation conditions used in this experiment (10^{-3} mbar D$_2$O at room temperature for 180 s). The broad band can be deconvoluted into three signal components with ν(OD) = 2737, 2745, and 2753 cm^{-1}, respectively, which are assigned to isolated and hydrogen-bond acceptor hydroxyls on the MgO surface (the corresponding OH bands are found at around 3750 cm^{-1}) [119]. Upon deposition of gold, the intensity of the OD signal at 2737 cm^{-1} is reduced (0.02 ML Au; Fig. 29a, middle) and by increasing the Au coverage further, this signal is almost completely depleted (0.2 ML Au; Fig. 29a, bottom). This result clearly indicates a very specific interaction of Au with hydroxylated MgO that involves only one particular surface hydroxyl site [116].

To obtain more detailed structural information about the specific hydroxyl site that is involved in the interaction with Au, it is instructive to inspect the OD-IR signals from MgO$_{hydr}$ as obtained following hydroxylation of MgO(001) at increasing D$_2$O partial pressure (Fig. 29b). The series of IR spectra shows, from top to bottom, an increasing hydroxyl coverage, in agreement with the water vapor pressure dependent hydroxylation activity of MgO, and corresponding shifts of the OD-IR frequencies, which indicate that different hydroxyl states (coordination,

H-bonding) are populated as a function of hydroxyl coverage. Of particular importance for the present discussion is the fact that hydroxyl species exhibiting frequencies of 2730–2740 cm^{-1}, i.e., in the range of the OD streching vibration that is depleted upon deposition of gold (Fig. 29a), appear exclusively after hydroxylation at low (10^{-4} mbar) water vapor pressure (Fig. 29b). It can be expected that under these conditions the MgO(001) terraces are not affected by water and hydroxylation occurs exclusively by water dissociation at defect sites. From STM investigations it is clear that step sites are the most abundant defect sites on MgO thin films as used in the present study (Fig. 29c). This leads us to propose that hydroxyl groups located at step sites such as the one sketched in Fig. 29d are involved in the nucleation of gold on MgO$_{hydr}$ [116].

In summary, the STM, XPS, and IRAS results presented in Figs. 27, 28, and 29 provide evidence for a selective chemical interaction between Au and hydroxyl groups on MgO. Hydroxyl groups were found to act as strong anchoring sites for Au and the spectroscopic results indicate that oxidized Au species are formed on MgO$_{hydr}$. The enhanced sinter stability of Au on MgO$_{hydr}$ may consequently be explained by the stronger interaction of Au with the MgO surface due to the formation of strong Au–O interfacial bonds. Though Au-oxide species are not stable at elevated temperature and decompose (see Fig. 28b; which shows that after 600 K annealing the Au particles are essentially neutral), this strong interfacial interaction is the main reason for the sinter resistance of Au on MgO$_{hydr}$.

4.2 Surface Science Approach to Supported Au Catalyst Preparation

The results presented in the previous section have shown that chemical modification of an oxide surface with hydroxyl groups can have a strong stabilizing effect on Au particles, which is an important criterion in Au-related catalysis. The strong impact of Au particle size on the catalytic activity of oxide-supported Au has first been demonstrated by Haruta. In the seminal work published in 1987 [112], catalysts active in CO oxidation were synthesized by coprecipitation from aqueous solutions containing HAuCl$_4$ and the nitrate of transition metals such as Fe, Co, or Ni. This type of preparation has later on been replaced by another procedure, which consists of suspending a metal oxide (the support) in a HAuCl$_4$ solution adjusted to a fixed pH in the range pH 7–10, aging of the solution for 1 h at 343 K, followed by washing with distilled water, drying, and calcination. This method is frequently termed deposition-precipitation (DP), although several conditions that are typical for DP, such as the gradual rise of pH and the preferential precipitation of the precursor at the interface, are not met as outlined by Louis et al. [120]. A critical factor for obtaining small Au nanoparticles is the removal of chlorine from the supported phase, as chlorine enhances the mobility of Au species during drying and calcination. This is achieved by replacement of the Cl ligands in AuCl$_4^-$ ions by hydroxo ligands through hydrolysis at neutral or alkaline pH such that Au-hydroxo complexes ($Au(OH)_4^-$ or $AuCl(OH)_3^-$) are the primary precursors interacting with the support, and by the washing step, which eliminates chlorine from the support surface.

The DP approach used for the preparation of supported Au catalysts is particularly well suited for studying the deposition of Au onto single-crystalline oxide supports from an Au precursor solution because it merely requires the oxide surface to be contacted with the precursor solution and does not rely on the presence of pores such as in incipient wetness impregnation or the simultaneous precipitation of the support and the active metallic phase as done in coprecipitation. We attempted to recreate a situation similar to a DP procedure reported by Haruta by exposing the surface of a 10 nm thin $Fe_3O_4(111)$ film grown on Pt(111) to an aqueous solution of $HAuCl_4$ adjusted to pH 10. $Fe_3O_4(111)$/Pt(111) was chosen because of its high stability in aqueous solutions, as previously shown in our studies of Pd deposition from both strongly acidic [121] and strongly alkaline precursor solutions [122], as well as because of its conducting properties, which allows STM to be straightforwardly applied for morphological characterization.

Figure 30a, b display the Au 4f and Cl 2p photoemission spectra from an Au-$Fe_3O_4(111)$ sample obtained by exposure of $Fe_3O_4(111)$ to 1 mM $HAuCl_4$, pH10, for 5 min. at room temperature, after individual stages of preparation. The "as deposited" state represents the raw catalyst, which has been dried at room temperature but not rinsed. The main Au component present at this stage of preparation exhibits an Au $4f_{7/2}$ binding energy of 87.1 eV, which is in line with values reported for strongly oxidized Au and $Au(OH)_3$. The hydroxo complex $[Au(OH)_4]^-$ is the main hydrolysis product of aqueous $HAuCl_4$ at pH 10 and the occurrence of the high binding energy Au 4f component in XPS strongly supports previous conclusions about the Au-hydroxo complex as the main adsorbing species during DP of Au. In addition, two Au $4f_{7/2}$ components at lower binding energy, at 85.5 and 84.6 eV, are present, which are attributed to oxidic and metallic Au resulting from partial decomposition of the Au-hydroxo precursor on the surface. As expected, a large amount of chloride is present on the raw catalyst.

Rinsing the surface with distilled water completely removed chloride from the surface, but led in addition to reduction of the Au precursor to metallic Au, which exhibits an Au $4f_{7/2}$ binding energy of 84.2 eV at this stage of preparation (Fig. 30a, b, "rinsed"). Subsequent heating to 600 K in UHV leads only to a small additional 0.1 eV shift of the Au $4f_{7/2}$ component to 84.1 eV (Fig. 30a, "annealed").

The XPS results provide already a good indication for the successful realization of Au deposition onto the $Fe_3O_4(111)$ surface from aqueous precursors. In order to obtain further proof, the annealed Au-$Fe_3O_4(111)$ model catalyst was investigated by STM. Figure 30c displays a corresponding image taken in air, wherein the bright features are assigned to Au nanoparticles. Closer inspection of this image reveals a bimodal particle size distribution, with the larger particles exhibiting a diameter of 6–8 nm, and some smaller ones, which are more abundant on the left side of the STM image, with a diameter of about 4 nm. For comparison, Fig. 30d shows an STM image of a $Fe_3O_4(111)$ surface with Au particles obtained by physical vapor

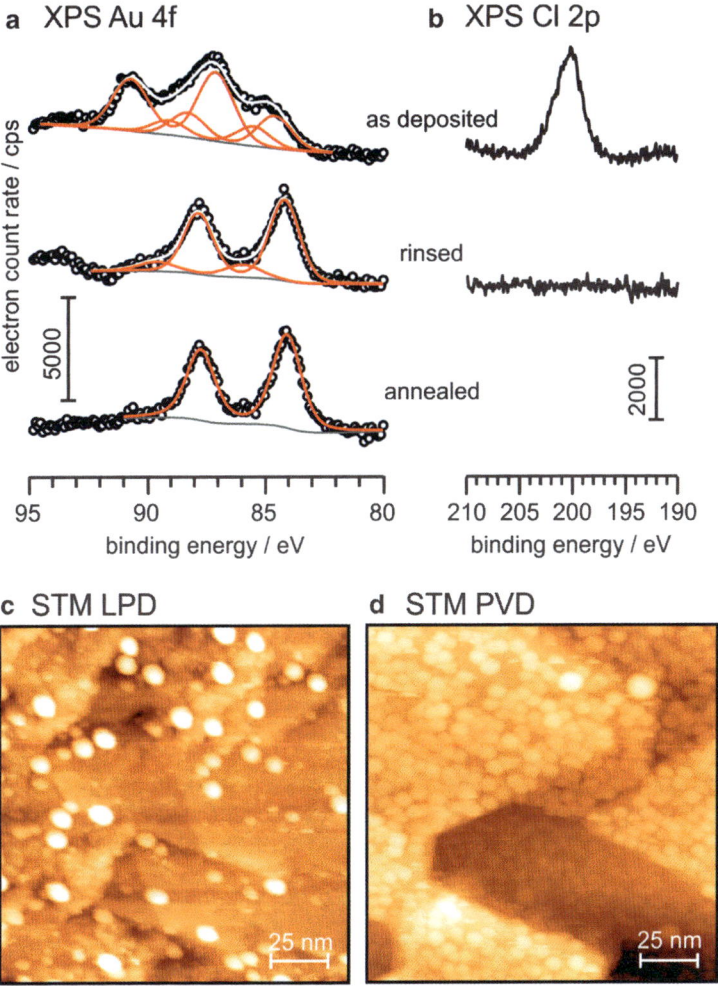

Fig. 30 Au 4f (**a**) and Cl 2p (**b**) photoemission spectra from an Au-Fe$_3$O$_4$(111) sample at different stages of preparation (from top to bottom: "as deposited" is after drying, "rinsed" is after rinsing with distilled water, "annealed" is after additional heating to 600 K in UHV). (**c**) STM image (150 × 150 nm) of the annealed Au-Fe$_3$O$_4$(111) sample prepared via aqueous Au precursor. (**d**) STM image (150 × 150 nm) of an annealed Au-Fe$_3$O$_4$(111) sample prepared by physical vapor deposition of Au in UHV

deposition of Au in UHV followed by annealing to 600 K. In this case, the Au particles are homogeneously distributed over the entire surface and the particle size distribution is narrow.

The chemical complexity of the system as well as the different interaction of Au with the chemically modified substrate is most probably responsible for the more heterogeneous Au particle distribution and morphology on the Au-Fe$_3$O$_4$(111)

sample prepared via the DP approach as compared the sample prepared in UHV. While more work is certainly necessary to understand the details of the metal-support interaction in this particular case, this work has demonstrated that progress is being made in applying surface science methodologies to study Au catalyst preparation procedures.

5 Synopsis

The growth of Au on ultrathin films and thicker, more bulk-like substrates by physical vapor deposition has been thoroughly studied and analyzed. The combination of imaging and spectroscopic techniques has allowed us derive a detailed picture of how clusters form from single Au atoms to clusters and further on to nanoparticles. The influence of charge of the particles and their relation to the adsorption of CO molecules have been discussed. By comparing three case studies, i.e., MgO, alumina, and iron oxide, the influence of the nature of the support, in particular, with respect to the charge, has been elucidated. While ultrathin MgO and alumina lead to electron transfer to the supported Au, iron oxide leads to transfer from the Au to the support. In order to proceed to more realistic supports we have modified the magnesia support chemically by hydroxylation and discuss the consequences for Au particle formation, as well as the impact of this modification on the particle charge. In the case of an iron oxide support it has even been possible to study Au particle deposition from solution and it is found that if certain conditions for reduction and calcination are used, the result is rather similar to a preparation of Au particles from physical vapor deposition. This study also shows that Michael Mingos comments on the comparability of molecular compounds and surface adsorbates have been close to the truth.

References

1. Mingos DMP (1984) Polyhedral skeletal electron pair approach. Acc Chem Res 17 (9):311–319. doi:10.1021/ar00105a003
2. Mingos DMP (1984) Gold cluster compounds. Gold Bull 17(1):5–12. doi:10.1007/BF03214670
3. Wade K (1971) The structural significance of the number of skeletal bonding electron-pairs in carboranes, the higher boranes and borane anions, and various transition-metal carbonyl cluster compounds. J Chem Soc D 0(15):792–793. doi:10.1039/C29710000792
4. Gillespie RJ, Hargittai I (1991) The VSEPR model of molecular geometry, 8th edn. Allyn & Bacon, Boston
5. Ertl G (2007) Reactions at surfaces: from atoms to complexity. http://www.nobelprize.org/nobel_prizes/chemistry/laureates/2007/ertl_lecture.pdf
6. Schauermann S, Nilius N, Shaikhutdinov S, Freund H-J (2012) Nanoparticles for heterogeneous catalysis: new mechanistic insights. Acc Chem Res. doi:10.1021/ar300225s
7. Pacchioni G, Freund H (2012) Electron transfer at oxide surfaces. The MgO paradigm: from defects to ultrathin films. Chem Rev 113(6):4035–4072. doi:10.1021/cr3002017

8. Risse T, Shaikhutdinov S, Nilius N, Sterrer M, Freund H-J (2008) Gold supported on thin oxide films: from single atoms to nanoparticles. Acc Chem Res 41(8):949–956
9. Freund H-J (2010) Model studies in heterogeneous catalysis. Chem Eur J 16(31):9384–9397. doi:10.1002/chem.201001724
10. Haruta M (2002) Catalysis of gold nanoparticles deposited on metal oxides. CatTech 6(3):102–115
11. Haruta M, Tsubota S, Kobayashi T, Kageyama H, Genet MJ, Delmon B (1993) Low-temperature oxidation of CO over gold supported on TiO_2, α-Fe_2O_3, and Co_3O_4. J Catal 144(1):175–192. doi:10.1006/jcat.1993.1322
12. Winkelmann F, Wohlrab S, Libuda J, Bäumer M, Cappus D, Menges M, Al-Shamery K, Kuhlenbeck H, Freund H-J (1994) Adsorption on oxide surfaces: structure and dynamics. Surf Sci 307–309 (Part 2):1148
13. Wichtendahl R, Rodriguez-Rodrigo M, Härtel U, Kuhlenbeck H, Freund H-J (1999) TDS study of the bonding of CO and NO to vacuum-cleaved NiO(100). Surf Sci 423(1):90–98
14. Wichtendahl R, Rodriguez-Rodrigo M, Härtel U, Kuhlenbeck H, Freund H-J (1999) Thermodesorption of CO and NO from vacuum-cleaved NiO(100) and MgO(100). Phys Status Solidi A 173(1):93–100. doi:10.1002/(sici)1521-396x(199905)173:1<93::aid-pssa93>3.0.co;2-4
15. Sterrer M, Risse T, Heyde M, Rust H-P, Freund H-J (2007) Crossover from three-dimensional to two-dimensional geometries of Au nanostructures on thin MgO(001) films: a confirmation of theoretical predictions. Phys Rev Lett 98(20):206103
16. Yulikov M, Sterrer M, Heyde M, Rust HP, Risse T, Freund H-J, Pacchioni G, Scagnelli A (2006) Binding of single gold atoms on thin MgO(001) films. Phys Rev Lett 96(14):146804
17. Nambu A, Graciani J, Rodriguez JA, Wu Q, Fujita E, Sanz JF (2006) N doping of TiO_2(110): photoemission and density-functional studies. J Chem Phys 125(9):094706
18. Rodriguez JA, Hanson JC, Kim J-Y, Liu G, Iglesias-Juez A, Fernández-García M (2003) Properties of CeO_2 and $Ce_{1-x}ZrxO_2$ nanoparticles: X-ray absorption near-edge spectroscopy, density functional, and time-resolved X-ray diffraction studies. J Phys Chem B 107 (15):3535–3543. doi:10.1021/jp022323i
19. Xu Y, Li J, Zhang Y, Chen W (2003) CO adsorption on MgO(001) surface with oxygen vacancy and its low-coordinated surface sites: embedded cluster model density functional study employing charge self-consistent technique. Surf Sci 525(1–3):13–23. doi:10.1016/s0039-6028(02)02566-9
20. Neyman KM, Ruzankin SP, Rösch N (1995) Adsorption of CO molecules on a MgO(001) surface. Model cluster density functional study employing a gradient-corrected potential. Chem Phys Lett 246(6):546–554. doi:10.1016/0009-2614(95)01150-X
21. Neyman KM, Roesch N (1992) Co bonding and vibrational-modes on a perfect Mgo (001) surface – Lcgto-Ldf model cluster investigation. Chem Phys 168(2–3):267–280
22. Pacchioni G (2000) Ab initio theory of point defects in oxide materials: structure, properties, chemical reactivity. Solid State Sci 2(2):161–179. doi:10.1016/s1293-2558(00)00113-8
23. Del Vitto A, Pacchioni G, Delbecq F, Sautet P (2005) Au atoms and dimers on the MgO(100) surface: a DFT study of nucleation at defects. J Phys Chem B 109(16):8040–8048
24. Street SC, Xu C, Goodman DW (1997) The physical and chemical properties of ultrathin oxide films. Annu Rev Phys Chem 48(1):43–68. doi:10.1146/annurev.physchem.48.1.43
25. Chambers SA (2000) Epitaxial growth and properties of thin film oxides. Surf Sci Rep 39 (5):105–180
26. Bäumer M, Freund H-J (1999) Metal deposits on well-ordered oxide films. Progr Surf Sci 61 (7–8):127–198. doi:10.1016/s0079-6816(99)00012-x
27. Giordano L, Cinquini F, Pacchioni G (2006) Tuning the surface metal work function by deposition of ultrathin oxide films: density functional calculations. Phys Rev B 73(4):045414
28. Pacchioni G, Giordano L, Baistrocchi M (2005) Charging of metal atoms on ultrathin MgO/Mo(100) films. Phys Rev Lett 94(22):226104
29. Bonzel HP (1988) Alkali-metal-affected adsorption of molecules on metal-surfaces. Surf Sci Rep 8(2):43–125

30. Diehl RD, McGrath R (1996) Structural studies of alkali metal adsorption and coadsorption on metal surfaces. Surf Sci Rep 23(2–5):43–171. doi:10.1016/0167-5729(95)00010-0
31. Kliewer J, Berndt R (2001) Low temperature scanning tunneling microscopy of Na on Cu (111). Surf Sci 477(2–3):250–258. doi:10.1016/S0039-6028(01)00891-3
32. Sterrer M, Risse T, Martinez Pozzoni U, Giordano L, Heyde M, Rust H-P, Pacchioni G, Freund H-J (2007) Control of the charge state of metal atoms on thin MgO films. Phys Rev Lett 98(9):096107
33. Sterrer M, Risse T, Giordano L, Heyde M, Nilius N, Rust HP, Pacchioni G, Freund H-J (2007) Palladium monomers, dimers, and trimers on the MgO(001) surface viewed individually. Angew Chem Int Ed 46(45):8703–8706
34. Wu SW, Ogawa N, Nazin GV, Ho W (2008) Conductance hysteresis and switching in a single-molecule junction. J Phys Chem C 112(14):5241–5244. doi:10.1021/jp7114548
35. Wu SW, Nazin GV, Chen X, Qiu XH, Ho W (2004) Control of relative tunneling rates in single molecule bipolar electron transport. Phys Rev Lett 93(23):236802
36. Jaeger RM, Kuhlenbeck H, Freund H-J, Wuttig M, Hoffmann W, Franchy R, Ibach H (1991) Formation of a well-ordered aluminium oxide overlayer by oxidation of NiA(110). Surf Sci 259(3):235–252. doi:10.1016/0039-6028(91)90555-7
37. Nilius N, Ganduglia-Pirovano VG, Bradzova V, Kulawik M, Sauer J, Freund H-J (2008) Counting electrons transferred through a thin alumina film into Au chains. Phys Rev Lett 100 (9):09802-1–4
38. Giordano L, Pacchioni G (2011) Oxide films at the nanoscale: new structures, new functions, and new materials. Acc Chem Res 44(11):1244–1252. doi:10.1021/ar200139y
39. Nilius N, Ganduglia-Pirovano MV, Brazdova V, Kulawik M, Sauer J, Freund H-J (2010) Electronic properties and charge state of gold monomers and chains adsorbed on alumina thin films on NiAl(110). Phys Rev B 81(4):045422
40. Kulawik M, Nilius N, Freund H-J (2006) Influence of the metal substrate on the adsorption properties of thin oxide layers: Au atoms on a thin alumina film on NiAl(110). Phys Rev Lett 96(3):036103
41. Hamers RJ (1989) Atomic-resolution surface spectroscopy with the scanning tunneling microscope. Annu Rev Phys Chem 40(1):531–559. doi:10.1146/annurev.pc.40.100189. 002531
42. Tersoff J, Hamann DR (1983) Theory and application for the scanning tunneling microscope. Phys Rev Lett 50(25):1998–2001
43. Kittel C (1996) Introduction to solid state physics, 7th edn. Wiley, New York
44. Schmid M, Shishkin M, Kresse G, Napetschnig E, Varga P, Kulawik M, Nilius N, Rust HP, Freund H-J (2006) Oxygen-deficient line defects in an ultrathin aluminum oxide film. Phys Rev Lett 97(4):046101–046104
45. Rienks EDL, Nilius N, Rust HP, Freund H-J (2005) Surface potential of a polar oxide film: FeO on Pt(111). Phys Rev B 71:2414041–2414044
46. Giordano L, Pacchioni G, Goniakowski J, Nilius N, Rienks EDL, Freund H-J (2008) Charging of metal adatoms on ultrathin oxide Films: Au and Pd on FeO/Pt(111). Phys Rev Lett 101(2):026102
47. Simic-Milosevic V, Heyde M, Nilius N, Koenig T, Rust HP, Sterrer M, Risse T, Freund H-J, Giordano L, Pacchioni G (2008) Au dimers on thin MgO(001) films: flat and charged or upright and neutral? J Am Chem Soc 130(25):7814–7815. doi:10.1021/ja8024388
48. Simic-Milosevic V, Heyde M, Lin X, König T, Rust H-P, Sterrer M, Risse T, Nilius N, Freund H-J, Giordano L, Pacchioni G (2008) Charge-induced formation of linear Au clusters on thin MgO films: scanning tunneling microscopy and density-functional theory study. Phys Rev B 78(23):235429
49. Nilius N (2009) Properties of oxide thin films and their adsorption behavior studied by scanning tunneling microscopy and conductance spectroscopy. Surf Sci Rep 64(12):595–659
50. Frondelius P, Häkkinen H, Honkala K (2007) Adsorption of small Au clusters on MgO and MgO/Mo: the role of oxygen vacancies and the Mo-support. New J Phys 9:339. doi:10.1088/1367-2630/9/9/339

51. Nilius N, Wallis TM, Ho W (2002) Development of one-dimensional band structure in artificial gold chains. Science 297:1853–1856
52. Wallis TM, Nilius N, Ho W (2002) Electronic density oscillations in gold atomic chains assembled atom by atom. Phys Rev Lett 89(23):236802
53. Nilius N, Wallis TM, Persson M, Ho W (2003) Distance dependence of the interaction between single atoms: gold dimers on NiAl(110). Phys Rev Lett 90(19):196103
54. Henry CR (1998) Surface studies of supported model catalysts. Surf Sci Rep 31:231–326
55. Ricci D, Bongiorno A, Pacchioni G, Landman U (2006) Bonding trends and dimensionality crossover of gold nanoclusters on metal-supported MgO thin films. Phys Rev Lett 97 (3):036106
56. Lin X, Nilius N, Freund H-J, Walter M, Frondelius P, Honkala K, Häkkinen H (2009) Quantum well states in two-dimensional gold clusters on MgO thin films. Phys Rev Lett 102 (20):206801-1–4
57. Lin X, Nilius N, Sterrer M, Koskinen P, Haekkinen H, Freund H-J (2010) Characterizing low-coordinated atoms at the periphery of MgO-supported Au islands using scanning tunneling microscopy and electronic structure calculations. Phys Rev B 81 (15):153406-1–4
58. Andersin J, Nevalaita J, Honkala K, Häkkinen H (2013) The redox chemistry of gold with high-valence doped calcium oxide. Angew Chem Int Ed 52(5):1424–1427. doi:10.1002/anie. 201208443
59. Walter M, Frondelius P, Honkala K, Häkkinen H (2007) Electronic structure of MgO-supported au clusters: quantum dots probed by scanning tunneling microscopy. Phys Rev Lett 99(9):096102
60. de Heer WA (1993) The physics of simple metal clusters: experimental aspects and simple models. Rev Mod Phys 65(3):611–676
61. Valden M, Lai X, Goodman DW (1998) Onset of catalytic activity of gold clusters on Titania with the appearance of nonmetallic properties. Science 281:1647–1650
62. Cabrera N, Mott NF (1948) Theory of the oxidation of metals. Rep Progr Phys 12:163
63. Molina LM, Hammer B (2005) Some recent theoretical advances in the understanding of the catalytic activity of Au. Appl Catal A 291(1–2):21–31. doi:10.1016/j.apcata.2005.01.050
64. Green IX, Tang WJ, Neurock M, Yates JT (2011) Spectroscopic observation of dual catalytic sites during oxidation of CO on a Au/TiO$_2$ catalyst. Science 333:736–739
65. Abbet S, Riedo E, Brune H, Heiz U, Ferrari AM, Giordano L, Pacchioni G (2001) Identification of defect sites on MgO(100) thin films by decoration with Pd atoms and studying CO adsorption properties. J Am Chem Soc 123(25):6172–6178
66. Lin X, Yang B, Benia HM, Myrach P, Yulikov M, Aumer A, Brown M, Sterrer M, Bondarchuk O, Kieseritzky E, Rocker J, Risse T, Gao H, Nilius N, Freund H-J (2010) Charge-mediated adsorption behavior of CO on MgO-supported Au clusters. J Am Chem Soc 132(22):7745–7749
67. Hashmi ASK, Hutchings GJ (2006) Gold catalysis. Angew Chem Int Ed 45(47):7896–7936. doi:10.1002/anie.200602454
68. Herzing AA, Kiely CJ, Carley AF, Landon P, Hutchings GJ (2008) Identification of active gold nanoclusters on iron oxide supports for CO oxidation. Science 321:1331–1335. doi:10. 1126/science.1159639
69. Chen MS, Goodman DW (2004) The structure of catalytically active Au on titania. Science 306:252–255
70. Ekardt W (ed) (1999) Metal clusters. Wiley, Chichester
71. Kreibig U, Vollmer W (eds) (1995) Optical properties of metal clusters, vol 25. Springer series in materials science. Springer, Berlin-New York
72. Wendt S, Sprunger PT, Lira E, Madsen GKH, Li Z, Hansen JØ, Matthiesen J, Blekinge-Rasmussen A, Lægsgaard E, Hammer B, Besenbacher F (2008) The role of interstitial sites in the Ti3d defect state in the band gap of titania. Science 320(5884):1755–1759. doi:10.1126/ science.1159846

73. Kim HY, Lee HM, Pala RGS, Shapovalov V, Metiu H (2008) CO oxidation by rutile $TiO_2(110)$ doped with V, W, Cr, Mo, and Mn. J Phys Chem C 112(32):12398–12408. doi:10.1021/jp802296g

74. Shapovalov V, Metiu H (2007) Catalysis by doped oxides: CO oxidation by $Au_xCe_{1-x}O_2$. J Catal 245(1):205–214. doi:10.1016/j.jcat.2006.10.009

75. Mammen N, Narasimhan S, Sd G (2011) Tuning the morphology of gold clusters by substrate doping. J Am Chem Soc 133(9):2801–2803. doi:10.1021/ja109663g

76. Wang JX, Lunsford JH (1986) Characterization of [Li+O] centers in lithium-doped magnesium oxide catalysts. J Phys Chem 90(22):5883–5887. doi:10.1021/j100280a084

77. Ito T, Wang J, Lin CH, Lunsford JH (1985) Oxidative dimerization of methane over a lithium-promoted magnesium oxide catalyst. J Am Chem Soc 107(18):5062–5068. doi:10.1021/ja00304a008

78. Pala RGS, Metiu H (2007) The structure and energy of oxygen vacancy formation in clean and doped, very thin films of ZnO. J Phys Chem C 111(34):12715–12722. doi:10.1021/jp073424p

79. Nolan M, Verdugo VS, Metiu H (2008) Vacancy formation and CO adsorption on gold-doped ceria surfaces. Surf Sci 602(16):2734–2742. doi:10.1016/j.susc.2008.06.028

80. Shao X, Prada S, Giordano L, Pacchioni G, Nilius N, Freund H-J (2011) Tailoring the shape of metal Ad-particles by doping the oxide support. Angew Chem Int Ed 50(48):11525–11527. doi:10.1002/anie.201105355

81. Stavale F, Shao X, Nilius N, Freund H-J, Prada S, Giordano L, Pacchioni G (2012) Donor characteristics of transition-metal-doped oxides: Cr-Doped MgO versus Mo-doped CaO. J Am Chem Soc 134(28):11380–11383. doi:10.1021/ja304497n

82. Stavale F, Nilius N, Freund H-J (2012) Cathodoluminescence of near-surface centres in Cr-doped MgO(001) thin films probed by scanning tunnelling microscopy. New J Phys 14:033006. doi:http://iopscience.iop.org/1367-2630/14/3/033006/

83. Prada S, Giordano L, Pacchioni G (2013) Charging of gold atoms on doped MgO and CaO: identifying the key parameters by DFT calculations. J Phys Chem C 117(19):9943–9951. doi:10.1021/jp401983m

84. Henderson B, Imbusch GF (eds) (1989) Optical spectroscopy of inorganic solids. Oxford University Press, Oxford

85. Lide DR (1996) CRC handbook of chemistry and physics. CRC Press Inc, Boca Raton

86. Benia HM, Myrach P, Gonchar A, Risse T, Nilius M, Freund H-J (2010) Electron trapping in misfit dislocations of MgO thin films. Phys Rev B 81:241415. doi:http://link.aps.org/doi/10.1103/PhysRevB.81.241415

87. Li B, Metiu H (2010) DFT studies of oxygen vacancies on undoped and doped La_2O_3 surfaces. J Phys Chem C 114(28):12234–12244. doi:10.1021/jp103604b

88. Hu Z, Li B, Sun X, Metiu H (2011) Chemistry of doped oxides: the activation of surface oxygen and the chemical compensation effect. J Phys Chem C 115(7):3065–3074. doi:10.1021/jp110333z

89. Shao X, Nilius N, Freund H-J (2012) Li/Mo codoping of CaO films: a means to tailor the equilibrium shape of Au deposits. J Am Chem Soc 134(5):2532–2534. doi:10.1021/ja211396t

90. Sterrer M, Heyde M, Novicki M, Nilius N, Risse T, Rust HP, Pacchioni G, Freund H-J (2006) Identification of color centers on MgO(001) thin films with scanning tunneling microscopy. J Phys Chem B 110(1):46–49. doi:10.1021/jp056306f

91. Myrach P, Nilius N, Levchenko SV, Gonchar A, Risse T, Dinse K-P, Boatner LA, Frandsen W, Horn R, Freund H-J, Schlögl R, Scheffler M (2010) Temperature-dependent morphology, magnetic and optical properties of Li-doped MgO. ChemCatChem 2(7):854–862. doi:10.1002/cctc.201000083

92. Mihaylov M, Knözinger H, Hadjiivanov K, Gates BC (2007) Characterization of the oxidation states of supported gold species by IR spectroscopy of adsorbed CO. Chem Ing Tech 79(6):795–806. doi:10.1002/cite.200700029

93. Sterrer M, Yulikov M, Risse T, Freund H-J, Carrasco J, Illas F, Di Valentin C, Giordano L, Pacchioni G (2006) When the reporter induces the effect: unusual IR spectra of CO on Au_1/MgO(001)/Mo(001). Angew Chem Int Ed 45(16):2633–2635. doi:10.1002/anie. 200504473

94. Yang B, Lin X, Gao H-J, Nilius N, Freund H-J (2010) CO adsorption on thin MgO films and single Au adatoms: a scanning tunneling microscopy study. J Phys Chem C 114 (19):8997–9001. doi:10.1021/jp100757y

95. Ho W (2002) Single-molecule chemistry. J Chem Phys 117(24):11033–11061

96. Lauhon LJ, Ho W (1999) Single-molecule vibrational spectroscopy and microscopy:CO on Cu(001) and Cu(110). Phys Rev B 60(12):R8525–R8528

97. Nilius N, Wallis TM, Ho W (2002) Vibrational spectroscopy and imaging of single molecules: bonding of CO to single palladium atoms on NiAl(110). J Chem Phys 117 (24):10947–10952

98. Wang H, Tobin RG, Lambert DK (1994) Coadsorption of hydrogen and CO on Pt(335): structure and vibrational Stark effect. J Chem Phys 101(5):4277–4287

99. Hammer B (2006) Special sites at noble and late transition metal catalysts. Top Catal 37(1):3–16. doi:10.1007/s11244-006-0004-y

100. Sicolo S, Giordano L, Pacchioni G (2009) CO Adsorption on one-, two-, and three-dimensional au clusters supported on MgO/Ag(001) ultrathin films. J Phys Chem C 113(23):10256–10263. doi:10.1021/jp9023266

101. Blyholder G (1964) Molecular orbital view of chemisorbed carbon monoxide. J Phys Chem 68:2772–2778

102. Hövel H, Barke I (2003) Large noble metal clusters: electron confinement and band structure effects. New J Phys 5(1):31

103. Benia HM, Lin X, Gao HJ, Nilius N, Freund H-J (2007) Nucleation and growth of gold on MgO thin films: a combined STM and luminescence study. J Phys Chem C 111(28):10528–10533

104. Fielicke A, von Helden G, Meijer G, Simard B, Rayner DM (2005) Gold cluster carbonyls: vibrational spectroscopy of the anions and the effects of cluster size, charge, and coverage on the CO stretching frequency. J Phys Chem B 109(50):23935–23940

105. Shaikhutdinov SK, Meyer R, Naschitzki M, Bäumer M, Freund H-J (2003) Size and support effects for CO adsorption on gold model catalysts. Catal Lett 86(4):211–219. doi:10.1023/a:1022616102162

106. Ruggiero C, Hollins P (1997) Interaction of CO molecules with the Au(332) surface. Surf Sci 377–379(0):583–586. doi:10.1016/S0039-6028(96)01451-3

107. Weststrate CJ, Lundgren E, Andersen JN, Rienks EDL, Gluhoi AC, Bakker JW, Groot IMN, Nieuwenhuys BE (2009) CO adsorption on Au(310) and Au(321): 6-fold coordinated gold atoms. Surf Sci 603(13):2152–2157. doi:10.1016/j.susc.2009.04.026

108. Lemire C, Meyer R, Shaikhutdinov SK, Freund H-J (2004) CO adsorption on oxide supported gold: from small clusters to monolayer islands and three-dimensional nanoparticles. Surf Sci 552(1–3):27–34. doi:10.1016/j.susc.2004.01.029

109. Nilius N, Rienks EDL, Rust H-P, Freund H-J (2005) Self-organization of gold atoms on a polar FeO(111) surface. Phys Rev Lett 95(6):066101

110. Meyer R, Lemire C, Shaikhutdinov SK, Freund H-J (2004) Surface chemistry of catalysis by gold. Gold Bull 37(1–2):72–124. doi:10.1007/bf03215519

111. Starr DE, Pazhetnov EM, Stadnichenko AI, Boronin AI, Shaikhutdinov SK (2006) Carbon films grown on Pt(111) as supports for model gold catalysts. Surf Sci 600(13):2688–2695. doi:10.1016/j.susc.2006.04.035

112. Haruta M, Kobayashi T, Sano H, Yamada N (1987) Novel gold catalysts for the oxidation of carbon-monoxide at a temperature far below 0 °C. Chem Lett 2:405–408. doi:10.1246/cl. 1987.405

113. Liu P, Kendelewicz T, Gordon GE, Parks GA (1998) Reaction of water with MgO(100) surfaces. Part I: synchrotron X-ray photoemission studies of low-defect surfaces. Surf Sci 412–13:287–314. doi:10.1016/s0039-6028(98)00444-0

114. Mejias JA, Berry AJ, Refson K, Fraser DG (1999) The kinetics and mechanism of MgO dissolution. Chem Phys Lett 314(5–6):558–563. doi:10.1016/s0009-2614(99)00909-4

115. Carrasco E, Brown MA, Sterrer M, Freund H-J, Kwapien K, Sierka M, Sauer J (2010) Thickness-dependent hydroxylation of MgO(001) thin films. J Phys Chem C 114 (42):18207–18214. doi:10.1021/jp105294e

116. Brown MA, Fujimori Y, Ringleb F, Shao X, Stavale F, Nilius N, Sterrer M, Freund H-J (2011) Oxidation of Au by surface OH: nucleation and electronic structure of gold on hydroxylated MgO(001). J Am Chem Soc 133(27):10668–10676. doi:10.1021/ja204798z

117. Brown MA, Carrasco E, Sterrer M, Freund H-J (2010) Enhanced stability of gold clusters supported on hydroxylated MgO(001) surfaces. J Am Chem Soc 132(12):4064–4065. doi:10. 1021/ja100343m

118. Veith GM, Lupini AR, Dudney NJ (2009) Role of pH in the formation of structurally stable and catalytically active TiO$_2$-supported gold catalysts. J Phys Chem C 113(1):269–280. doi:10.1021/jp808249f

119. Chizallet C, Costentin G, Che M, Delbecq F, Sautet P (2007) Infrared characterization of hydroxyl groups on MgO: a periodic and cluster density functional theory study. J Am Chem Soc 129(20):6442–6452. doi:10.1021/ja068720e

120. Zanella R, Delannoy L, Louis C (2005) Mechanism of deposition of gold precursors onto TiO$_2$ during the preparation by cation adsorption and deposition-precipitation with NaOH and urea. Appl Catal A 291(1–2):62–72. doi:10.1016/j.apcata.2005.02.045

121. Wang HF, Ariga H, Dowler R, Sterrer M, Freund H-J (2012) Surface science approach to catalyst preparation – Pd deposition onto thin Fe$_3$O$_4$(111) films from PdCl$_2$ precursor. J Catal 286:1–5. doi:10.1016/j.jcat.2011.09.026

122. Wang HF, Kaden WE, Dowler R, Sterrer M, Freund H-J (2012) Model oxide-supported metal catalysts – comparison of ultrahigh vacuum and solution based preparation of Pd nanoparticles on a single-crystalline oxide substrate. Phys Chem Chem Phys 14(32):11525–11533. doi:10.1039/c2cp41459g

Struct Bond (2014) 162: 139–230
DOI: 10.1007/430_2014_140
© Springer International Publishing Switzerland 2014
Published online: 20 July 2014

Gas Phase Formation, Structure and Reactivity of Gold Cluster Ions

Athanasios Zavras, George N. Khairallah, and Richard A.J. O'Hair

Abstract With the advent of electrospray ionisation (ESI) and matrix-assisted laser desorption ionisation (MALDI), mass spectrometry (MS) is now routinely used to establish the molecular formulae of gold nanoclusters (AuNCs). ESI-MS has been used to monitor the solution phase growth of AuNCs when gold salts are reduced in the presence of phosphine or thiolate ligands. Beyond this analytical role, over the past 2 decades MS-based methods have been employed to examine the fundamental properties and reactivities of AuNC ions. For example, ion mobility and spectroscopic measurements may be used to assign structures; thermochemical data provides important information on ligand binding energies; unimolecular chemistry can be explored; and ion–molecule reactions with various substrates can be used to probe catalysis by AuNC ions. MS can also be used to monitor and direct the synthesis of AuNC bulk material either by guiding solution phase synthesis conditions or by soft landing a beam of mass-selected (i.e. monodisperse) AuNC ions onto a surface. This review showcases all areas in which mass spectrometry has played a role in AuNC science.

Keywords Catalysis · Gas-phase reactions · Gold nanoclusters · Mass spectrometry · Spectroscopy · Synthesis

Contents

1 Introduction 141
   1.1 Why Are Clusters Interesting? 141
   1.2 Why Are Gold Clusters Interesting? 142
   1.3 Definition of a Gold Cluster Ion 142
   1.4 Scope of the Review 144

A. Zavras, G.N. Khairallah, and R.A.J. O'Hair (✉)
School of Chemistry, University of Melbourne, Parkville, VIC 3010, Australia
e-mail: rohair@unimelb.edu.au

2   MS-Based Analysis of Gold Nanoclusters Synthesised in the Condensed Phase 145
    2.1   The Need for Net Charge to Allow for Analysis of Gold Nanoclusters via MS ... 145
    2.2   Top-Down and Bottom-Up Approaches for the Synthesis of Gas-Phase Gold
          Nanoclusters 147
    2.3   Thiolate Ligands 150
    2.4   Phosphine Ligands 152
    2.5   MS-Based Fragmentation Methods for the Production of Gas-Phase Gold
          Cluster Ions 161
    2.6   Formation of Carbides, Hydroxides, Phosphides and Tellurides from
          Miscellaneous Top-Down Approaches 164
3   Tools for Determining Structures and Their Application to Gold Cluster Ions 165
    3.1   MS-Based Fragmentation Methods 165
    3.2   Ion Mobility 170
    3.3   IR Spectroscopy 172
    3.4   UV–Vis Spectroscopy 173
    3.5   Trapped Ion Electron Diffraction 175
    3.6   Photoelectron Spectroscopy 175
    3.7   Thermochemistry 179
4   Reactivity of Gold Cluster Ions 180
    4.1   Reactions of Bare Gold Cluster Ions: Overview 183
    4.2   Ligated Gold Cluster Ions 192
    4.3   Gold-Containing Bimetallic Cluster Ions 199
    4.4   Catalysis by Gold Cluster Ions 202
5   From the Gas Phase to Materials 206
    5.1   MS-Directed Synthesis of Gold Clusters 206
    5.2   MS-Selected Deposition of Gold Cluster Ions 207
6   Conclusions 211
References 212

Abbreviations

2D	Two-dimensional
3D	Three-dimensional
AuNP	Gold nanoparticle
BINAP	(±)-2,2′-Bis(diphenylphosphino)-1,1′-binapthylene
BTBC	Borane *tert*-butylamine complex
CID	Collision-induced dissociation
ClAuPPh$_3$	Chlorotriphenylphosphinegold(I)
DFT	Density functional theory
DLS	Direct light scattering
DMA	Differential mobility analysis
DMG	*N,N*-Dimethylglycine
DMSO	Dimethylsulfoxide
dppe	1,2-Bis(diphenylphosphino)ethane
dppm	Bis(diphenylphosphino)methane
dppp	1,3-Bis(diphenylphosphino)propane OR 1,5-Bis(diphenylphosphino)pentane
EA	Electron affinity
EI	Electron ionisation

EID	Electron-induced dissociation
ESI	Electrospray ionisation
eV	Electron volt
FAB	Fast atom bombardment
FT	Fourier transform
ICP	Inductively coupled plasma
ICR	Ion cyclotron resonance
IMS	Ion mobility spectrometry
IR	Infrared
K	Kelvin
MALDI	Matrix-assisted laser desorption ionisation
MPC	Monolayer-protected cluster
MS	Mass spectrometry
MS^n	Tandem mass spectrometry
NC	Nanocluster
nm	Nanometre
NMR	Nuclear magnetic resonance
PD	Photodissociation
PES	Photoelectron spectroscopy
RRKM	Rice–Ramsperger–Kassel–Marcus
SAM	Self-assembled monolayer
SEC	Size exclusion chromatography
SIMS	Secondary ion mass spectrometry
SORI	Sustained off-resonance irradiation
THF	Tetrahydrofuran
TIED	Trapped ion electron diffraction
TOF	Time of flight
TS	Transition state
UV–Vis	Ultraviolet visible

1 Introduction

1.1 Why Are Clusters Interesting?

As Castleman and Jena have previously noted [1], clusters are interesting since they (1) 'bridge phases as well as disciplines', including the studies of (a) the environment, materials science and biology [2] and (b) physics and chemistry [3] and (2) 'have come to symbolize a new embryonic form of matter that is intermediate between atoms and their bulk counterpart' [1]. For some time the motivation to study clusters was to model bulk behaviour by assuming that their properties vary smoothly as some power law until they reach the bulk limit. It has now become well appreciated that for certain cluster size regimes, the properties of materials can change in a highly non-monotonic fashion [3]. Thus the new motivation has become to control the size and shape of these cluster materials so that their properties can be controlled for novel material and catalytic applications.

1.2 Why Are Gold Clusters Interesting?

There are at least two reasons why there has been intense interest in gold clusters: (1) they serve as models for structure and bonding of cluster compounds [4–9], a topic further elaborated in [10] ('Theoretical studies on gold clusters and nanoparticles'). (2) Seminal work by Bond [11], Hutchings [12], Haruta [13] and others [14–16] in the 1970s and 1980s in the field of heterogeneous catalysis led to a paradigm shift by showing that gold, the noblest of all metals, can be turned into a highly active catalyst simply by reducing its dimensions to a few nanometres [13]. Examples of catalytic reactions that were discovered include hydrogenation of alkenes (Eq. (1)) [11], hydrochlorination of acetylene (Eq. (2)) [12] and oxidation of CO (Eq. (3)) [13]. Finally, it has been recognised for some time that homogenous catalysis may give rise to nanoparticles that maybe the actual catalysts [17]. Indeed a recent report by Corma et al. [18] has highlighted that such nanoclusters can be highly reactive catalysts for the ester-assisted addition of water to alkynes (Eq. (4)), with turnover numbers of ten million [19].

$$\begin{array}{c} R \\ \diagup \\ H \end{array} = \begin{array}{c} H \\ \diagup \\ H \end{array} \quad \xrightarrow[\text{Au cat.}]{H_2} \quad \begin{array}{c} R \quad H \\ \diagup \diagdown H \\ H \diagdown \diagup H \\ H \end{array} \tag{1}$$

$$H-C\equiv C-H \quad \xrightarrow[\text{Au cat.}]{HCl} \quad \begin{array}{c} H \quad H \\ \diagup \diagdown \diagup \\ H \quad Cl \end{array} \tag{2}$$

$$C=O \quad \xrightarrow[\text{Au cat.}]{O_2} \quad O=C=O \tag{3}$$

$$\begin{array}{c} O \\ \parallel \\ Cl \diagup \diagdown R \end{array} + \begin{array}{c} OH \\ \diagup \diagup \\ \end{array} \quad \xrightarrow[\text{Au NCs}]{H_2O\ (1\text{eq.})} \quad \begin{array}{c} O \\ \diagdown \diagup O \diagdown R \\ \parallel \\ O \end{array} \tag{4}$$

1.3 Definition of a Gold Cluster Ion

A diverse range of scientists have been involved in the study of clusters, which has given rise to a 'Tower of Babel' of other terms for clusters, including colloids, Q-particles, quantum dots, nanoparticles, grains, aerosols, hydrosols, dust, foam, non-covalent complexes, supramolecular complexes, protein assemblies, etc. Johnston [20] suggests that the earliest scientific reference to clusters may have been made by Robert Boyle in 1661, when he wrote of '...minute masses or clusters ... as were not easily dissipable into such particles as compos'd them' [21]. Since then, the word 'cluster' has been commonly used by physicists, chemists and mass spectrometrists. Each group has their own definition of a cluster (The old IUPAC mass spectrometry definition of a cluster is as follows: An ion formed by the combination of more ions or atoms or molecules of a chemical species often in association with a second species: [22]),[1,2] and these definitions can be further expanded to include the types of chemical bonding that 'holds the cluster together' [23]. The most recent recommended IUPAC definition of a *cluster ion* in mass spectrometry is an 'Ion formed by the combination via non-covalent forces of two or more atoms or molecules of one or more chemical species with an ion' [24], and thus this definition encompasses a wide range of ions formed from chemical and biological sources and also across all the types of bonding.

Further distinctions have been made in the literature based on size and other criteria:

1. An encyclopaedia article [25] has classified clusters as

   (a) 'Microclusters having from 3 to 10–13 atoms. Concepts and methods of molecular physics are applicable';
   (b) 'Small clusters have from 10 to 13 to about 100 atoms. Many different geometrical isomers exist for a given cluster size with almost the same energies. Molecular concepts lose their applicability';
   (c) 'Large clusters have from 100 to 1,000 atoms. A gradual transition is observed to the properties of the solid state. Small nanoparticles and nanocrystals have at least 1,000 atoms. These bodies display some of the properties of the solids state'.

2. Kreibig and Vollmer have defined metal clusters by the number, N, of atoms per cluster as being:

   (a) 'very small clusters' fall in the range $2 < N \leq 20$;
   (b) 'small clusters' are those where $20 \leq N \leq 500$;
   (c) 'larger clusters' span a substantial range of sizes: $500 \leq N \leq 10^7$ [26].

3. Ott and Finke have made an important distinction between transition-metal nanoclusters and colloids: 'nanoclusters are expected to be smaller (1–10 nm)

---

[1] The IUPAP does not appear to have a formal definition of a cluster. Johnston [20] suggests the following 'physics' definition of clusters: I will take the term cluster to mean an aggregate of a countable number (2–10n, where n can be as high as 6 or 7) of particles (i.e. atoms or molecules). The constituent particles may be identical, leading to homo-atomic (or homo-molecular) clusters, Aa, or they can be two or more different species – leading to hetero-atomic (hetero-molecular) clusters AaBb.

[2] The IUPAC inorganic chemistry definition of cluster is as follows: A number of metal centres grouped close together which can have direct metal bonding interactions or interactions through a bridging ligand, but are not necessarily held together by these interactions. See IUPAC. Compendium of Chemical Terminology, 2nd ed. (the "Gold Book"). Compiled by A. D. McNaught and A. Wilkinson. Blackwell Scientific Publications, Oxford (1997). XML online corrected version: http://goldbook.iupac.org (2006) created by M. Nic, J. Jirat, B. Kosata; updates compiled by A. Jenkins. ISBN 0-9678550-9-8. doi:10.1351/goldbook.

with near-monodisperse size distributions ($\leq$15%), while colloids are often >10 nm with much broader size distributions' [27].

Since there are limits on many mass analysers used in mass spectrometry-based approaches to the study of gold cluster ions, in this chapter we take as our definition of a gold cluster ion:

A cluster containing 2 or more gold atoms and up to 1,000 atoms that has an overall negative or positive charge. The cluster can be either a 'bare gold cluster' or a 'ligated gold cluster'.

1.4 Scope of the Review

There have been a number of general reviews on gas-phase cluster ions [28–32] as well as reviews on gold cluster ions [33, 34]. This review focuses on literature reports that utilise mass spectrometry (MS): (1) as a key analytical tool to examine the formation and reactions of gold clusters; (2) as basic methods to study the structure, properties and reactivity of gold cluster ions in the gas phase and (3) as a way of soft landing gold cluster ions onto surfaces to generate catalysts.

All mass spectrometers consist of: (1) an ionisation source to generate ions in or to transfer ions to the gas phase, (2) a mass analyser to separate ions according to their mass to charge ratio (m/z), (3) a detector to 'count' the number of ions at each m/z value and (4) a data processor/recorder [35]. The development of matrix-assisted laser desorption (MALDI) and electrospray ionisation (ESI) has revolutionised the analysis and study of inorganic and organometallic complexes and nanoclusters via mass spectrometry [36]. MALDI and ESI have been coupled to a range of mass analysers, and the resultant mass spectrometers are commonly used in the analysis of metal clusters. The mode of operation and the 'figures of merit' of the various types of mass analysers available are not discussed here – readers are referred to an excellent review [37].

There are many potential combinations of ionisation source and mass analyser that can be used for fundamental studies, but these are not discussed in any detail here. The most widely used mass spectrometers in gold cluster ion studies over the past decades have been ion cyclotron resonance (ICR) mass spectrometers (including their Fourier transform variants) [38–40], triple quadrupole mass spectrometers [41, 42], flowing afterglow reactors [43] and ion-trap mass spectrometers [44].

While computational chemistry plays a key role in supporting many experimental studies on the structure and reactivity of gold clusters, papers solely dealing with theoretical calculations are largely neglected except where they shed important insights into prior experimental work.

The structure and reactivity of neutral gold clusters have also been studied in the gas phase (see Sect. 3.3). Readers interested in this topic should also refer to recent work on the absorption of CO onto silver-doped gold clusters [45] and the oxidation of neutral gold carbonyl clusters by O_2 and N_2O [46] together with the reviews on neutral metal clusters by Knickelbein [47] and Shi and Bernstein [48].

Finally, while there are a growing number of reports on the use of gold nanoparticles as a co-additive in the analysis of low molecular weight compounds by laser desorption ionisation MS [49–53], these are not reviewed here.

2 MS-Based Analysis of Gold Nanoclusters Synthesised in the Condensed Phase

Historically, mass spectrometry-based analysis of gold cluster compounds has lagged behind other methods such as X-ray crystallography due to traditional ionisation methods such as electron ionisation or chemical ionisation being incompatible with compounds of low volatility. It appears that the first analytical studies of gold cluster compounds had to wait until the advent of fast atom bombardment (FAB) [54], which allowed transfer of gold cluster cations such as $[Au_6(PPh_3)_6]^{2+}$ to the gas phase [55].

2.1 The Need for Net Charge to Allow for Analysis of Gold Nanoclusters via MS

While MALDI-MS and ESI-MS have been a boon for the analysis of metal complexes and clusters, these ionisation methods are challenged by compounds that have no net charge. For inorganic and organometallic complexes, a number of strategies have been developed to overcome this limitation: (1) neutral metal halides can have a halide anion replaced with a neutral ligand such as pyridine, resulting in a MS detectable cationic complex [36]; (2) neutral ligands can be swapped for related fixed charge analogues [56]; (3) addition of acids or bases can allow protonation or deprotonation of coordinated ligands, thereby providing a positive or negative charge [36]; (4) oxidation to form a charged complex in which the metal centre is in a higher oxidation state [36]; (5) addition of a metal ion such as Ag^+ to form a coordination complex [36] and (6) carbonyl derivatisation of neutral metal carbonyls by alkoxides, azides and hydrides [36].

Gold NCs that have been synthesised in solution and then subjected to analysis via ESI or MALDI can be considered to have the following general formula given in Eq. (5):

$$\left[Au_a^{(0)} Au_b^{(+1)} L_c A_d^{(-1)} - nH^+ \right]^z \tag{5}$$

where L is a neutral ligand such a phosphine or bisphosphine and A is an anionic ligand such as a thiolate or halide.

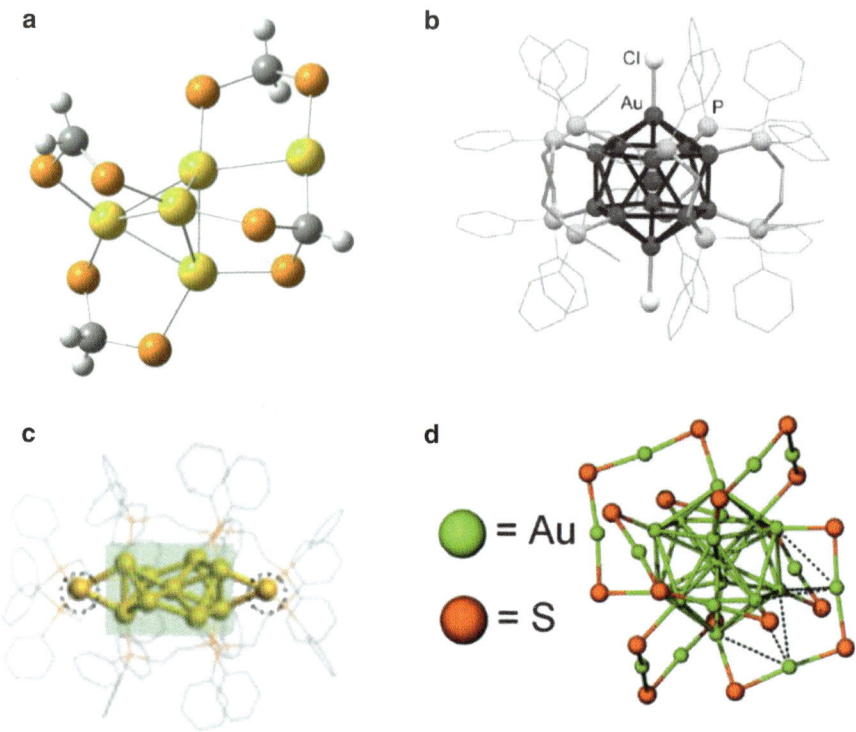

Fig. 1 Examples of charged AuNCs that fit the $[Au^{(0)}{}_a Au^{(+1)}{}_b L_c A^{(-1)}{}_d - nH^+]^z$ formulation and that have been analysed by both MS and X-ray crystallography: (**a**) $[Au_5(dppm)_4 - H^+]^{2+}$ (where $a = 2$; $b = 3$; $c = 4$; $n = 1$; $z = +2$) [57]; (**b**) $[Au_{13}(dppe)_5Cl_2]^{3+}$ (where $a = 8$; $b = 5$; $c = 5$; $d = 2$; $z = +3$) [58]; (**c**) $[Au_{11}(dppe)_6]^{3+}$ (where $a = 8$; $b = 3$; $c = 6$; $z = +3$) [59] and (**d**) $[Au_{25}(SCH_2CH_2Ph)_{18}]^+$ (where $a = 6$; $b = 19$; $d = 18$; $z = +1$) [60]. Non-coordinating counter ions are omitted for clarity. Figures reproduced from (**a**) [57]; (**b**) [58]; (**c**) [59]; (**d**) [60]

Thus the net charge of the NC is given by Eq. (6):

$$z = b(+1) + d(-1) - n \qquad (6)$$

Note that either L or A can be deprotonated to add a negative charge to the NC.

Figure 1 highlights some of the isolated charged AuNCs that have been structurally characterised by both X-ray crystallography and MS. The NC $[Au_5L^1{}_4-H]^{2+}$ (Fig. 1a) is interesting as it possesses a tetrahedral gold core with an exopolyhedral $Au(CH(PPh_2)_2)$ moiety formed via deprotonation of the dppm at the methylene carbon.

Some of the strategies listed above for the manipulation of inorganic and organometallic compounds that have no net charge have also been applied to AuNCs that have no net charge. Most examples are for AuNCs with thiolate ligands: (1) thiolate ligand possessing acidic protons can be further deprotonated,

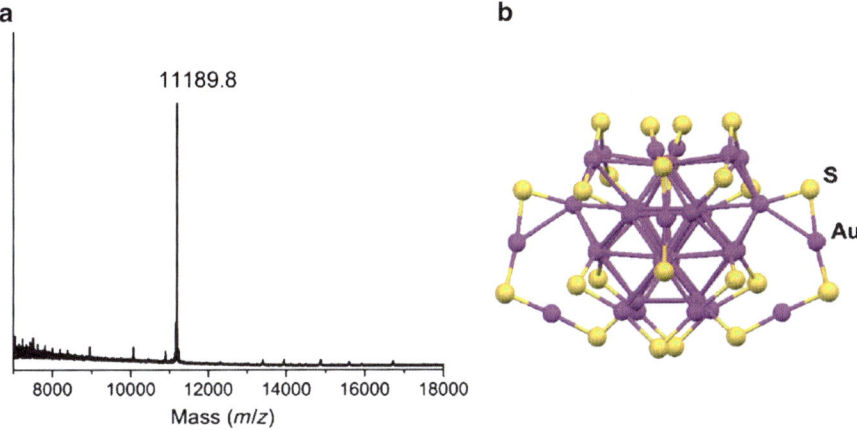

Fig. 2 Example of a neutral AuNC that has been analysed via MS: (**a**) ESI-MS Cs^+ adduct of $Au_{36}(SPh-tBu)_{34}$; (**b**) crystal structure highlighting the $Au_{36}S_{24}$ framework. Figure reproduced from [61]

as highlighted for glutathione ligands, which produce anionic NCs; (2) ESI-MS of caesium acetate-doped solutions of $Au_{36}(SPh-tBu)_{24}$ give rise to cationic caesium adducts (Fig. 2a) [61].

The recent gold cluster literature can largely be divided into two classes based on the capping ligands used:

1. *Thiolate ligands*: these anionic ligands often produce neutral clusters that make them challenging to analyse via mass spectrometry. Nonetheless, there are quite a few examples of MALDI/MS and ESI-MS being used to examine their formation and size distributions. Since this area has been reviewed in 2010 [33], only subsequent work is briefly described below.
2. *Phosphine ligands*: these neutral ligands allow the ready identification of charged cationic gold nanoclusters via ESI-MS. Clusters capped by either monophosphines or bisphosphines (monodentate or bidentate phosphine ligands) have been studied. The latter exhibit interesting size-selectivity effects, as discussed in detail in the following sections.

2.2 Top-Down and Bottom-Up Approaches for the Synthesis of Gas-Phase Gold Nanoclusters

The techniques developed to study bare and ligated gold cluster ions in the gas phase fall into the same two categories, Fig. 3, as identified by Ott and Finke [27] for solution phase studies of clusters:

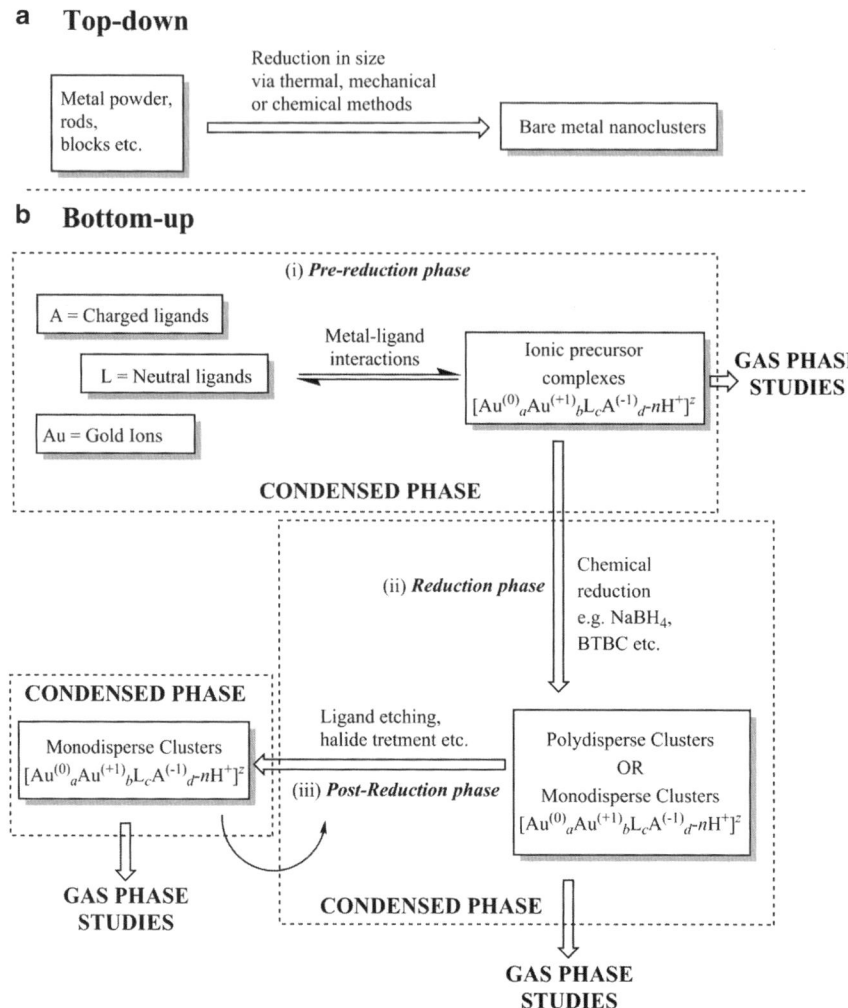

Fig. 3 The two general techniques used to synthesise nanoclusters are (**a**) the 'top-down' approach and (**b**) the 'bottom-up' approach

1. The 'top-down' approach utilises bulk material (metal powder, rods, blocks, etc.) which is broken down by physical methods to smaller architectures on the nanometre scale. Methods that have been applied to the synthesis of gold nanoclusters following the 'top-down' approach include lithographic techniques [62, 63], mechanical methods such as ball milling [64], thermal methods such as laser ablation vaporisation [65–67] and chemical methods such as ligand etching [68–70]. Issues concerning the top-down approach include difficulties with reproducibility, polydispersity and the need to require mechanical equipment and instrumentation which may add time and costs to the synthesis.

De Heer has reviewed a range of top-down methods available for the production of gas-phase metal cluster ions [71]. Of these, the laser vaporisation cluster source, also known as the 'Smalley source', has been commonly used for the production of gas-phase clusters for over 30 years [72]. Indeed this method has been most widely used to generate bare gold cluster ions in the gas phase. Readers interested in the design and operational aspects of these laser vaporisation cluster source should refer to an excellent recent review by Duncan [72].

2. The 'bottom-up' approach involves chemical methods whereby a metal salt or complex, in the presence of an appropriate chelating ligand, is reduced to form ligand-protected metal nanoclusters. Reduction commonly occurs via treatment with a chemical reducing agent such as sodium borohydride, borane *tert*-butylamine complex and citrate [73]. The addition of chemical reducing agents to gold ions in solution provides a source of electrons whereby neutral gold atoms are formed. When the onset of reduction occurs in the presence of an appropriate chelating ligand, such as a phosphine or thiolate, the aggregation of neutral gold atoms to bulk material is perturbed by the properties of the ligand in a given environment. Size selectivity of a ligand is an inherent property that can be optimised experimentally by varying factors such as the temperature, stir rate, concentration, solvent, etc.

Gold cluster synthesis via the bottom-up method requires a deep understanding of the assembly processes, governed by the interactions of the atomic and molecular components, that leads to the formation of stable and monodisperse nanoclusters [74–77]. This can lead to the rational design and fine tuning of the AuNCs architecture. Although the bottom-up method is seen as the inverse method to the top-down approach, it may also be used in addition to top-down techniques [78] in particular via processes that involve ligand etching. There is a growing awareness that there are three phases associated with AuNC synthesis via the bottom-up method and that ESI-MS can be used to examine ionic species present during these phases (Fig. 3b):

1. *Pre-reduction phase*: To provide insight into the fundamental processes that govern the formation of nanoclusters it is important to establish the identity and relative abundance of ionic complexes that exist in solution before the initiation of chemical reduction. These complexes are the 'molecular building blocks' that can then interact to form particles of well-defined stoichiometry upon reduction. ESI-MS has been used to examine the identity and relative abundance of the cationic precursors that spontaneously assemble in solution prior to reduction.

2. *Reduction phase*: A key step in the synthesis of AuNCs in the condensed phase involves the use of chemical reducing agents. Reducing agents which are routinely used for AuNCs are sodium borohydride ($NaBH_4$), a fast reducing agent, and borane *tert*-butylamine complex (BTBC), a comparatively slow reducing agent. Subsequently, ESI-MS can be used to monitor the abundance and identity of AuNC ions present in solution over time from the onset of reduction. This provides a tool to monitor reactive cluster intermediates,

potentially revealing the reaction mechanisms that result in the formation of the final product distribution of AuNCs.

3. *Post-reduction phase*: ESI-MS can be used to monitor gold cluster reactions after the exhaustion of chemical reducing agents. Processes such as ligand etching and exchange will govern the formation or transformation of clusters to reach the final thermodynamically stable product(s). These products can be further transformed via the addition of other reagents such as HCl.

The bottom-up approach has been widely used to synthesise ligand-protected gold NCs. There are numerous reports which have examined a range of ligands, solvents, concentrations and many other experimental conditions. Here we solely focus on literature reports in which ESI-MS and MALDI-MS have been used as a key tool to monitor the formation and reactions of gold nanoclusters.

2.3 Thiolate Ligands

A great deal of attention has been invested into the formation of monodisperse thiolate-protected nanoclusters, and several reviews described their synthesis and properties [79–83]. In 1994, Brust [84] first reported the synthesis of less than 5 nm-sized alkanethiolate-protected gold nanoclusters by using sodium borohydride as a reducing agent. It was noted that these nanoclusters should exhibit intriguing properties due to quantum confinement effects. Since then, the synthetic method of Brust has been refined and optimised by numerous researchers. Here we consider literature reports which have dealt with the three phases of nanocluster growth described in Fig. 3.

2.3.1 Pre-reduction Phase

Briñas et al. [85] used direct light scattering (DLS), UV–Vis and size exclusion chromatography (SEC) studies to monitor how the size of thiolate-protected gold nanoparticles is dependent on the pH-induced size of the gold(I)–thiolate precursor complex formed upon addition of glutathione to an aqueous solution of HAuCl$_4$. Lower pH levels resulted in the formation of larger gold(I)–thiolate precursor complexes, which in turn gave rise to larger thiolate-protected gold nanoparticles upon reduction with sodium borohydride. In contrast, higher pH levels resulted in the formation of smaller gold(I)–thiolate complexes which resulted in the formation of smaller thiolate-protected gold nanoclusters upon addition of NaBH$_4$.

Simpson et al. [86] used a modified Brust method [84] to study the identity and structure of gold(I)–thiolate precursors formed upon mixing N-(2-mercapto-propionyl)-glycine (tiopronin) and HAuCl$_4$. MALDI-TOF-MS analysis revealed the sole presence of a tetramer $[Au_4(Tio)_4 + Na_x\text{-}H_{x-1}]^+$ ($x = 1$–4). A cyclic structure was proposed for this tetramer (Fig. 4a) due to the absence of other cluster

Fig. 4 Proposed structures for the gold(I)–thiolate precursor showing (**a**) a discrete tetrameric gold(I)–thiolate and (**b**) a straight chain gold(I)–thiolate polymer

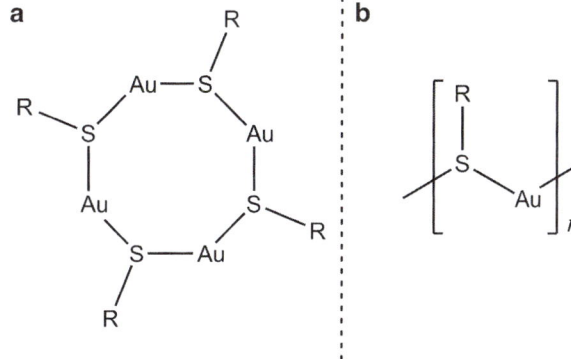

Fig. 5 MALDI-TOF-MS of AuNPs using the DCTB matrix operating at threshold laser fluence. Figure reproduced from [87]

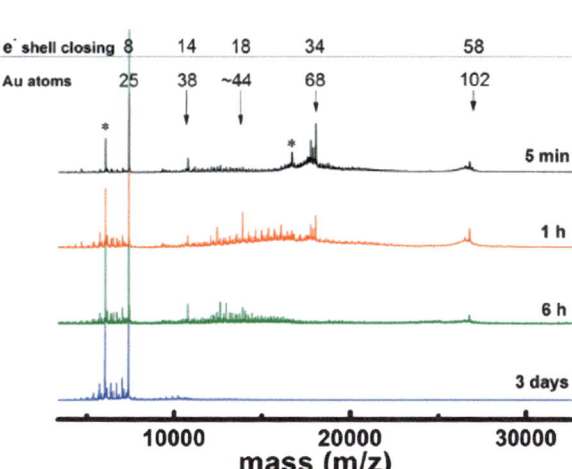

stoichiometries, which would otherwise indicate an open chain polymer structure (Fig. 4b).

2.3.2 Reduction Phase

Few reports have appeared on the use of MS to examine the evolution of thiolate-protected gold nanoclusters during the reduction phase. A rare example probed the size evolution to monodisperse $Au_{25}(SCH_2CH_2Ph)_{18}$ using MALDI-TOF-MS (Fig. 5) [87]. At approximately 5 min after the addition of sodium borohydride to a stirring solution of gold/phenylethanethiol (1:6) in THF, polydisperse ligand-protected AuNPs with the gold cores of Au_{25}, Au_{38}, Au_{44}, Au_{68} and Au_{102} are observed. The Au_{68} ligand-protected cluster seems to act as an intermediate during the reduction phase, where its disappearance occurs between 1 and 6 h. The Au_{102} ligand-protected cluster persists until at least 6 h.

2.3.3 Post-reduction Phase

Mass spectrometry has been used to monitor ligand exchange reactions of thiolate-protected gold nanoclusters. For example, Spivey et al. [88] used MALDI-TOF-MS to show that treatment of glutathione-protected nanoclusters with dodecanethiol ($SC_{12}H_{25}$) results in the transformation to monodisperse $Au_{38}(SC_{12}H_{25})_{12}$. Bare Au_{38} clusters, formed via reduction of $Au_{38}(SC_{12}H_{25})_{12}$ by H_2, were attached to TiO_2 supports, and their role in catalysing CO oxidation was investigated.

Whetten et al. have used mass spectrometry to assign the number of ligands and gold atoms of synthesised thiolate-protected gold nanoclusters. For instance, they used ESI-MS to identify the various charge states of $[Au_{144}Cl_{60}]^z$. This cluster was synthesised from the well-known thiolate-protected cluster $[Au_{144}(SR)_{60}]^z$ via ligand exchange with chloride [89]. They concluded that the $z = 2$ and 4^+ charge states yielded a highly symmetric (I_h) cluster. Mass spectrometry was also used in their study of the $Au_{67}(SR)_{35}$ cluster via MALDI-TOF [90] as well as the study of $[Au_{25}(SC_6H_{13})_{18}]^x$ cluster via ESI-MS [91]. In this latter study, they concluded that geometric rather than electronic factors are responsible for the stability of $[Au_{25}(SC_6H_{13})_{18}]^x$ ($x = -1, 0, +1$).

Recently, Jin et al. [92] used ESI-MS and UV–Vis spectroscopy to monitor the solution phase reactivity of monodisperse $Au_{38}(PET)_{24}$, (PET = phenylethanethiol), with an excess of 4-*tert*-butylbenzenethiol (TBBT) and sampled at various time intervals, Fig. 6. The excess addition of the bulkier ligand TBBT ultimately results in the formation of monodisperse $Au_{36}(TBBT)_{24}$.

Given the data in Fig. 6 the evolution of the thiol-induced $Au_{36}(TBBT)_{24}$ from $Au_{38}(PET)_{24}$ was divided into four stages (Fig. 7): (1) ligand exchange reactions of PET for TBBT occur, (2) ligand exchange reaction continues together with a structural distortion of the cluster core as observed by the optical spectra, (3) disproportionation reaction occurs as identified by ESI-MS whereby 2 equiv. of $Au_{38}(TBBT)_m(PET)_{24-m}$ give $Au_{36}(TBBT)_m(PET)_{24-m}$ and $Au_{40}(TBBT)_{m+2}$ $(PET)_{24-m}$, and (4) size focusing occurs, resulting in the formation of monodisperse $Au_{36}(TBBT)_{24}$.

2.4 Phosphine Ligands

Gold nanoclusters protected by phosphine ligands have been studied in the condensed phase since the pioneering work of Malatesta in the 1960s [93–97]. These early studies were motivated by the desire to develop models for cluster bonding and thus a key aim was the isolation of crystalline material suitable for X-ray diffraction studies [57, 98–108]. A significant achievement was the isolation and characterisation of a triphenylphosphine monolayer-protected clusters (MPCs) of monodisperse $Au_{55}(PPh_3)_{12}Cl_6$ by Schmid [109–112]. In recent years ESI-MS has been used to monitor the role of bis(phosphino)alkane ligands of the type

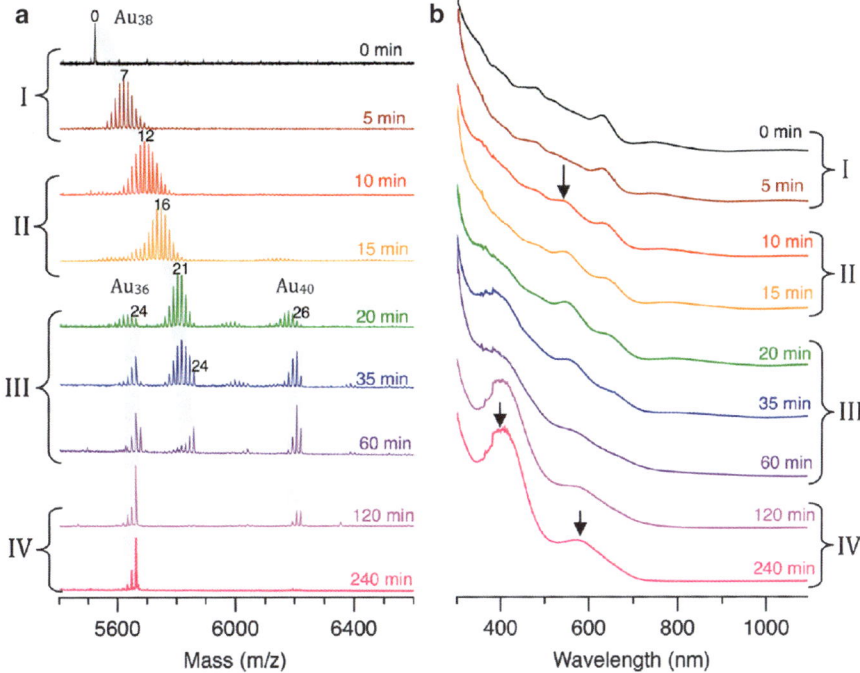

Fig. 6 (**a**) Time-dependent ESI-MS reaction of $Au_{38}(PET)_{24}$ with excess TBBT. The doubly charged region is shown. The three grey shadows indicate three groups of peaks: (*left*) $Au_{36}(TBBT)_m(PET)_{24-m}$, (*middle*) $Au_{38}(TBBT)_m(PET)_{24-m}$, and (*right*) $Au_{40}(TBBT)_m(PET)_{24-m}$. The numbers on top of the mass peaks indicate the number of TBBT ligands (m) exchanged onto the cluster; (**b**) corresponding UV–Vis spectra at different times taken parallel to the ESI-MS data. Figure reproduced from [92]

$Ph_2P-(CH_2)_n-PPh_2$, where $n = 1$–6, in promoting the formation of 'size-selected' monodisperse clusters. Table 1 summarises the types of bis(phosphino)alkane-protected gold clusters that have been observed by ESI-MS. From Table 1 it is evident that a variety of gold core nuclearities are observed, depending on the bisphosphine ligand used. This size selectivity correlates to the variation in the alkane chain of $Ph_2P-(CH_2)_n-PPh_2$, where $n = 1$–6. In the next sections, literature reports which have dealt with the three phases of bisphosphine-protected nanoclusters growth are discussed.

2.4.1 Pre-reduction Phase

Colton et al. [143] were the first to use ESI-MS to monitor the ligand exchange reactions of $ClAuPPh_3$ with various monodentate and bidentate phosphine ligands added at various concentrations. The aims of their study were to examine (1) the identity and abundance of gold(I) complexes that undergo ligand exchange and

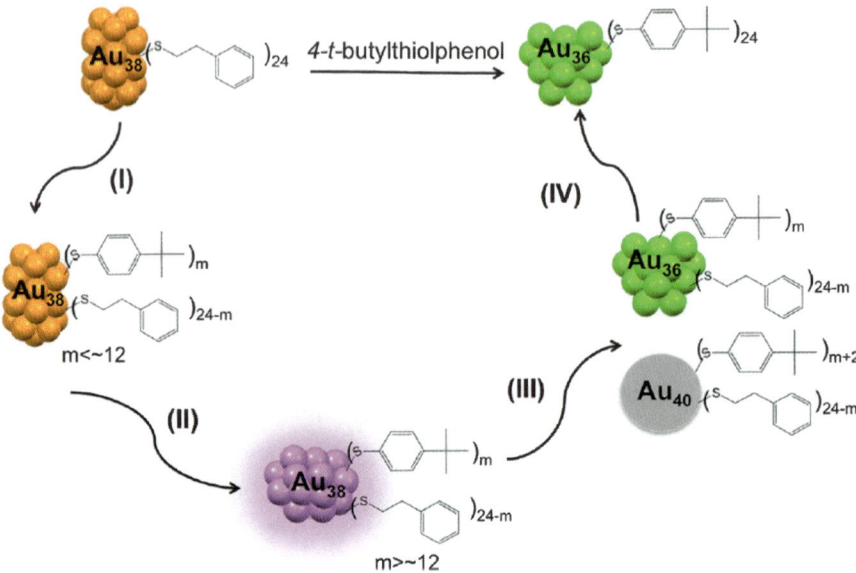

Fig. 7 'Reaction pathway for conversion of $Au_{38}(PET)_{24}$ to $Au_{36}(TBBT)_{24}$. Stage I, ligand exchange; Stage II, structure distortion; Stage III, disproportionation; Stage IV, size focusing'. Figure reproduced from [92]

Table 1 A comparison of the types of gold nanoclusters observed by ESI-MS which are formed via reduction of Au salts in the presence of bisphosphine ligand as a function of the ligand

Cluster	L^1	L^2	L^3	L^4	L^5	L^6
$[Au_5L^n_4 - H^+]^{2+}$	✔a					
$[Au_6L^n_3]^{2+}$			✔e	✔b	✔b	✔b
$[Au_6L^n_4]^{2+}$			✔d, f, g			
$[Au_8L^n_4]^{2+}$			✔f, g	✔b	✔b, g, h	✔b, i
$[Au_9L^n_4]^{2+}$					✔g	
$[Au_9L^n_4]^{3+}$	✔a	✔c				
$[Au_9L^n_5]^{3+}$	✔a	✔c				
$[Au_{10}L^n_4]^{2+}$	✔a		✔f	✔b	✔b, g–h	✔b, i
$[Au_{10}L^n_5]^{2+}$					✔h	✔i
$[Au_{11}L^n_5]^{3+}$	✔a	✔c	✔b, d–g	✔b		
$[Au_{11}L^n_6]^{3+}$	✔a					
$[Au_{12}L^n_6]^{4+}$			✔f			
$[Au_{13}L^n_6]^{3+}$	✔a					
$[Au_{14}L^n_7 - H^+]^{3+}$	✔a					

The letters a–i are used to denote the following references: (a) [113], (b) [114], (c) [58], (d) [115], (e) [116], (f) [117–139], (g) [140], (h) [141] and (i) [142]. A ✔ indicates that the ion was observed (reported). A blank box indicates that the ion was not observed (not reported). L^n denotes the bisphosphine ligands, Ph_2P-$(CH_2)_n$-PPh_2, where $n = 1$–6. Table reproduced from [113]

Fig. 8 Representations of (**a**) singly charged mononuclear bis-ligated gold(I) complex $[AuL_2]^+$ and (**b**) doubly charged dinuclear bis-ligated gold(I) complex $[Au_2L_2]^{2+}$

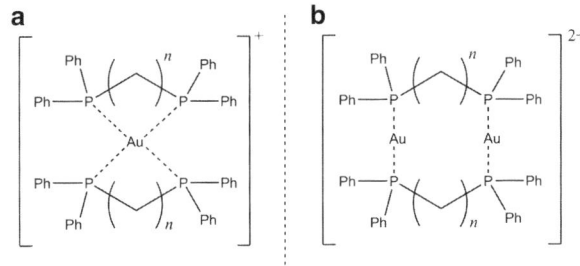

(2) the maximum number of ligands associated with gold(I) ions. For monodentate phosphines they observed the formation of $[Au(PR_3)_n]^+$, where $n = 2$ and 3 but not 4. The studies provided evidence on the instability of a three coordinate monodentate phosphine complex, $[Au(PPh_3)_3]^+$ consistent with previous solution phase NMR experiments [144, 145].

Subsequently Hudgens et al. [146] set out to systematically investigate the relationship between the types of AuNPs formed and the cationic ligand complexes of gold that exist in solution prior to treatment with the reducing agent. The series of bisphosphine ligands, $Ph_2P-(CH_2)_n-PPh_2$ where $n = 1–6$, were proposed to form the mononuclear, bis-ligated complex shown in Fig. 8a. It was suggested that the ligand in the $[AuL_2]^+$ complex acts as a trap, protecting the bis-ligated gold(I) via the steric bulk of the ligand. This ultimately hinders the reduction and post-reduction nucleation process as supported by previous studies [116]. The bridged dinuclear species, Fig. 8b, may be further stabilised by aurophilic interactions [147, 148].

Control of the pre-reduction phase is important in order to tailor the synthesis of AuNCs [113]. The undesirable $[Au(PPh_3)_2]^+$, known to be a precursor to colloid formation, could be minimised by the addition of 2 equiv. of the bis(diphenyl-phosphino)methane ligand. Reduction of the now enriched cationic precursors to nanocluster resulted in a polydisperse solution as monitored via ESI-MS and UV–Vis spectroscopy. Several of these clusters had not been previously described (Table 1).

In efforts to investigate how bidentate phosphine ligands control the 'size selectivity' of gold NC formation, ESI-MS was used in a series of experiments to examine the role of ligand concentration relative to the gold salt concentration. The incremental addition of 1,6-bis(diphenylphosphino)hexane, (L^6), to a methanolic solution containing 10.0 mg of $AuClPPh_3$ was monitored via ESI-MS to observe the identity and abundance of cationic gold-ligand complexes that form prior to the addition of sodium borohydride [142].

The fractional total ion current of each 'pre-reduction' complex was plotted, Fig. 9, as a function of the various ratios of $[L^6]/[PPh_3]$, where $0 \leq [L^6]/[PPh_3] \leq 16$. The aim was to monitor how the complexes formed during the 'pre-reduction' phase determine the fate of the identity and abundance of AuNPs in solution. It was concluded that the ultimate formation of bisphosphine-protected clusters

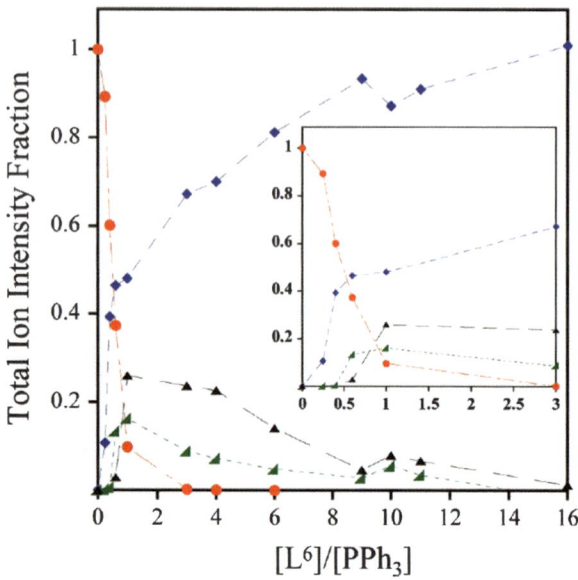

Fig. 9 Fractional total ion current measured via ESI-MS as a function of the ligand to AuClPPh$_3$ ratio, [L^6]/[PPh$_3$]. Triphenylphosphine ligands are increasingly replaced on the metal complexes as L^6 is added incrementally to a solution comprising 10.0 mg of AuClPPh$_3$ dissolved in chloroform/methanol (50:50). The symbols are assigned as follows: *red circles*, [Au(PPh$_3$)$_2$]$^+$; *blue diamonds*, [Au$_2$L$^6_2$Cl]$^+$; *green wedges*, [Au$_2$L$^6_2$Cl]$^+$; and *black triangles*, [Au$_2$L$^6_3$Cl]$^+$. The *insert* displays the fractional total ion current for $0 \leq$ [L^6]/[PPh$_3$] ≤ 3. Figure reproduced from [142]

observed via ESI-MS could be 'tuned' by the nature of the gold-ligand complex distribution prior to reduction. Figure 10 summarises the fractional product distribution of AuNPs as a function of [L^6]/[PPh$_3$] and will be discussed in greater detail in Sect. 2.4.3.

2.4.2 Reduction Phase

In 2006 Wang et al. [149] used ESI-MS and UV–Vis spectroscopy to monitor the formation of AuNCs upon the slow reduction of AuClPPh$_3$ in chloroform by the addition of 5 equiv. of BTBC. A range of PPh$_3$-protected AuNC intermediates were observed by examining aliquots of the reaction at various time intervals. Their studies were extended to the reduction of AuClPPh$_3$ in chloroform mixed with equimolar bisphosphine ligand Ph$_2$P-(CH$_2$)$_n$-PPh$_2$, = 3–6. It was concluded that the bidentate ligands showed exceptional size selectivity, Fig. 11, compared to the monodentate triphenylphosphine ligand. Monodisperse [Au$_{11}$(L^n)$_5$]$^{3+}$ was selectively synthesised when $n = 3$.

The formation of gold nanoclusters described initially by Wang et al. was later refined by Hudgens et al. [140], in which bisphosphine (L^3)-protected gold

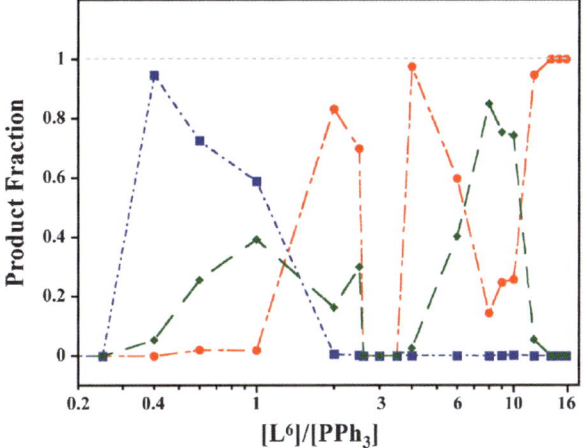

Fig. 10 Fractional product distribution as a function of $[L^6]/[PPh_3]$ as determined from fractional ion intensities measured with ESI-MS. The symbols are assigned as follows: *red circles*, $[Au_8L^6_4]^{2+}$; *blue squares*, $[Au_9L^6_4Cl]^{2+}$; and *green diamonds*, $[Au_{10}L^6_x]^{2+}$ ($x = 4$ and 5). See [142] for more detail. Figure reproduced from [142]

nanoclusters were formed and the substitution of the solvent from neat chloroform to a (50:50) methanol/chloroform solution assisted in both the stabilisation of the clusters over time and the ESI ion current. They observed that using neat chloroform resulted in the decomposition of clusters over time and led to the formation of a red precipitate which was soluble in methanol. Additionally, when the synthesis was carried out in neat chloroform alone, the Au_{11} cluster was formed in relatively low abundance. In contrast, when the (50:50) methanol/chloroform solvent system was used, the $[Au_{11}(L^3)_5]^{3+}$ was present in greater intensity and the additional clusters $[Au_6(L^3)_3]^{3+}$ and $[Au_8(L^3)_4]^{3+}$ were identified as minor components. This refined solvent system was also applied to a synthesis of $Et_2P(CH_2)_3P(Et)_2$-protected AuNCs [150] which were also studied via density functional theory (DFT) calculations.

The reaction mechanisms that govern the formation of gold nanoclusters protected by 1,3-bis(diphenylphosphino)propane were further elaborated by Hudgens et al. [117], who recorded ESI-MS at various time intervals prior to and after the addition of sodium borohydride.

Tsukuda et al. [151] have also shown that the dinuclear system $Au_2(BINAP)X_2$, X=Cl or Br, can be treated with $NaBH_4$ to form monodisperse undecagold clusters using either of the bidentate ligand (±)-2,2′-bis(diphenylphosphino)-1,1′-binapthylene (BINAP), as revealed via ESI-MS, which gave the ions as $[Au_{11}(BINAP)_4X_y]^{z+}$ (X=Cl or Br; y, z = 1,2; 2,1), Fig. 12.

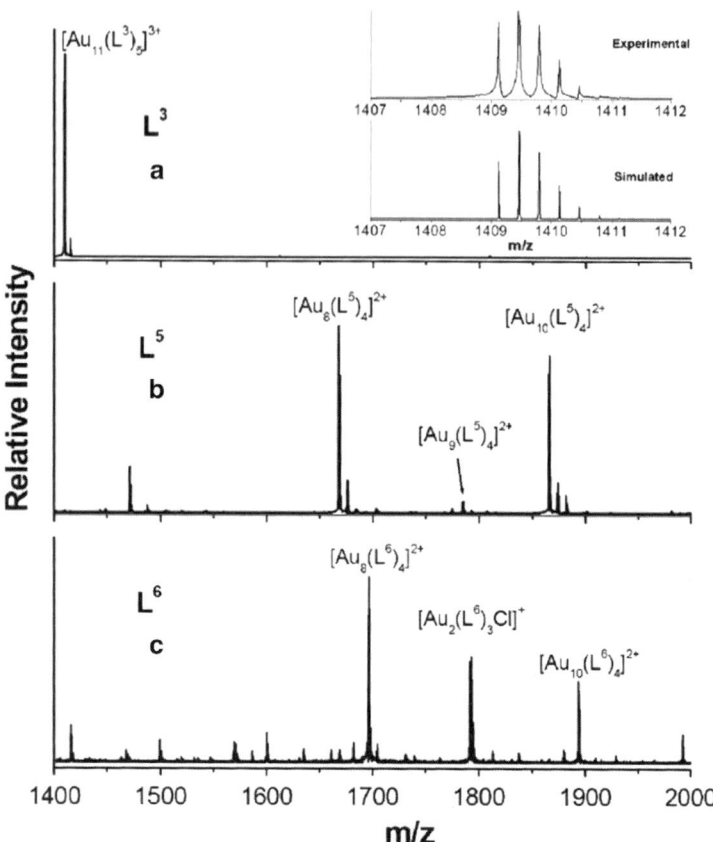

Fig. 11 High-resolution positive ion ESI-MS characterisation of Au clusters stabilised with the indicated diphosphine ligand, L^n, where $L=Ph_2P-(CH_2)_n-PPh_2$: (**a**) L^3 ($n = 3$); (**b**) L^5 ($n = 5$); (**c**) L^6 ($n = 6$). The inset in (**a**) shows the isotopic pattern of the $[Au_{11}(L^3)_5]^{3+}$ peak and the simulated isotopic pattern using the natural isotopic abundances of Au, P, C and H. Figure reproduced from [149]

2.4.3 Post-reduction Phase

Several studies have examined post-synthetic modification of gold nanoclusters [152–157]. Konishi et al. [58] have used ESI-MS to show that the polydisperse solution of gold nanoclusters formed upon reduction of the dinuclear complex $Au_2(dppe)Cl_2$ by $NaBH_4$ (Fig. 13a) transforms of the monodisperse cluster $[Au_{13}L_5Cl_2]^{3+}$ after treatment with HCl for 3 h (Fig. 13b).

ESI-MS is a valuable tool to monitor the reactions that occur after chemical reduction [142], which is known to be complete within hours after the addition sodium borohydride [158]. In Fig. 14, the initial distribution of $[Au_8L^6_4]^{2+}$ remains persistent over at least 6 days after the addition of sodium borohydride. After 14 days, a new peak corresponding to $[Au_9L^6_4Cl]^{2+}$ is observed.

Fig. 12 Positive ion ESI-MS of BINAP-protected AuNCs. The bottom panels show the experimental and simulated isotopic patterns for each of the main peaks.
Figure reproduced from [151]

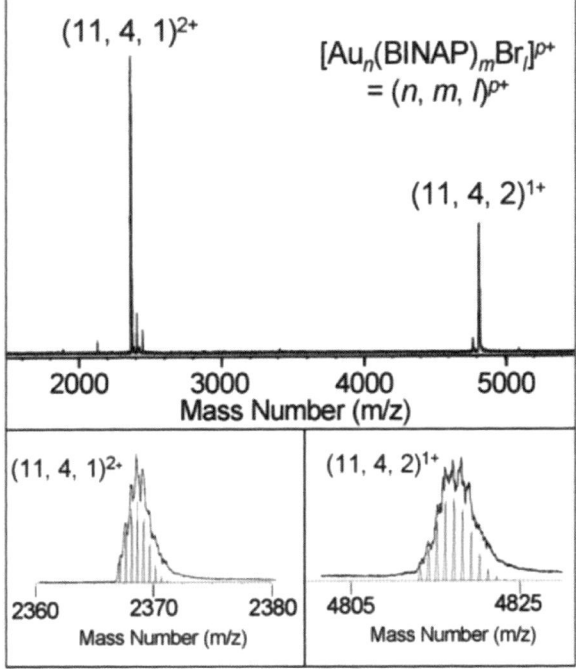

Fig. 13 Positive ion ESI-MS spectra of (**a**) polydisperse gold nanoclusters protected by (dppe), (**b**) after treatment of the polydisperse cluster with HCl for 3 h and (**c**) post-purification of the monodisperse cluster.
Figure reproduced from [58]

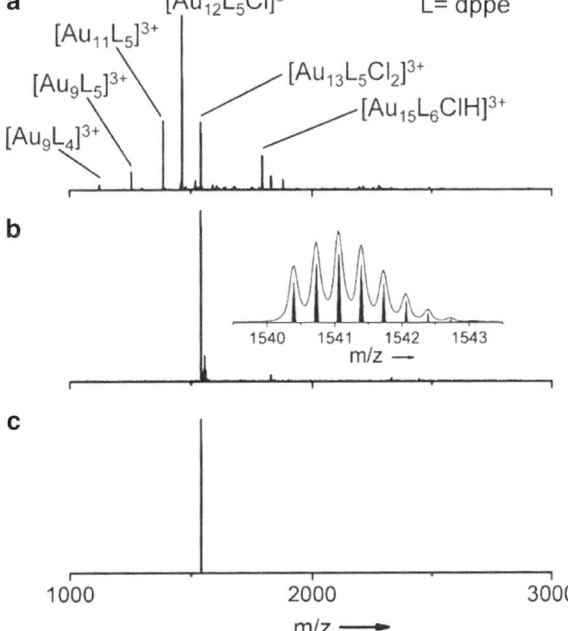

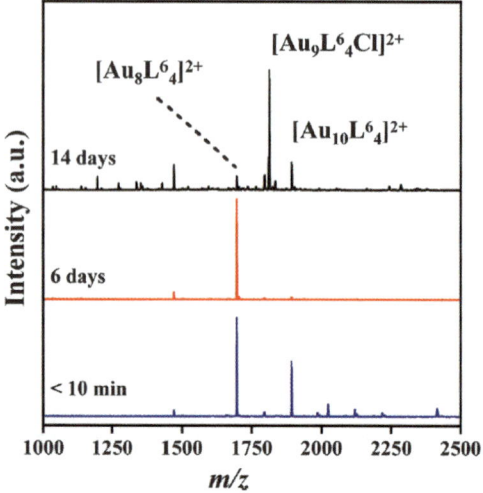

Fig. 14 Time resolved ESI-MS for nanocluster synthesis from solutions containing $[L^6]/$
$[PPh_3] = 1$. The sequence shows that $[Au_9L^6_4Cl]^{2+}$ cluster growth is preceded by the formation
of monodisperse $[Au_8L^6_4]^{2+}$ clusters through degradation of larger clusters. Figure reproduced
from [142]

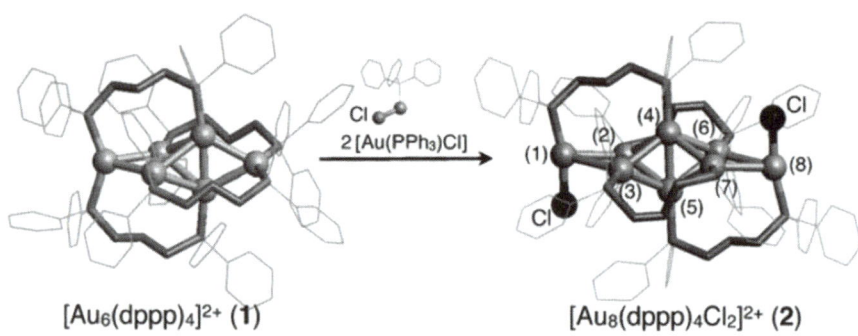

Fig. 15 X-ray crystal structures of the cations of $[Au_6(dppp)_4](NO_3)_2$ (**1**) and $[Au_8(dppp)_4Cl_2]$
$(PF_6)_2$ (**2**). dppp = 1,3-Bis(diphenylphosphino)propane. Hydrogen atoms omitted for clarity.
Figure reproduced from [159]

Kamei et al. [159] used ESI-MS to monitor the reactions of two AuNCs
$[Au_6(dppp)_4](NO_3)_2$ ((**1**) in Fig. 15) [102] and $[Au_9(PPh_3)_8](NO_3)_2$ [160]. ESI-MS
revealed that $[Au_6(dppp)_4]^{2+}$ (Fig. 16a) transforms to monodisperse
$[Au_8(dppp)_4Cl_2]^{2+}$ upon treatment with 20 equiv. of ClAuPPh_3. The subsequent
addition of an alcoholic solution of $NaPF_6$ resulted in the precipitation of crude
$[Au_8(dppp)_4Cl_2](PF_6)_2$ which was recrystallised in dichloromethane to yield crystals suitable for X-ray crystallography ((**2**) in Fig. 15). ESI-MS was also used to

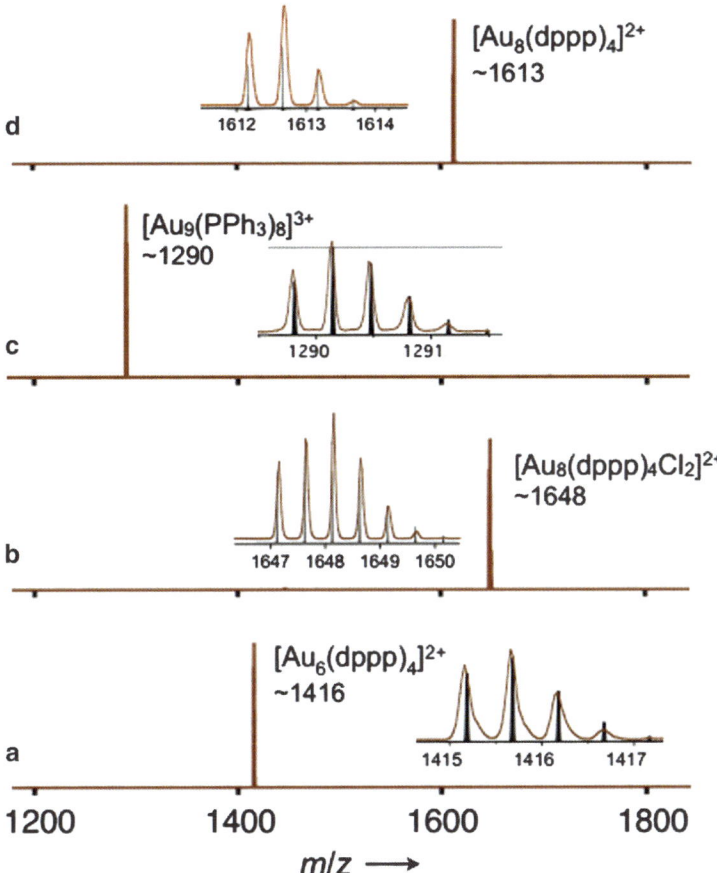

Fig. 16 ESI-MS of (**a**) $[Au_6(dppp)_4](NO_3)_2$, (**b**) the product of reacting $[Au_6(dppp)_4](NO_3)_2$ [102] in methanol with 2 molar equivalents of $AuClPPh_3$ in chloroform, (**c**) $[Au_9(PPh_3)_8](NO_3)_3$ [160] and (**d**) the product of reacting $[Au_9(PPh_3)_8](NO_3)_3$ in dichloromethane with 6 equiv. of dppp also in dichloromethane. dppp $= 1,3$-bis(diphenylphosphino)propane. Figure reproduced from Supplementary Material of [159]

monitor the ligand etching of $[Au_9(PPh_3)_8]^{3+}$ (Fig. 16c) into monodisperse $[Au_8(dppp)_4]^{2+}$ upon addition of 6 equiv. of dppp.

2.5 MS-Based Fragmentation Methods for the Production of Gas-Phase Gold Cluster Ions

Fragmentation of mass-selected gold cluster ions provides (1) structural information and (2) a gas-phase top-down approach to the 'synthesis' of new gold cluster

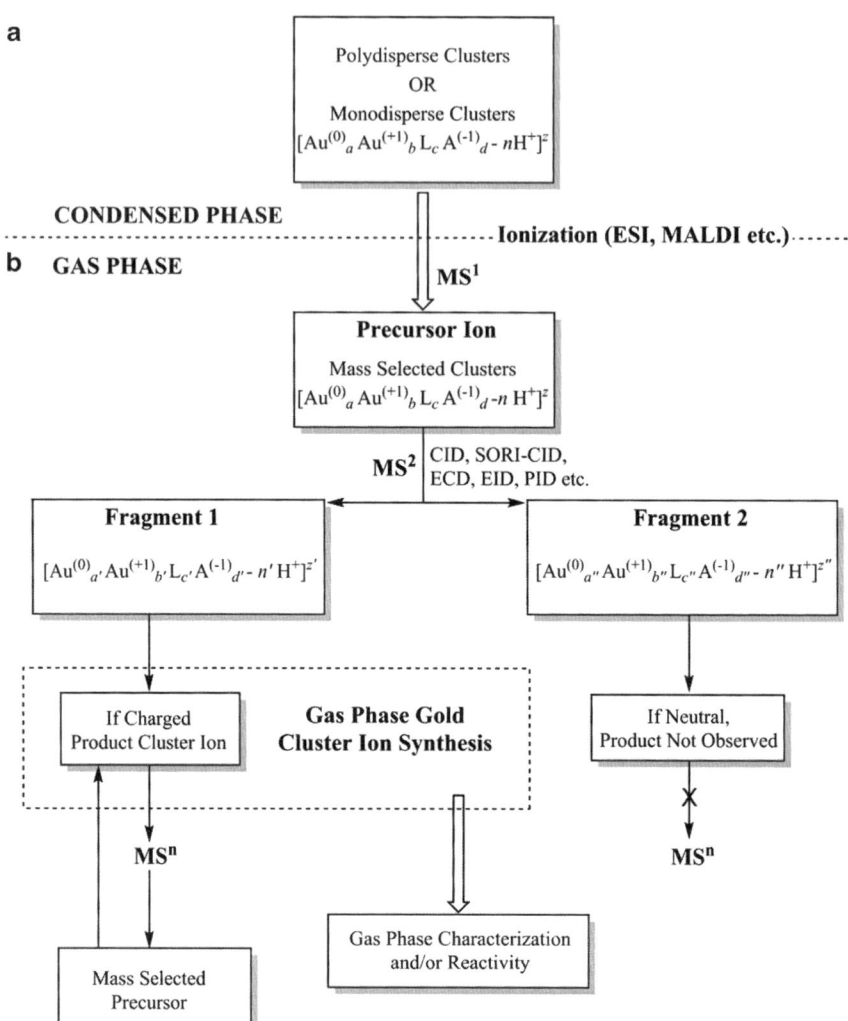

Fig. 17 Combination of solution phase synthesis, ESI and multistage mass spectrometry experiments for the gas phase 'synthesis' of gold cluster ions. (**a**) condensed phase clusters transferred to the gas phase and (**b**) fragmentation (via CID, SORI-CID, ECD, EID, PID, etc.) produces new cluster ions

ions (Fig. 17), whose reactivity can be examined using instruments with multistage mass spectrometry (MS^n) capabilities. Several examples where this approach has been adopted are discussed later in Sect. 3.1.

As discussed further in Sect. 3.1, fragmentation of a gas-phase cluster ion can be induced in several ways. The subsequent fragmentation of the activated mass-selected precursor ion can occur via three general channels:

Table 2 The general formula $[Au^{(0)}_a Au^{(+1)}_b L_c A^{(-1)}_d - nH^+]^z$ (Eq. (5)) can be tabulated and used to account for the fragmentation channels of a mass-selected gold cluster ion

	a	b	c	d	n	z
Mass-selected precursor ion $[Au^{(0)}_a Au^{(+1)}_b L_c A^{(-1)}_d - nH^+]^z$	a	b	c	d	n	z
Product ion $[Au^{(0)}_a Au^{(+1)}_b L_c A^{(-1)}_d - nH^+]^z$	a'	b'	c'	d'	n'	z'
Loss from precursor to form product $[Au^{(0)}_a Au^{(+1)}_b L_c A^{(-1)}_d - nH^+]^z$	a''	b''	c''	d''	n''	z''

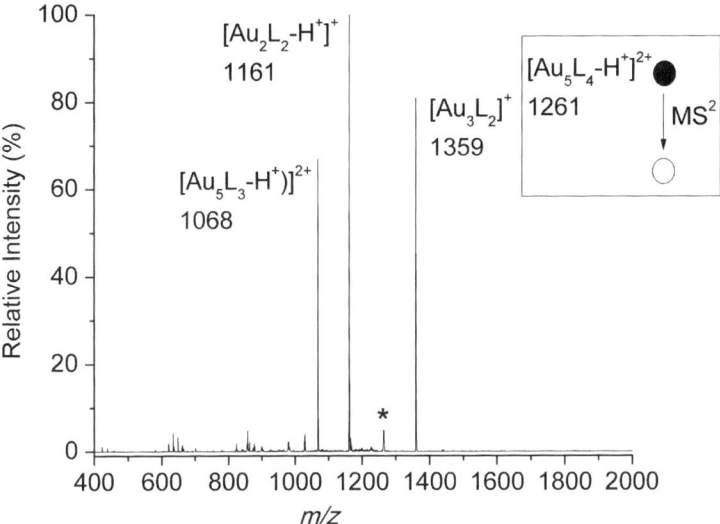

Fig. 18 Linear ion trap low-energy collision-induced dissociation of $[Au_5L_4-H^+]^{2+}$. The most intense peak in the cluster is represented by the m/z value. *Asterisk* refers to the mass-selected precursor ion peak $[Au_5L_4-H^+]^{2+}$. Figure adapted from [113]

1. *Ligand loss* occurs where no dissociation of the gold cluster core remains inert.
2. *Core fission* occurs when the cluster core undergoes fragmentation. As a subset of core fission, the precursor ion can also split into complimentary ion pairs.
3. *Ligand activation* occurs when the protecting ligand fragments undergo fragmentation via loss of functional groups or rearrangement.

Table 2 summarises how precursor ion fragmentation can be accounted for using the general formula $[Au^{(0)}_a Au^{(+1)}_b L_c A^{(-1)}_d-nH^+]^z$, (Eq. (5)), for gold cluster ions. For instance, the fragmentation of $[Au_5L^1_4-H^+]^{2+}$, where L = dppm, has been studied [113] via collision-induced dissociation (CID) in a LTQ FT hybrid linear ion-trap mass spectrometer. Mass selection of $[Au_5L^1_4-H^+]^{2+}$ and subsequent ion activation, Fig. 18, results in the mass-selected precursor, $[Au_5L^1_4-H^+]^{2+}$ having two main fragmentation channels. A charge separation fragmenting into the complimentary ion pairs $[Au_3L^1_2]^+$ (Table 2, $a' = 2$, $b' = 1$, $c' = 2$, $d' = 0$, $n' = 0$)

and $[Au_2L^1{}_2\text{-}H^+]^+$ (Table 2, $a'' = 0$, $b'' = 2$, $c'' = 2$, $d'' = 0$, $n'' = 1$) and a ligand loss to generate $[Au_5L^1{}_3\text{-}H^+]^{2+}$.

2.6 Formation of Carbides, Hydroxides, Phosphides and Tellurides from Miscellaneous Top-Down Approaches

Top-down approaches that utilise particle or photon-surface interactions have been used to 'synthesise' a range of clusters in which gold is combined with other elements. Most of these studies have only used MS to determine the cluster ion stoichiometries, and thus their structures remain unknown. Exceptions are Wang's photoelectron spectroscopy studies on gold cluster anions containing other elements, which are discussed further in Sect. 3.6.

Gold carbide clusters, $Au_xC_y{}^+$, have been formed in several ways, including the reactions of 'hot' laser-desorbed gold cluster cations with ethene and alkanes [65], laser vaporisation of gold-coated carbon rods [161] and via impact of the $C_{60}{}^-$ anion at keV energies onto gold surfaces [162]. The latter approach produces both gold carbide cluster cations and anions. The structures of these gold carbide cluster ions have intrigued scientists and theoretical modelling has been employed to examine potential isomers [161–164]. Many of the isomers found consist of gold atoms coordinated to the end of carbon chains, and thus these do not possess gold–gold bonds.

Caesium ion bombardment at keV energies of gold surfaces prepared by vapour deposition of gold onto silver or glass substrates produces a wide range of anions containing a combination of different numbers of gold, oxygen or hydrogen [165]. The source of oxygen and hydrogen was assumed to be due to a surface layer of adsorbed H_2O and O_2 or their derived products. Anions with both even and odd electron counts were observed with nearly similar abundances for the oligomeric series $Au_xO_{2x}H_y{}^-$ and $Au_xO_{2x+1}H_z{}^-$ (where $x = 2$ or 3). Although the structures of these clusters were not determined, the authors made analogies to the known structures of neutral gold fluorides.

Finally laser ablation of nanogold mixtures with either red phosphorus or tellurium has been used to produce a range of gold phosphide [166] and gold telluride [167] cluster ions. A rich set of gold phosphide cations and anions were identified including $AuP_n{}^+$ ($n = 1$, 2–88 (even numbers)); $Au_2P_n{}^+$ ($n = 1$–7, 14–16, 21–51 (odd numbers)); $Au_3P_n{}^+$ ($n = 1$–6, 8, 9, 14); $Au_4P_n{}^+$ ($n = 1$–9, 14–16); $Au_5P_n{}^+$ ($n = 1$–6, 14, 16); $Au_6P_n{}^+$ ($n = 1$–6); $Au_7P_n{}^+$ ($n = 1$–7); $Au_8P_n{}^+$ ($n = 1$–6, 8); $Au_9P_n{}^+$ ($n = 1$–10); $Au_{10}P_n{}^+$ ($n = 1$–8, 15), $Au_{11}P_n{}^+$ ($n = 1$–6) and $Au_{12}P_n{}^+$ ($n = 1, 2, 4$); $AuP_n{}^-$ ($n = 4$–6, 8–26, 30–36 (even numbers), 48); $Au_2P_n{}^-$ ($n = 2$–5, 8, 11, 13, 15, 17); $Au_3P_n{}^-$ ($n = 6$–11, 32); $Au_4P_n{}^-$ ($n = 1, 2, 4, 6, 10$);

$Au_6P_n^-$; and $Au_7P_n^-$. Several new Au_mTe_n ($m = 1–11$; $n = 1–4$) clusters were identified.

3 Tools for Determining Structures and Their Application to Gold Cluster Ions

Ever since the development of soft ionisation methods such as FAB, ESI and MALDI, mass spectrometry has been applied to the analysis of gold cluster compounds. The primary concern has generally been the assignment of cluster formulae. For larger clusters where the stoichiometries of gold to ligand are unknown, or which are polydisperse mixtures, or mixtures containing different ligands, this can be challenging. Various approaches have been adopted to overcome these problems. For example, Maity et al. have used [168] 50:50 phenylacetylene and para-tolylacetylene mixtures to assign the stoichiometry of organogold clusters protected by phenylacetylene.

In this section we do not review all of the analytical applications of MS in gold cluster chemistry. Rather we highlight the types of tools that have been developed to study the structures of both bare and ligated cluster ions in the gas phase. Quantum chemical calculations are an essential adjunct to experimental techniques, which allow the gas-phase structures of many clusters to be obtained. A recent interesting review by Kappes et al. highlights the techniques used to date to determine the structures of bare gold clusters [169]. The experimental techniques discussed below include (1) fragmentation methods (identify atom connectivity), (2) ion mobility spectrometry (IMS) (identify the collision cross section), (3) infrared multiphoton dissociation (IRMPD) (identify IR absorption), (4) UV–Vis spectroscopy (identify the UV–Vis absorption spectra to investigate the clusters structure), (5) trapped ion electron diffraction (TIED) (identify molecular scattering) and (6) photoelectron spectroscopy (PES) (identify the electron binding energy).

3.1 MS-Based Fragmentation Methods

The most commonly used approach in mass spectrometry to gain structural information is the use of activation methods to induce fragmentation of a mass-selected precursor ion. Activation can occur, for example, via collision-induced dissociation (CID, discussed in Sect. 3.1.1), surface-induced dissociation (SID), laser-based activation methods such as photodissociation (PD, discussed in Sect. 3.1.1) and ion–electron interactions (discussed in Sect. 3.1.2). When tunable lasers are used, this allows the spectroscopy of AuNCs to be examined, and this is discussed further in Sects. 3.3 and 3.4 below.

3.1.1 Collision-Induced Dissociation and Photodissociation of Gold
Cluster Ions in the Gas Phase

Bare Gold Cluster Cations and Anions

Schweikhard's group has widely studied the fragmentation reactions of bare gold cluster anions and cations in a Penning ion trap using a range of activation techniques including CID [170–178], PD [176, 179–195] and electron-induced dissociation (EID) [196–201]. The latter studies are discussed in Sect. 3.1.2 below.

The collision of positively charged gold clusters, Au_n^+ ($2 \leq n \leq 23$ atoms), with rare gases, results in dissociation via loss of a single gold atom, (Eq. (7), $z = +1$), or two gold atoms. The loss of two atoms occurs most probably through the emission of a dimer, (Eq. (8), $z = +1$), rather than by a sequential evaporation. The minimum kinetic energies of clusters required to induce dissociation exhibit a pronounced odd–even effect. Clusters with an even number of delocalised electrons are more stable than those with an odd number. This was observed for Au_n^+, $n \leq 15$. When $15 \leq n \leq 23$, monomer loss is the main fragmentation pathway suggesting that the binding energies of all cluster cations in this range lie above the dimer binding energy:

$$[Au_n]^z \rightarrow [Au_{n-1}]^z + Au \tag{7}$$

$$\rightarrow [Au_{n-2}]^z + Au_2 \tag{8}$$

$$[Au_n]^z \rightarrow [Au_n]^{(z+1)} + e^- \tag{9}$$

Ervin's group have examined the fragmentation patterns and extracted the bond dissociation energies of small gold anionic clusters, Au_n^-, using the energy-resolved threshold collision-induced dissociation ($n = 2$–7) as well as the time-resolved photodissociation lifetime ($n = 6, 7$) techniques [176]. In both cases the main fragmentation channels are found to be loss of Au atom (Eq. (7), $z = -1$), which is favoured for odd electron anions (i.e. when $n = $ even) or Au_2 (Eq. (8), $z = -1$), which dominates for even electron ions (i.e. when $n = $ odd). The dissociation energies extracted from the two techniques were in good agreement and show an even–odd alternation for the loss of an atom from gold cluster anions, $D_0(Au_{n-1}^- -Au)$. Finally, photoelectron loss (Eq. (9)) appears to be important for $[Au_6]^-$ and was inferred via depletion of the signal for the cluster anion. Table 3 summarises the fragmentation methods used and the key observations in the study of bare gold cluster ions.

Ligated Gold Cluster Cations

One of the first studies in which a ligated gold cluster cation was subjected to CID appears to be that of Wang et al. [202], who studied the gas-phase fragmentation of $[Au_{20}(PPh_3)_8]^{2+}$ under conditions of sustained off-resonance irradiation collision-induced dissociation (SORI-CID) in an FT-ICR. Losses of $nPPh_3$ ligands (where

Table 3 Summaries of studies on the fragmentation of bare gold cluster cations and anions

Au_n^z	(n)	Method	Key observation(s)	References
Au_n^+	2–23	CID	CID for $n \leq 15$ results in dimer evaporation. All larger clusters fragment via monomer evaporation	[170]
Au_n^+	2–23	CID	Study of binding energies for the dissociation of clusters. Even electron clusters were found to be more stable than odd electron clusters	[171]
Au_{15}^{2+}	15	CID	Fragments into the complimentary ion pairs Au_3^+ and Au_{11}^+	[172]
Au_n^+	3–23	PD	Undergo dimer evaporation	[179]
Au_n^+	9, 21	PD	Dissociation energies determined	[180]
Au_n^{2+}	12	CID	Sequential dissociation was used to circumvent interference with singly charged clusters. Fragmentation into the complimentary ions pairs Au_3^+ and Au_9^+ was observed for Au_{12}^{2+}	[173]
Au_n^{2+}	7–35	CID	Fission occurs for small clusters at $n = 15$ (Eq. (8), $z = +2$). Larger clusters undergo gold atom evaporation (Eq. (7), $z = +2$)	[174]
Au_n^{3+}	19–35	CID	Fission occurs for small clusters at $n = 25$ (Eq. (8), $z = +3$) and larger clusters exhibit gold atom evaporation (Eq. (7), $z = +3$)	[174]
Au_n^-	16–30	EID	First observation of gold cluster dianions $n = 20$–30	[196]
Au_n^-	12–28	EID	Formation of dianions. Strong odd–even effect observed	[197]
Au_n^-	2–21	CID	Neutral monomer and dimer evaporation	[175]
Au_n^+	12–72	EID	Odd–even effects. For larger clusters magic numbers are observed	[198]
Au_n^-	6, 7	PD	Determination of dissociation energies	[176]
Au_n^-	2–7	CID	Determination of dissociation energies	[176]
Au_n^{2+}	7–35	CID	Odd–even effect disappears for $n < 11$	[177]
Au_n^{3+}	19–35	CID	First studies for the CID of Au_n^{3+}. The dissociation energy as a function of cluster size is smaller than Au_n^{2+}. No odd–even effects and no particularly stable clusters were observed	[177]
Au_n^+	7–15	PD	The energy dependence for monomer and dimer evaporation has been calculated for the size-selected clusters decay pathway	[181]
Au_n^+	2–27	PD	Au_9^+ shows anomalous behaviour to its odd neighbours with less likelihood of dimer evaporation. This is due to electron shell closing at $n = 8$	[182]
Au_n^+	14, 16	PD	Determination of dissociation energies	[183]

(continued)

Table 3 (continued)

Au_n^z	(n)	Method	Key observation(s)	References
Au_n^+	14–24	PD	Determination of dissociation energies	[184]
Au_n^-		EID	Production of dianionic and trianionic clusters	[199, 200]
Au_n^-	25	EID	Threshold for production of Au_{25}^{2-} is determined	[201]
Au_n^-	7	PD	Photodissociation with a green laser pointer shows decay into Au_6^- and Au_5^-	[185]
Au_n^{2-}	29	PD	Decay observed by PD	[185]
Au_n^+	9, 11, 13, 15	PD	Determination of dissociation energy via dimer evaporation	[186]
Au_n^+	17–21	PD	Excitation energies far above the clusters dissociation thresholds have been used to induce multisequential fragmentation by the evaporation of neutral monomers	[187]
Au_n^-	11–40 51–70	Low-energy electrons	Production of dianions and trianions	[188]
Au_n^+	3–21	PD	Fragmentation pathways monitored as a function of cluster size. Monomer and dimer evaporation	[189]
Au_n^+	5, 8, 12	PD	Isomers can be distinguished by their distinct decay rates	[190]
Au_n^+	30	PD	The determination of relative dissociation energies	[191]
Au_n^+	7–27	PD	The monomer-dimer branching ratios as a function of the excitation energy and dissociation energy were determined	[192]
Au_n^{2-}	29	PD	Dissociation products determined	[193]
Au_n^-	14, 17	PD	Study of decay rates for cluster anions	[194]
Au_n^{2-}	35, 40, 45, 50	PD	Photoexcited gold cluster dianions resulted in neutral atom evaporation (Eq. (7), $z = -2$) and electron emission (Eq. (9), $z = -2$)	[195]
Au_n^{2-}	21–31	CID	Singly charged clusters have been observed at low collision energy, indicating the emission of one electron Eq. (9)	[178]

$n = 1$–4) were observed, with no core fission occurring under these low-energy collision conditions. These results highlight the relative structural stability of the Au_{20} core, which was predicted from theoretical calculations to be tetrahedral [203].

The CID of bisphosphine-protected gold nanoclusters where the protecting ligand was Ph_2P-$(CH_2)_n$-PPh_2 ($n = 1$–6) has been studied (Sect. 2.5 above). This

Table 4 Types of reactions and their branching ratios (BR) observed in the ECD spectra of multiply charged dppm-capped gold nanoclusters

Ion	BR (Eq. (10))	BR (Eq. (11))	BR (n)
$[Au_2L^1_2]^{2+}$			100%
$[Au_5L^1_4 - H^+]^{2+}$		100%	
$[Au_9L^1_4]^{3+}$	89.0%	6.9%	4.1%
$[Au_9L^1_5]^{3+}$	87.7%	12.3%	
$[Au_{10}L^1_4]^{2+}$	100.0%		
$[Au_{11}L^1_4]^{3+}$	100.0%		
$[Au_{11}L^1_5]^{3+}$	100.0%		
$[Au_{11}L^1_6]^{3+}$	41.6%	58.4%	
$[Au_{13}L^1_5]^{3+}$	100.0%		
$[Au_{13}L^1_6]^{3+}$	99.3%	0.7%	
$[Au_{14}L^1_7 - H^+]^{3+}$		100.0%	

No ionic fragments are observed for Au_8 and Au_6 clusters. Blank cells indicate that the channel was not observed

approach not only provides structural information but also allows for the synthesis of new metal clusters in the gas phase (Sect. 2.5).

3.1.2 Electron Capture Dissociation and Electron-Induced Dissociation

Multiply charged gold cluster ions present the opportunity to investigate fragmentation induced by one electron reduction. In solution the electron source for reduction of Au(I) to Au(0) originates from the chemical reducing agent (e.g. $NaBH_4$, BTBC and other boranes). Additionally, the reducing environment persists after the initial formation of gold nanoclusters during the reduction phase. Mass selection of reactive intermediates of the general equation $[Au^{(0)}_a Au^{(+1)}_b L_c A^{(-1)}_d - nH^+]^z$ where $z \geq 2$ could provide useful information regarding the electronically driven fragmentation processes that are required for the formation of subsequent clusters, whether they are poly- or monodispersed. Compared to the condensed phase, gas-phase single-electron electrochemical reactions of gold cluster ions are free of competing neutral radical Au(0) atoms and small clusters.

Three types of reactions were observed in the electron capture dissociation (ECD) studies of multiply charged dppm-protected gold clusters [113] (Table 4): (1) charge reduction Eq. (10), (2) charge reduction and ligand loss Eq. (11) and (3) charge reduction and ligand activation Eq. (12):

$$\left[Au_xL_y\right]^{n+} + e^- \rightarrow \left[Au_xL_y\right]^{(n-1)+\bullet} \tag{10}$$

$$\left[Au_xL_y\right]^{n+} + e^- \rightarrow \left[Au_xL_{y-1}\right]^{(n-1)+\bullet} + L \tag{11}$$

$$\left[Au_xL_y\right]^{n+} + e^- \rightarrow \left[Au_xL_{y-1}(CH_2PPh_2)\right]^{(n-1)+} + Ph_2P^\bullet \tag{12}$$

Compared to the CID fragmentation studies, no core fission was observed, suggesting the preservation of aurophilic interactions in the reduced cluster cations.

Bare gold cluster cations, Au_n^+ ($n = 12$–72), generated via laser vaporisation using a Smalley-type source were mass selected and isolated in a Penning trap [198]. The mass-selected clusters were then bombarded with an electron beam (typically ca. 150 eV for 1.2 s) generated from a rhenium filament. Electron impact resulting in the dissociation and further ionisation of mass-selected gold clusters reveals that small gold clusters, Au_n^{z+} ($n \leq 30$) where $z = 1$ or 2, exhibit an odd–even alternation in their abundance with preference to even electron clusters. For Au_n^+, n is generally an odd number and for Au_n^{2+} n is generally an even number.

3.2 Ion Mobility

The mobility of gas-phase ions through a gas and under the influence of an electric field is a useful property to exploit in order to develop techniques to separate nanoparticles and nanoclusters based on their size, shape and charge [204]. When coupled to mass spectrometry, termed IMS [205], it allows for accurate measurements of the particle mass and therefore its composition. The marriage of this technique with theoretical calculations has proven to be very powerful in determining the structures of gas-phase cluster ions [206]. The use of IMS is gaining grounds in the analysis of clusters. Weis has reviewed the use of IMS in fundamental studies of metal clusters, including gold clusters [207], and a chapter in this book, by A. Fielicke et al., highlights work done on bare AuNC ions. Here we briefly review IMS studies on bare AuNCs (Sect. 3.2.1) and ligated AuNCs (Sect. 3.2.2).

3.2.1 Bare Gold Nanoclusters

In early work, laser ablation coupled with IMS and density functional theory calculations was used to determine the structures of bare gold anions and cations, Au_n^- and Au_n^+ (n up to 13) [208, 209]. Cationic clusters were found to have planar structures for $n \leq 3$–7 and form 3D structures at $n \geq 8$. Anionic clusters on the other hand are planar for $n \leq 3$–11, for $n = 12$ both a planar and a three-dimensional structure seem to coexist, and for $n = 13$, only a three-dimensional structure is observed (Fig. 19). Kappes also studied mixed silver–gold clusters $Ag_mAu_n^+$ ($m + n$) <6. Triangular, rhombus (or Y-shaped) and connected triangles (or trigonal bipyramid) were observed as the main structures for trimers, tetramers and pentamers, respectively.

3.2.2 Ligated Gold Nanoclusters

Ion mobility was also used to study ligated gold clusters. Lenggoro et al. [210] have reported the use of ES-DMA (electrospray–differential mobility analysis) to study AuNPs. This technique allows the segregation of the nanoclusters by size

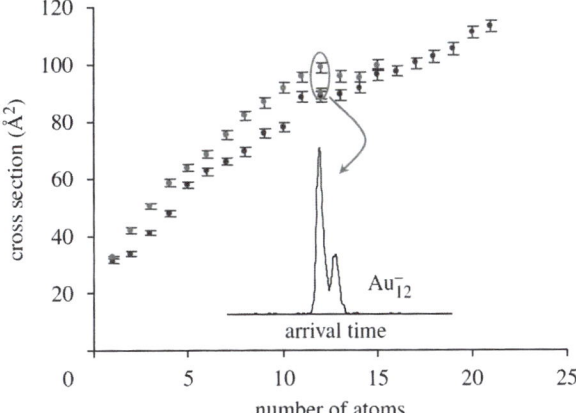

Fig. 19 IMS determined experimental cross section for gold cluster cations (*black circles*) Au_n^+ and anions (*grey circles*) Au_n^- at 300 K. For the same n value, the anions clearly have a larger cross section. For $n = 12$; a bimodal arrival time is observed indicating the presence of 2 isomers. Figure reproduced from [209]

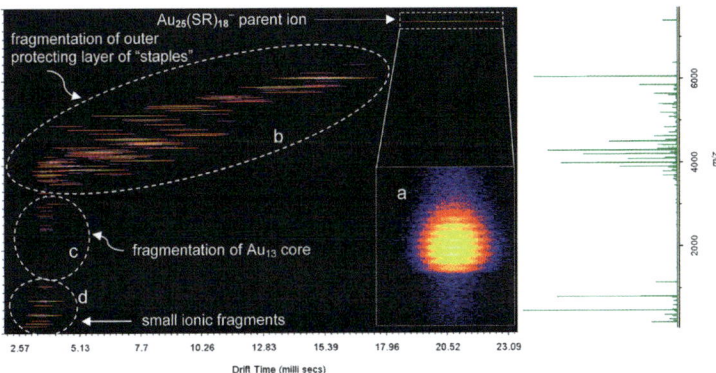

Fig. 20 IM-MS/MS driftscope plot showing m/z (*y* axis) versus drift time (*x* axis) for the analysis of $Au_{25}(SCH_2CH_2Ph)_{18}$. Inset *a* shows an expansion of the isotopic distribution of the parent ion, $Au_{25}(SCH_2CH_2Ph)_{18}^-$. The CID of $Au_{25}(SCH_2CH_2Ph)_{18}$ results in the fragmentation of the [-SR-Au-SR-Au-SR-] staples as shown by the slant oval *b*. Fragmentation of the $Au_{13}(SR)_m(S)_n$ core is shown by the vertical oval *c*, and the small ionic fragments are shown by *d*. Figure reproduced from [216]

[211]. Zachariah et al. have studied gold nanorods [212], aggregation and kinetics of colloidal gold [211] and the conformation of bovine serum albumin on AuNPs [213]. Hackley et al. reported recently the use of a new hyphenated technique where ES-DMA was coupled to ICP-MS. They show that this technique can be used to segregate gold nanoparticles based on their size and their elemental composition [214, 215]. ES-DMA and variants were also coupled directly to mass spectrometry (ESI-IMS) and used to decipher the structures of ligated gold nanoparticles. Dass

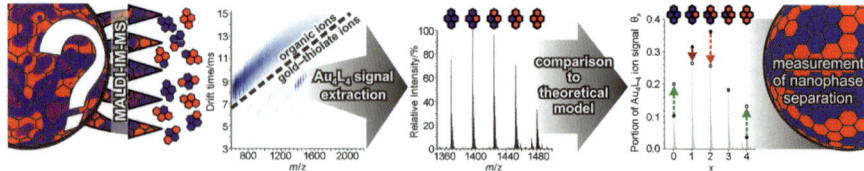

Fig. 21 Workflow for MALDI-IM-MS experiments conducted by Cliffel et al. on mixed-ligand AuNPs with unknown levels of nanophase separation. The MALDI process leads to the fragmentation and the gold–thiolate ions undergo gas-phase separation from organic ions via IM. The Au_4L_4 ion species re-extracted from the data by software, and their abundances are compared to a theoretical model. Deviations indicate nanophase separation in the AuNP monolayer. Figure reproduced from [219]

et al. used ESI-IMS to study the structures of $Au_{25}(SCH_2CH_2Ph)_{18}$ and its fragments upon CID [216] (Fig. 20). As seen in Fig. 20, their results show that the Au_{25} exists in the gas phase as one structural isomer, with an Au_{13} as the core and six staples (-RS-Au-SR-Au-SR). In addition, the IMS results show formation of a series of eight bands that are related to the fragmentation of the outer protecting 'staples' shell, each containing an increasing number of Au/SmRn products (Fig. 20). Fragments relating to the breakdown of the Au_{13} core can also be observed, along with other low-mass fragments.

McLean et al. used MALDI-IMS [217] for the analysis of low-mass Au–thiolate fragments derived from thiolate-protected AuNPs. Positive fragment ions generated were identified as identical to the gold–thiolate precursor complex, and hence, the authors postulate that a reversal of the reduction of the precursor complex has occurred due to the energy provided by the ionisation method (i.e. MALDI). On the other hand, negative fragment ions were found to be similar to capping structural motifs that are well established in the literature [218].

Cliffel et al. also used MALDI-IMS to study the mixed thiolate ligands on AuNPs and measured the phase segregation in the protecting monolayer. This is achieved when fragmentation of the Au–thiolate complex occurs during the MALDI process from the AuNP surface. These Au–thiolate separate via ion mobility in the gas phase from organic ions. Au_4L_4 species were identified and then analysed. They claim this as a novel strategy for the analysis of nanophase separation on nanoparticles (Fig. 21) [217, 219].

3.3 IR Spectroscopy

Infrared (IR) spectroscopy is increasingly becoming a widely used technique aimed at determining gas-phase structures of metal clusters amongst others [220]. For instance, Armentrout et al. determined the structures of the dehydrogenated products of methane achieved by group 5 transition metals [221], and Asmis et al. have been determining the structures of transition-metal oxides [222, 223].

Table 5 Gas-phase IR spectroscopy of ligated neutral and anionic gold clusters. Experimentally observed absorption bands and structures are listed

Gold cluster	Absorption wave number(s) (cm^{-1})	Ligand	Structure	References
Au_4^0	1,502	O_2	2D distorted Y-shaped	[226]
Au_7^0	1,063	O_2	2D centred hexagone	[226]
Au_9^0	1,064	O_2	3D bicapped trigonal prism	[226]
Au_{11}^0	1,058	O_2	–	[226]
Au_{21}^0	1,069	O_2	–	[226]
Au_2^0	ca. 57	Kr	–	[224]
Au_3^0	ca. 95	Kr	Obtuse-angled isosceles triangle	[224]
Au_4^0	ca. 150	Kr	2D rhombus and Y-shaped	[224]
Au_7^0	165, 186, 201	Kr	2D structure	[224]
Au_{19}^0	149, 167	Kr	Truncated pyramid	[225]
Au_{20}^0	148	Kr	Tetrahedral	[225]
Au_4^-	ca. 1,060	O_2	2D distorted Y-shape	[232]
Au_6^-	ca. 1,060	O_2	2D D_{3h}	[232]
Au_8^-	1,043, 1,059, 1,102	O_2	Planar edge-capped hexagon	[232]
Au_{10}^-	ca. 1,060	O_2	Planar – contains a 7 Au hexagonal motif	[232]
Au_x^- ($x = 12$–20)	ca. 1,060	O_2	–	[232]

Several gas-phase IR spectroscopic studies aimed at understanding the structures of small gold cluster anions, cations and neutrals were reported [224–232] and are the subject of another chapter in this book by A. Fielicke et al. Table 5 above is a summary of the absorption wave numbers observed for selected neutral and anionic gold clusters.

3.4 UV–Vis Spectroscopy

Gas-phase UV–Vis spectroscopic studies of gold clusters have been rarely reported. The first report on neutral and cationic clusters of the form Au_n ($n = 7, 9, 11, 13$) and Au_n^+ ($n = 6$–13) appeared around 20 years ago [233]. The authors highlighted the odd–even alterations in the UV–Vis spectra; however, they concluded that little structural information could be extracted from these spectra. Thus, only Au_{13} was reported to be consistent with an icosahedral structure. Recently, the UV–Vis spectrum of the glutathione-protected gold cluster $[Au_{25}(SG)_{18}$–$6H]^{7-}$ was reported [234]. Although no structural information was extracted, the authors have shown that the UV–Vis spectrum for this AuNC is similar in both the gas and solution phases.

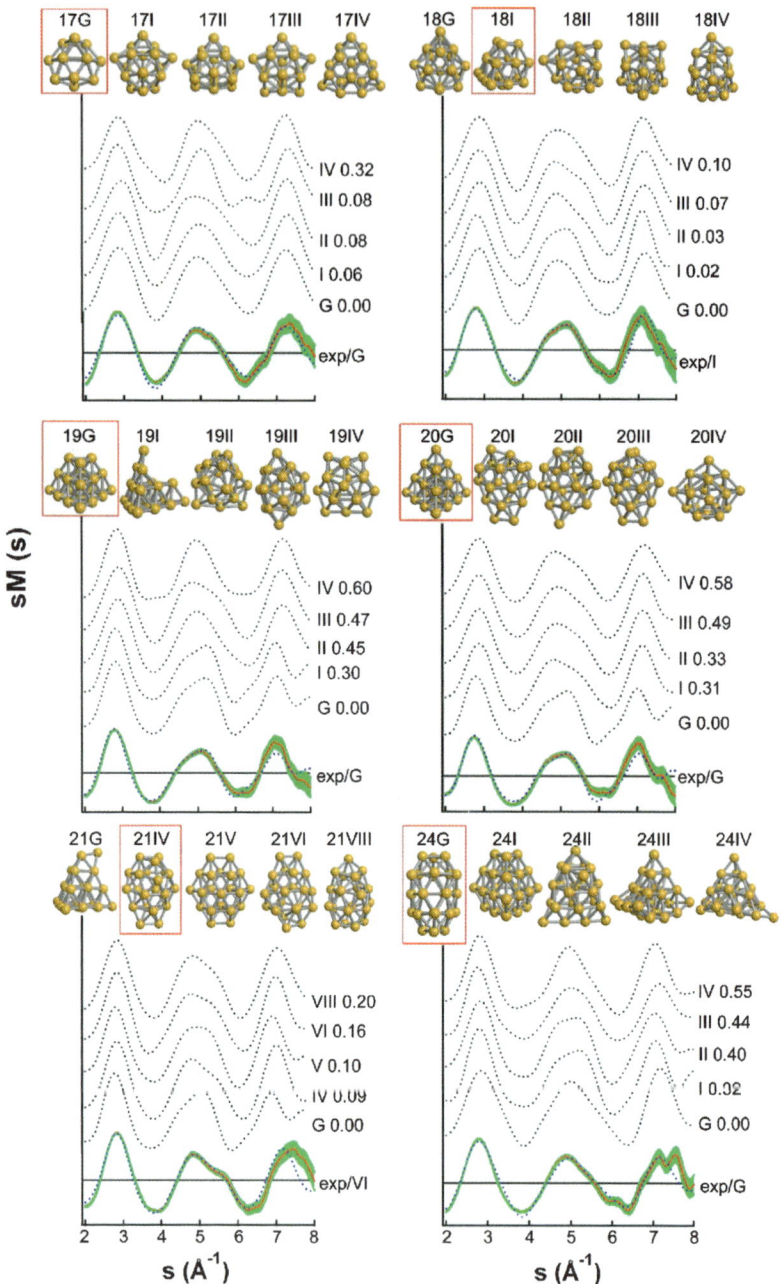

Fig. 22 Gas-phase electron diffraction of gold cluster anions. Calculated isomeric structures for Au$_n^-$ clusters with $11 \leq n \leq 16$ are shown at the top of each panel with their corresponding diffraction patterns ($sM(s)$ vs s (A^{-1})) depicted below (*dotted curves*). Experimental diffraction $sM(s)$ (*solid curve, red*) along with the best fit (*dotted curve, blue*) are shown at the bottom. The grey (*online green*) shading shows the data uncertainty $\pm\sigma$. The energy of the isomer relative to the ground state (GS) is shown on the right for each corresponding curve. The diffraction intensity has arbitrary units. The isomers that constitute the best fit are marked by a red frame in each panel. Figure reproduced from [236]

3.5 Trapped Ion Electron Diffraction

TIED was also used in the structural determination of gold cluster ions. The analysis of the diffraction measurements is appropriate since it directly relates to the spatial arrangement of the scattering atoms. This technique however suffers from limitations due to, for example, the atomic scattering intensities which limit the lower end of detection and the inability to distinguish between isomers belonging to the same structural family.

Gold cluster anions were studied by Schooss and Kappes et al. [235] as well as by Parks et al. [236]. The work of Kappes et al. was recently reviewed [169]. Briefly, the structures of the anions Au_n^- ($n = 11–20$) have been determined by TIED and compared to theoretically calculated structures. The smallest cluster studied, Au_{11}^-, possessed a flat C_s structure as was predicted by theory [237]. The structures of Au_n^- ($n = 11, 13$) were reported to be in agreement with IMS studies (see before) and PES (Sect. 3.6). Au_n^- ($n = 16–20$) possess a cagelike structural motif, in agreement with previous PES studies. Parks et al.'s work on gold cluster anions Au_n^- ($n = 11–24$) reported [236] structures that were also compared to DFT calculations. For Au_n^- ($n = 11–12$), structures were found that match those determined via IMS [209]. For Au_{13}^-, however, a mixture of isomers was reported, in disagreement with the previously mentioned TIED work. In addition the 3D structures reported for Au_n^- ($n = 14, 15$) were in stark contrast to the previously mentioned TIED work. Structures for Au_n^- ($n = 16–20$) were in agreement with other TIED reports highlighting their cagelike structures (Fig. 22). Au_{21}^- and Au_{24}^- were reported to possess an elongated cage and a single-wall tube-like structure, respectively (Fig. 22).

Structures for gold cluster cations, $Au_{11–20}^+$, were also studied by TIED. The 3D structures of $Au_{11–13}^+$ were found to be in close agreement to those reported by IMS. $Au_{14–17}^+$ were found to have layered structures, whereas a change in the structural motif to decorated cagelike structures is observed for $Au_{18–20}^+$ with Au_{20}^+ possessing 2 isomers.

In many cases, the minimum-energy structures obtained by calculations were not consistent with the experimental scattering function. It was rather calculated structures with higher energies that fitted the experimental data.

3.6 Photoelectron Spectroscopy

Another technique used in determining the structures of anionic clusters is PES. We should note that experiments conducted in the last century generated useful data, however failed to yield detailed structural information. In all cases, an odd–even oscillation in the values of the electron affinity was observed [233, 238]. Bare anionic gold clusters were widely studied in particular by Lai-Sheng Wang et al. who have recently reviewed their wide body of work (see [239] with references therein) and included some comparisons to other reports. Briefly, the structures of the anionic clusters Au_n^- ($n = 3$–15) were reported as 2D flat structures for $n \leq 12$ [240]. This was surprising since metal clusters are expected to form 3D structures in this size range. For $n = 12$, a mixture of isomers (2D and 3D) has been reported, and hence this size represents the transition from 2D to 3D structures for gold cluster anions. In the case of $n = 13$–15, 3D structures are expected. Mixtures of isomeric structures were identified for most of the cluster sizes for $n = 7$–15.

Argon tagging and oxygen titration were used to shed more light on the structures of Au_n^- [241]. For instance, they were used to resolve the issue of isomeric Au_{10}^- where four distinct isomers were found to coexist and they were readily distinguished (Fig. 23). It should be noted that the formation of a superoxo was observed and confirmed via PES, upon adsorption of O_2 onto gold cluster anions (Au_n^-) with an even number of gold atoms [242, 243]. In addition, substitution in Au_n^- by isoelectronic Cu or Ag was also used as another method that provide information on the various isomers [244].

Structures of the gold clusters, Au_n^-, in the range $n = 16$–19 [245] were also reported (Fig. 24). The major isomers found for $n = 16$–18 were dominated by hollow cage motifs, with the Au_{16}^- structure being the most interesting. This highly symmetrical cage structure (T_d symmetry) shows, amongst others, a break from the odd–even trend observed for coinage metal cluster ions. Au_{19}^- structure was resolved and assigned as a single isomer with a pyramidal structure. A transition from cage structure to pyramidal is observed at Au_{18}^-.

A tetrahedral symmetric structure for Au_{20}^- was reported (Fig. 24) [203]. This highly symmetric cluster is special, possessing a large HUMO–LUMO gap of 1.77 eV reminiscent of C_{60}^-. Excluding Au_2 and Au_6, this HUMO–LUMO gap is the largest amongst all known coinage metal clusters [238]. The high symmetry of this structure led several researchers to attempt the synthesis of ligand-protected golden pyramids with an Au_{20} core. For example, the isolation of the thiolate-capped $Au_{20}(SCH_2CH_2Ph)_{16}$ cluster has recently been reported [246].

The experimental vertical detachment energies and adiabatic detachment energies measured for Au_n^- ($n = 2$–20) gold cluster anions via PES are listed in Table 6.

Few experimental structures of the anions in the range Au_n^- $n = 21$–35 have been proposed [247–249] to range from pyramidal to tubular to core–shell. Au_{21}^- possesses a pyramidal structure, whereas $Au_{22,23}^-$ were reported to have mixed isomers: pyramidal and fused planar. Au_{24}^- was found to be tubular and Au_{25}^- with a core–shell structure. This latter cluster is the smallest anionic gold cluster with an atom in the core. Larger clusters were reported to have low-symmetry core–shell structures.

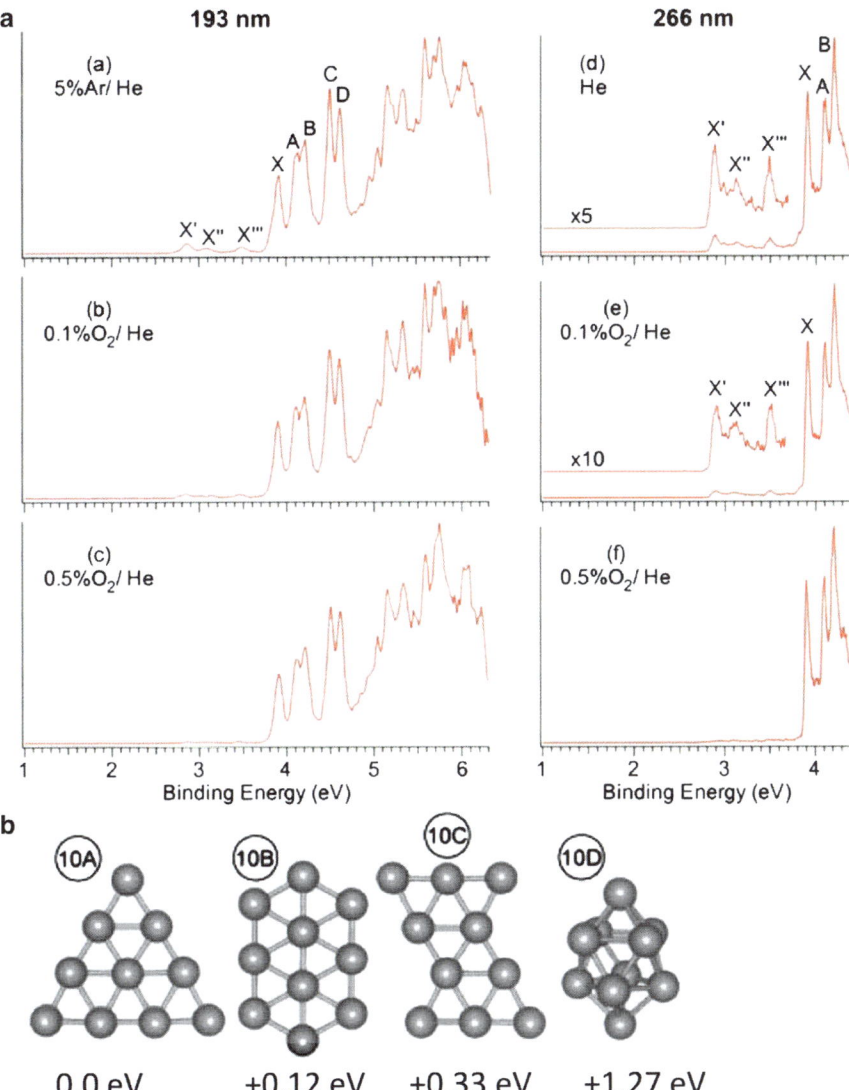

Fig. 23 (**a**) Comparison of the 193 nm and 266 nm PES spectra of Au$_{10}^-$ produced using a 5% Ar–He (*a*) or pure He (*d*) as carrier gases with those using 0.1% O$_2$–He (*b, e*) and 0.5% O$_2$–He (*c,f*) as carrier gases. Note that the weak features (X′, X″, X‴) are titrated out with increasing O$_2$ concentration. A-D are spectral features due to the D$_{3h}$ global minimum structure. (**b**) Calculated minimum-energy structures for Au$_{10}^-$ (10A is the global minimum). Figures reproduced from (**a**) [240] and (**b**) [241]

Fig. 24 Lowest-energy
structures of the Au$_n^-$
clusters for $n = 16$–20 as
determined via DFT
calculations.
Figure reproduced from
[245]

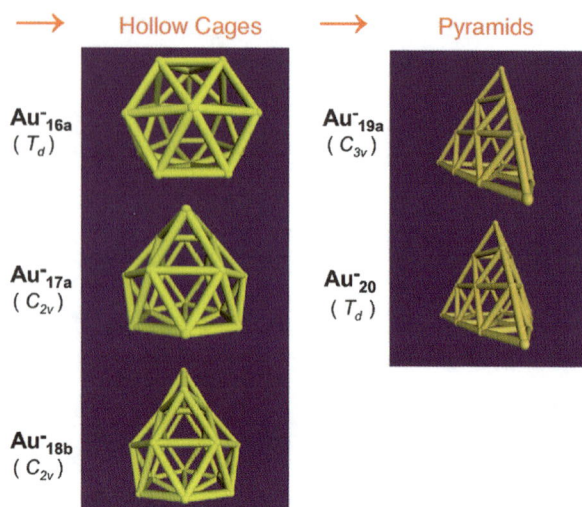

Table 6 Experimental vertical detachment energies (VDE) and adiabatic detachment energies
(ADE) for Au$_n^-$ ($n = 2$–20)

Cluster Au$_n^-$	Experimental VDE (eV)	Experimental ADE (eV)	References
2	2.01	1.92	[240]
3	3.88	3.88	[240]
4	2.75	2.70	[240]
5	3.09	3.06	[240]
6	2.13	2.06	[240]
7	3.46	3.40	[240]
8	2.79	2.73	[240]
9	3.83	3.81	[240]
10	3.91	3.89	[240]
11	3.80	3.76	[240]
12	3.06	3.03	[240]
13	3.94	3.91	[240]
14	3.00	2.94	[240]
15	3.63	–	[245]
16	4.03	–	[245]
17	4.08	–	[245]
18	3.32	–	[245]
19	3.74	–	[245]
20	2.751	2.745	[203]

Table 7 Key gas-phase bond energies derived from guided ion beam studies of Au^+ reacting with substrates

Substrate	Reaction products	Bond energy derived	Bond energy eV (kJ mol^{-1})	References
H_2 (D_2)	AuH^+ (AuD^+) + H (D)	Au^+-H	2.13 ± 0.11 (206 ± 11)[a]	[258]
CH_4	$AuH^+ + CH_3$	Au^+-H	1.94 ± 0.08 (187 ± 8)[b]	[260]
CH_4	$AuCH_3^+ + H$	Au^+-CH_3	1.91 ± 0.13 (184 ± 13)	[260]
CH_4	$AuCH_2^+ + H_2$	Au^+-CH_2	3.75 ± 0.05 (362 ± 5)	[260]
O_2	$AuO^+ + O$	Au^+-O	$>0.42 \pm 0.20$ ($>41 \pm 19$)[c]	[259]
N_2O	$AuO^+ + N_2$	Au^+-O	$>0.55 \pm 0.07$ ($>53 \pm 7$)[c]	[259]
N_2O	$AuN_2^+ + O$	Au^+-N_2	$>0.30 \pm 0.04$ ($>29 \pm 4$)[c]	[259]

[a]Average from measurements of both H_2 and D_2, adjusted for the zero point energy difference
[b]Lower limit of bond energy
[c]Lower limit due to impulsive behaviour

3.7 Thermochemistry

There is a rich history on the use of mass spectrometry-based methods to measure the binding energies of substrates to metal ions and metal cluster ions using gas-phase ion–molecule reactions [31, 250–257].

Much of the earlier literature dealt with the thermochemistry associated with bare metal atomic cations [253]. Recent guided ion beam measurements from Armentrout et al. [258–260] have examined the reactions of Au^+ with H_2, D_2, HD, O_2, N_2O and CH_4 and have provided important Au^+–X bond energies, which are listed in Table 7.

Two key methods have been used to measure the binding energies of substrates to gold cluster ions: (1) equilibrium measurements, whereby an equilibrium between the cluster ion, $Au_n^{+/-}$, and the neutral, L, at a known concentration and temperature is established (Eq. (13)). Measurement of the relative ion signals for the cluster ion, $Au_n^{+/-}$, and its complex, $Au_n(L)^{+/-}$, combined with the known concentration of the neutral allows the equilibrium constant to be determined, from which the binding energy can be calculated. (2) Kinetic measurements, whereby the rate of reaction between the cluster ion, $Au_n^{+/-}$, and the neutral, L, at a known concentration and temperature is determined (Eq. (14)). The initially formed energised adduct undergoes collisional stabilisation with the bath gas, which is typically helium in ion-trapping instruments. Treatment of the kinetic data using theoretical approaches such as Rice–Ramsperger–Kassel–Marcus (RRKM) theory can allow the binding energy to be estimated:

$$Au_n^{+/-} + L \rightleftharpoons Au_n(L)^{+/-} \qquad (13)$$

$$Au_n^{+/-} + L \rightarrow \left[Au_n(L)^{+/-}\right]^* \overset{He}{\rightarrow} Au_n(L)^{+/-} \tag{14}$$

Table 8 lists the gas-phase binding energies of gold cluster ions towards a range of substrates [261, 262]. The motivation of many of these studies was to provide a firm thermochemical understanding of key steps associated with possible catalytic cycles for the processing of substrates. Thus the reader is directed to related studies on ion–molecule reactions (Sect. 4.1) and catalytic cycles (Sect. 4.3).

An examination of the data in Table 8 reveals some interesting trends:

1. The thermochemistry associated with room temperature CO adsorption on isolated gold cluster cations exhibits a pronounced cluster size effect for the adsorption energies for the first CO molecule binding to the gold cluster ions. The binding energies decrease with increasing cluster size from 1.09 eV for $n = 6$ to less than 0.65 eV or $n = 26$. Exceptions were local maxima of between 0.73 and 0.75, which were found for $n = 30$, 31, and 48, 49. The atom-by-atom variations were suggested to arise from different binding sites on the gold clusters, consistent with DFT calculations on the smallest cluster sizes $n = 3$–9 and $n = 20$.
2. Not surprisingly, the binding energies also depend on the nature of the substrate. Thus CO binds more strongly than CH_4 to the gold cluster cations.
3. The magnitude of the cluster charge is important. Thus CO binds more strongly to the cluster cations than the cluster anions.

The combination of experiment and DFT calculations provides valuable insight into how binding energies change upon sequential addition of CO and how this influences the geometry of the gold cluster core. A case in point is the binding of CO to the homo and hetero, M gold and silver cluster cations, $M_{5-x}M'_x{}^+$ (where M=Au and M'=Ag) [267]. At room temperature, $Au_5{}^+$ rapidly absorbed four CO molecules, with absorption of a fifth CO only occurring at temperatures lower than 250 K. This is consistent with the results of DFT calculations (Fig. 25), which reveal that the 'bow-tie' structure for the bare $Au_5{}^+$ cluster can readily bind four CO molecules at each of the four corner atoms. Binding of a fifth CO requires the gold cluster core to undergo a rearrangement to a structure described as a 'side-capped tetrahedron'. Related structural transitions were determined for the mixed gold–silver clusters $Au_3Ag_2{}^+$, $Au_2Ag_3{}^+$ and $AuAg_4{}^+$.

4 Reactivity of Gold Cluster Ions

The gas-phase reactions of gold cluster anions and cations with single and multiple neutral substrates have been widely studied over the past 2 decades. Much of this work has been inspired by Haruta's discovery of CO oxidation by gold clusters

Table 8 Gas-phase binding energies of substrates to gold cluster ions

Au cluster ion	Substrate	Binding energy eV (kJ mol^{-1})	Method	Reference
Au$^+$	CO	2.08 ± 0.15 (201 ± 14)	a	[261, 262]
Au$_5^+$	CO	1.03 ± 0.1 (99 ± 10)	c	[263]
Au$_5^+$	CO	1.04 ± 0.1 (100 ± 10)	c	[263]
Au$_6^+$	CO	1.09 ± 0.1 (105 ± 10)	c	[263, 264]
Au$_7^+$	CO	1.07 ± 0.1 (103 ± 10)	c	[263]
Au$_8^+$	CO	0.89 ± 0.1 (86 ± 10)	c	[263]
Au$_9^+$	CO	0.85 ± 0.1 (82 ± 10)	c	[263]
Au$_{10}^+$	CO	0.90 ± 0.1 (87 ± 10)	c	[263]
Au$_{11}^+$	CO	0.90 ± 0.1 (87 ± 10)	c	[263]
Au$_{12}^+$	CO	0.87 ± 0.1 (84 ± 10)	c	[263]
Au$_{13}^+$	CO	0.86 ± 0.1 (83 ± 10)	c	[263]
Au$_{14}^+$	CO	0.86 ± 0.1 (83 ± 10)	c	[263]
Au$_{15}^+$	CO	0.88 ± 0.1 (85 ± 10)	c	[263]
Au$_{16}^+$	CO	0.86 ± 0.1 (83 ± 10)	c	[263]
Au$_{17}^+$	CO	0.80 ± 0.1 (77 ± 10)	c	[263]
Au$_{18}^+$	CO	0.85 ± 0.1 (82 ± 10)	c	[263]
Au$_{19}^+$	CO	0.84 ± 0.1 (81 ± 10)	c	[263]
Au$_{20}^+$	CO	0.78 ± 0.1 (75 ± 10)	c	[263]
Au$_{21}^+$	CO	0.81 ± 0.1 (78 ± 10)	c	[263]
Au$_{22}^+$	CO	0.71 ± 0.1 (69 ± 10)	c	[263]
Au$_{23}^+$	CO	0.66 ± 0.1 (64 ± 10)	c	[263]
Au$_{24}^+$	CO	0.70 ± 0.1 (68 ± 10)	c	[263]
Au$_{25}^+$	CO	0.70 ± 0.1 (68 ± 10)	c	[263]
Au$_{26-29}^+$	CO	< 0.65 (<63)	c	[263]
Au$_{30}^+$	CO	0.73 ± 0.1 (70 ± 10)	c	[263]
Au$_{31}^+$	CO	0.75 ± 0.1 (72 ± 10)	c	[263]
Au$_{32}^+$	CO	0.69 ± 0.1 (67 ± 10)	c	[263]
Au$_{33-40}^+$	CO	< 0.65 (<63)	c	[263]
Au$_{41}^+$	CO	0.65 ± 0.1 (63 ± 10)	c	[263]
Au$_{42-47}^+$	CO	< 0.65 (<63)	c	[263]
Au$_{48}^+$	CO	0.74 ± 0.1 (71 ± 10)	c	[263]
Au$_{49}^+$	CO	0.74 ± 0.1 (71 ± 10)	c	[263]
Au$_{50-65}^+$	CO	< 0.65 (<63)	c	[263]
Au$_2^+$	CH$_4$	0.65 ± 0.03 (63 ± 3)	d, e	[265]
Au$_2^+$	CH$_4$	0.91 ± 0.04 (88 ± 4)	d, f	[265]
Au$_3^+$	CH$_4$	0.40 ± 0.06 (39 ± 6)	d, e	[265]
Au$_3^+$	CH$_4$	0.72 ± 0.07 (69 ± 7)	d, f	[265]
Au$_4^+$	CH$_4$	0.36 ± 0.04 (35 ± 4)	d, e	[265]
Au$_4^+$	CH$_4$	0.64 ± 0.04 (62 ± 4)	d, f	[265]
Au$_5^+$	CH$_4$	0.32 ± 0.09 (31 ± 9)	d, e	[265]
Au$_5^+$	CH$_4$	0.57 ± 0.09 (55 ± 9)	d, f	[265]
Au$_6^+$	CH$_4$	0.20 ± 0.02 (19 ± 2)	d, e	[265]
Au$_6^+$	CH$_4$	0.41 ± 0.03 (40 ± 3)	d, f	[265]
Ag$_2$Au$^+$	CO	0.81 ± 0.1 (78 ± 10)	d	[266]
Ag$_2$Au(CO)$^+$	CO	0.77 ± 0.1 (74 ± 10)	d	[266]
Ag$_2$Au(CO)$_2^+$	CO	0.73 ± 0.1 (70 ± 10)	d	[266]

(continued)

Table 8 (continued)

Au cluster ion	Substrate	Binding energy eV (kJ mol^{-1})	Method	Reference
AgAu$_4$$^+$	CO	1.01 ± 0.1 (97 ± 10)	c	[264]
Ag$_2$Au$_3$$^+$	CO	0.78 ± 0.1 (75 ± 10)	c	[264]
Ag$_3$Au$_2$(CO)$_2$$^+$	CO	0.52 ± 0.1 (50 ± 10)	b	[267]
Ag$_3$Au$_2$(CO)$_3$$^+$	CO	0.47 ± 0.1 (45 ± 10)	b	[267]
Ag$_3$Au$_2$CO)$_4$$^+$	CO	0.33 ± 0.1 (32 ± 10)	b	[267]
Ag$_4$Au$^+$	CO	0.77 ± 0.1 (74 ± 10)	b	[267]
Ag$_4$Au(CO)$^+$	CO	0.73 ± 0.1 (70 ± 10)	b	[267]
Ag$_4$Au(CO)$_2$$^+$	CO	0.55 ± 0.1 (53 ± 10)	b	[267]
Ag$_4$Au(CO)$_3$$^+$	CO	0.05 ± 0.1 (5 ± 10)	b	[267]
Ag$_4$Au(CO)$_4$$^+$	CO	0.07 ± 0.1 (7 ± 10)	b	[267]
Ag$_4$Au(CO)$_5$$^+$	CO	0.07 ± 0.1 (7 ± 10)	b	[267]
AgAu$_5$$^+$	CO	0.96 ± 0.1 (93 ± 10)	c	[264]
Ag$_2$Au$_4$$^+$	CO	0.92 ± 0.1 (89 ± 10)	c	[264]
Ag$_3$Au$_3$$^+$	CO	0.77 ± 0.1 (74 ± 10)	c	[264]
Au$_2$$^-$	CO	0.18 ± 0.02 (17 ± 2)	d, e	[268]
Au$_2$$^-$	CO	0.38 ± 0.04 (37 ± 4)	d, f	[268]
Au$_3$$^-$	CO	0.28 ± 0.04 (27 ± 4)	d, e	[268]
Au$_3$$^-$	CO	0.46 ± 0.05 (44 ± 5)	d, f	[268]
Au$_3$(CO)$^-$	CO	0.28 ± 0.1 (27 ± 10)	d	[266]
Au$_2$$^-$	O$_2$	0.60 ± 0.10 (58 ± 10)	d, e	[268]
Au$_2$$^-$	O$_2$	0.93 ± 0.10 (90 ± 10)	d, f	[268]
AgAu$^-$	O$_2$	1.10 ± 0.15 (106 ± 14)	d, e	[268]
AgAu$^-$	O$_2$	1.59 ± 0.20 (153 ± 19)	d, f	[268]

[a]Bracketing method
[b]Equilibrium measurement
[c]Radiative association
[d]Kinetic measurement with RRKM modelling
[e]Modelled with a tight TS
[f]Modelled with a loose TS

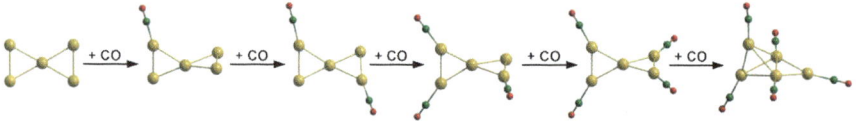

Fig. 25 DFT-calculated changes to the gold core structure upon sequential binding of CO to Au$_5$$^+$. Figure reproduced from [267]

(Eq. (3)). By examining the role of charge and cluster size, a considerable body of reactivity patterns has been accumulated. Reactions of bare gold cluster ions are discussed in Sect. 4.1 and those of ligated clusters in Sect. 4.2, while mixed metal clusters are discussed in Sect. 4.3. Finally complete catalytic cycles are described in Sect. 4.4.

Table 9 Gas-phase products and rates constants for the reactions of gold dimers with neutral substrates

Substrate	Au_2^+ rate[a]	Au_2^+ Product(s)	Au_2 rate[b]	Au_2 product(s)[a]	Au_2^- rate[c]	Au_2^- product(s)
O_2	NR[c]	NR[b]	NR[b]	NR[b]	83 ± 8[c]	$Au_2(O)_2^-$
N_2O	0.24[d]	$NO^+ + Au_2N$	NR[b]	NR[b]	NS	NS
N_2	NS	NS	NR[b]	NR[b]	NS	NS
CH_4	2.9[e]	$Au_2CH_4^+$	NR[b]	NR[b]	NS	NS
CO	NR	NR	2.6 ± 0.9	$Au_2(CO)$	5.1 ± 1[c]	$Au_2(CO)^-$
NH_3	60[b]	$Au(NH_3)^+$	22 ± 4	$Au_2(NH_3)$	NS	NS
C_2H_4	NS	NS	230 ± 50	$Au_2(C_2H_4)$	NR	NR
H_2	NS	$Au_2(H_2)_3^{+f}$	1.4 ± 0.3	$Au_2(H)_2$	NS	NS

[a]Second order, rate constant with units of 10^{-11} cm^3 s^{-1}
[b]Data collected in a fast-flow reactor at room temperature under thermalised conditions [270]. Termolecular rates reported units of 10^{-30} cm^6 s^{-1}
[c]Data collected in a variable-temperature ion trap [268, 271]
[d]Data collected in an ion trap [272]
[e]Data collected in an ion trap at 300 K [265]
[f]Data collected in an ion trap at 100 K [271]
NR = no reaction observed, which sets a limit on the bimolecular rate constants of less than 5×10^{-15} cm^3 s^{-1} at 6 Torr He
NR no reaction, NS not studied

4.1 Reactions of Bare Gold Cluster Ions: Overview

The total electron count on bare gold clusters can be manipulated in two ways:

1. Clusters of gold atoms which are neutral, cationic or anionic have different electron counts. Apart from the role of total electron count, the presence or absence of a charge influences the collisional rate constant. Thus the collisional rate constants of singly charged ions are substantially larger than those of analogous neutrals.

   One of the first studies to have systematically examined the gas-phase reactions of a range of neutral substrates (D_2, CH_4 and O_2) with gold clusters Au_n ($n = 2$–25), in their cationic, neutral and anionic forms, revealed that the cluster charge and electron count can have a pronounced effect on their reactivities [269]. For the reactions with D_2, the reactivity follows the order Au_n^+ (reactive for $n = 2$–15) > Au_n (only $n = 3$ and 7 reactive) >> Au_n^- (nonreactive). For O_2, the reactivity follows the order Au_n^- (clusters where n is even are reactive) >> Au_n^+ (only $n = 10$ reactive).

   To further highlight the role of electron count and charge on reactivity, we have summarised from the literature, data for the reactions of the gold dimer cation, neutral and anion with a range of neutral substrates (Table 9).

2. Pioneering work by the groups of Cox and Ervin on the gas-phase reactivity of small gold clusters with substrates has shown that the addition or removal of single Au atom can have a profound effect on reactivity [273, 274]. Figure 26 shows that the relative reactivities of bare anionic gold clusters with both carbon

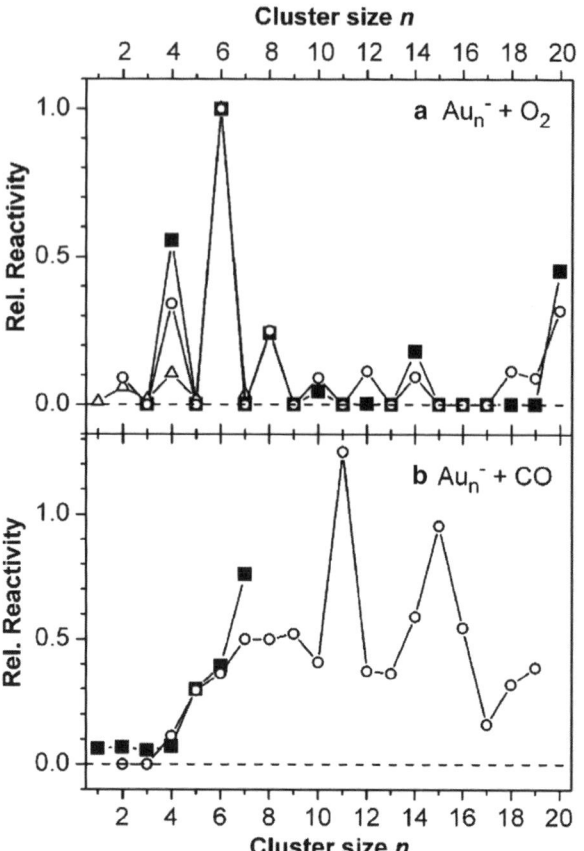

Fig. 26 Literature data on the relative reactivity of gold cluster anions, Au_n^-, in the adsorption reaction of one O_2 or one CO molecule, respectively, as a function of the cluster size n. (**a**) Reactions of Au_n^- with O_2: (*filled square*) data from [274], (*open triangle*) data from [273], and (*circle*) data from [275]. For comparison, all data are normalised to the reactivity of Au_6^-. (**b**) Reactions of Au_n^- with CO: (*filled square*) data from [273] and (*circle*) data from [276]. Again, the data shown have been normalised to the reactivity of Au_6^- towards O_2. Figure reproduced from [34]

monoxide and oxygen are dependent on cluster size. It is evident that there is a 'saw' pattern for O_2, whereby the clusters with even numbers of gold atoms and which have open-shell electronic structures are more reactive. It has been suggested that this effect arises from the interaction of the cluster's HOMO with the unpaired electron in the O_2 π^* antibonding orbital. This interaction is attractive for the anionic clusters with an even number of gold atoms as they possess an odd number of electrons, and thus the HOMO contains an unpaired electron.

4.1.1 Reactions of Bare Gold Cluster Ions with a Single Substrate

Reactions with CO

The adsorption of CO to gold cluster cations and anions has been widely studied under a range of experimental conditions. The thermochemistry for binding of the

Table 10 Bimolecular collision rates for the reactions of Au_x^-; $x \leq 7$ with CO

Cluster	k_{exp}[a]	k_{theor}[b]	Efficiency (%)[c]
Au^-	0.20	6.74	0.3
Au_2^-	0.22	6.53	0.3
Au_3^-	0.18	6.46	0.3
Au_4^-	0.23	6.42	0.4
Au_5^-	0.96	6.40	1.5
Au_6^-	1.26	6.38	2.0
Au_7^-	2.43	6.38	3.8

[a]k_{exp} Experimental bimolecular rate constant
[b]Calculated theoretical collision rate [277]
[c]$k_{exp}/k_{theor} \times 100$
Table adapted from [273]

first CO was described in Sect. 3.7. Multiple CO adsorptions have been observed for gold cluster anions and cations (Eq. (15)):

$$Au_x^{-/+} + yCO \rightarrow Au_n(CO)_y^{-/+} \qquad (15)$$

Ervin and Lee studied the reactions of gold cluster anions (Au_x^-; $x \leq 7$) with O_2 and CO in a flow tube reactor [273]. Unlike in the case of O_2 (see below), no even–odd alteration of the reactivity was observed with CO, and gold clusters Au_x^- with $x > 4$ were more reactive than the smaller ones ($x \leq 4$) (Table 10).

Wallace and Whetten have published several studies on the reaction of CO with size-selected gold clusters at room temperature [276, 278–280]. For instance, in a study aimed at understanding the size-dependent reactivity of Au_x^- ($x = 4$–19) [280], they concluded that initial products formed seem to correspond mainly to clusters with a gold electron shell filling at 8, 14, 18 and 20 electrons (i.e. Au_5CO^-, $Au_{11}CO^-$, $Au_{15}CO^-$, $Au_{15}(CO)_2^-$). When the concentration of CO is increased, cluster saturation is observed with 4–8 molecules of CO adsorbed (e.g. $Au_5(CO)_4^-$, $Au_8(CO)_5^-$, $Au_9(CO)_6^-$, $Au_{12}(CO)_8^-$). In another study, Whetten and Wallace reported that Au_2^- and Au_3^- were unreactive [276] at room temperature. This contradicted Ervin's results that these small clusters react with CO, albeit slowly [273]. This was later explained when they reported that preadsorbed water on Au_2^- and Au_3^- allowed the subsequent adsorption of the carbonyl and displacement of water [278].

The temperature-dependent reaction kinetics of the bare gold cluster anions Au_x^- ($x = 1$–3) to CO (and O_2 discussed below) adsorption have been studied by Bernhardt et al. in a variable-temperature ion trap [268]. At room temperature Au_x^- ($x = 1$–3) was found to be unreactive with CO, as previously described by Whetten et al. [280]. For an ion-trap temperature of 250 K, the gold clusters Au_x^- ($x = 2, 3$) are able to adsorb only one molecule of CO. At 100 K $Au_x(CO)_y^-$ ($x, y = 2,2; 3,2$) are also detected, and Au^- remains unreactive even at lower temperatures. Kinetic measurements reveal that $Au_2(CO)^-$ and $Au_2(CO)_2^-$ reach an equilibrium when reaction times are extended. In contrast $Au_3(CO)_2^-$ is almost the sole product of allowing $Au_3(CO)^-$ to react with CO (Eq. (15)) over extended reaction times.

This binding can be reversible, as demonstrated by the fact that certain cluster ions are metastable, undergoing CO loss (Eq. (16)) [279]:

$$Au_x(CO)_y^{-/+} \rightarrow Au_x(CO)_{y-z}^{-/+} + zCO \qquad (16)$$

Reactions with O_2

O_2 is more selective in its reactions with gold cluster ions. Small gold cluster cations are essentially unreactive, while anions exhibit a pronounced odd–even effect for x in adding O_2 (Eq. (17)) [273, 275]. Cluster anions with an unpaired electron (x = even) are highly reactive, and it has been suggested that these undergo electron transfer to form a superoxide anion bound to Au_x (see Sect. 3.6 and A. Fielicke et al.'s chapter in this book):

$$Au_x^- + O_2 \rightarrow Au_x(O_2)^- \qquad (17)$$

Reactions with N_2

The reactions of Au_x^+ (x = 3 and 5) with N_2 have been studied in a variable-temperature ion trap [281]. The addition of multiple N_2 to Au_x^+ is very sensitive to the temperature. At room temperature no $Au_x(N_2)_y^+$ are observed. For Au_3^+ the ions Au_3^+ and $Au_3(N_2)_3^+$ are observed at 200 K, while at 100 K only $Au_3(N_2)_3^+$ is found. In the case of Au_5^+ both $Au_5(N_2)_3^+$ and $Au_5(N_2)_4^+$ are observed at 200 K, and only the latter ion is observed at 100 K. These nitrogen adducts facilitate absorption of H_2 and O_2, as discussed below.

Reactions with H_2

Three studies of the reactions of gold cluster ions with hydrogen have been reported using different instruments, which highlight the role of reaction conditions (pressure and temperature). Cox's group used a fast-flow reactor to examine the reactions of gold cluster cations and anion with deuterium. They found that while small ($n < 15$) gold cations react readily with D_2 via addition of D_2, no reactions were observed for the anions [269]. Sugawara et al. have noted that in the lower-pressure regime of an FT-ICR mass spectrometer, no reaction occurs for gold cluster cations Au_n^+ (n = 1–12) with H_2 [67]. Under variable-temperature ion-trap conditions, where the hydrogen adducts can be collisionally cooled with the helium bath gas [271], Au_x^+ (x = 2–7) exhibit interesting temperature-dependent reactivity patterns towards molecular hydrogen. At 300 K, only Au_5^+ adsorbs up to 3 molecules of H_2. Lowering the temperature to 200 K results in a dramatic change of reactivity. Although Au_2^+ remains unreactive, the other cluster cations adsorb the following numbers of hydrogen molecules: Au_3^+ and Au_4^+ up to three; Au_5^+ and Au_6^+ up to four; and Au_7^+ up to two. Further lowering the temperature to 100 K leads to a

Fig. 27 Cluster size-dependent hydrogen saturation at $T_R = 100$ K and minimum-energy structures of the investigated cluster sizes. 'Corner' site atoms are indicated by filled circles. For the case of Au_6^+, the two lowest-energy isomers are displayed with the nontriangular incomplete hexagonal (ih) isomer being 0.19 eV higher in energy than the triangular (t) one. Figure reproduced from [271]

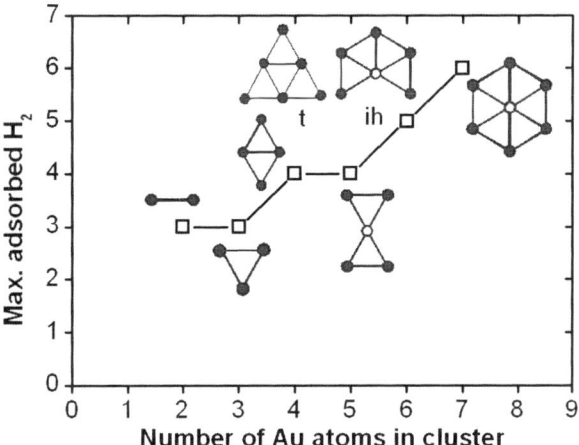

size-dependent hydrogen saturation, which nicely matches the numbers of corner sites available for the lowest-energy structures of the clusters as shown in Fig. 27.

Reactions with N_2O

The reactions of small gold cluster cation with N_2O have been studied in an ion-trap mass spectrometer [272]. Au_2^+ reacts with N_2O to yield NO^+ as the major product (Eq. (18)). In contrast, Au_3^+ reacts via adduct formation to give $Au_3N_2O^+$ (Eq. (19)) and via O atom abstraction to yield Au_3O^+ (Eq. (20)). The former channel operates for Au_4^+:

$$Au_2^+ + N_2O \rightarrow NO^+ + Au_2N \tag{18}$$

$$Au_3^+ + N_2O \rightarrow Au_3N_2O^+ \tag{19}$$

$$Au_3^+ + N_2O \rightarrow Au_3O^+ + N_2 \tag{20}$$

Reactions with NH_3 and CH_3NH_2

Pronounced odd versus even electron effects have been observed in the reactions of gold cluster cations, Au_x^+, with ammonia and methylamine [282]. Slow association reactions are observed for $x = 1$ and 3 (Eq. (21)), whereas Au atom displacement was observed for $x = 2$ and 4 (Eq. (22)). The primary product for $x = 4$ undergoes further cluster degradation via loss of Au_2 (Eq. (23)):

$$Au_x^+ + RNH_2 \rightarrow Au_x(RNH_2)^+ \tag{21}$$

$$Au_x^+ + RNH_2 \rightarrow Au_{x-1}(RNH_2)^+ + Au \tag{22}$$

$$Au_3(RNH_2)^+ + RNH_2 \rightarrow Au(RNH_2)_2^+ + Au_2 \tag{23}$$

Reactions with the Sulphur Compounds H_2S and CH_3SSCH_3

Sugawara et al. examined the reactions of gold cluster cations Au_n^+ ($n = 1$–12) with H_2S in an FT-ICR mass spectrometer [67]. Four types of primary reaction channels were observed: (1) adduct formation (Eq. (24)), a reaction that occurs for clusters where $n = 9$, 11 and 12; (2) HS abstraction (Eq. (25)), a reaction that only occurs for $n = 2$; (3) sulphuration to yield Au_nS^+ (Eq. (26)), which occurs for $n = 4$–8 and 10 and (4) Au displacement (Eq. (27)), a reaction that only occurs for $n = 2$. Au_3^+ was unreactive towards H_2S and even n cluster cations were found to be more reactive than adjacent odd n clusters. Subsequent sulphuration reactions of Au_nS^+ proceeded to give $Au_nS_m^+$ but stopped at $Au_nS_{m+x}H_2^+$ when H_2 loss did not occur. The maximum number of sulphur atoms, m, observed in $Au_nS_m^+$ increased with the cluster size up to $n = 7$ and 8 (where $m = 5$), while the sulphuration reaction stopped at early stages for $n > 9$:

$$Au_x^+ + H_2S \rightarrow Au_x(SH_2)^+ \tag{24}$$

$$\rightarrow Au_x(SH)^+ + H \tag{25}$$

$$\rightarrow Au_xS^+ + H_2 \tag{26}$$

$$\rightarrow Au_{x-1}(SH_2)^+ + Au \tag{27}$$

Using FT-ICR mass spectrometry, Höckendorf et al. have shown that small odd electron gold cluster anions Au_x^- ($x = 2$ and 4) react with dimethyldisulphide via S–S (Eq. (28)) and S–C bond activation (Eq. (29)) [283]. Au_2^- also reacts via displacement of a gold atom (Eq. (30)) or anion (Eq. (31)):

$$Au_x^- + CH_3SSCH_3 \rightarrow Au_x(SCH_3)^- \tag{28}$$

$$\rightarrow Au_x(SSCH_3)^- \tag{29}$$

$$\rightarrow Au_{x-1}(CH_3SSCH_3)^- + Au \tag{30}$$

$$\rightarrow Au^- + Au_{x-1}(CH_3SSCH_3) \tag{31}$$

Reactions with H_2 and CH_4

The co-adsorption of hydrogen and methane on small bare cationic gold clusters Au_x^+ ($x = 3, 5$) has been proposed to occur on the same adsorption site of the gold cluster (i.e. the same gold atom) [284], termed permissive co-adsorption. For these experiments Bernhardt et al. used a variable-temperature ion trap. For short reaction times, at 300 K, $Au_3CH_4^+$ is observed as a single product. Increasing the reaction time to 1 and 2 s results in formation of $Au_3(CH_4)_2(H_2)_2^+$. When the ion-trap temperature is reduced to 200 K, the formation of the CH_4 and H_2 adsorption with the Au_3^+ cluster is accelerated. At a reaction time as short as 0.1 s, minute amounts of $Au_3(CH_4)_2(H_2)_2^+$ are observed, and the dominant peaks correspond to $Au_3(CH_4)_x(H_2)_y^+$, $(x, y = 3, 0; 3, 3)$. At reaction times greater than 0.5 s, the only product observed is $Au_3(CH_4)_3(H_2)_3^+$.

Mass-selected Au_5^+ stored in an ion trap under similar conditions reacts in a similar way to Au_3^+. At an ion-trap temperature of 300 K and a reaction time of 0.1 s, only one CH_4 molecule is adsorbed. Increasing the reaction time from 0.5 to 2 s results in the addition of a second adsorbate being H_2 to give $Au_5(CH_4)(H_2)^+$. Reducing the temperature to 200 K enhances the reactivity of the cluster considerably. At 0.1 s $Au_5(CH_4)_2^+$ is the dominant peak. Increasing the reaction time to 0.5 s shows $Au_5(CH_4)_3(H_2)^+$ as the dominant peak with $Au_5(CH_4)_2(H_2)^+$ and $Au_5(CH_4)_3(H_2)_4^+$. Further increasing the reaction time from 0.5 up to 2 s results in mainly $Au_5(CH_4)_4(H_2)_4^+$.

Reactions with H_2 and $CH_3CH=CH_2$

The mass-selected bare gold cluster cations, Au_3^+ and Au_5^+, were reacted with propylene ($CH_3CH=CH_2$) comparable to the reactions discussed above, however replacing CH_4 with $CH_3CH=CH_2$. Both clusters appear to immediately react with $CH_3CH=CH_2$. At a reaction time of 0.1 s and an ion-trap temperature of either 300 K or 200 K, the main peak is solely $Au_3(CH_3CH=CH_2)_3^+$ for the mass-selected trimer. In comparison, the Au_5^+ cluster at 300 K forms predominantly $Au_5(CH_3CH=CH_2)_4^+$ and at 200 K $Au_5(CH_3CH=CH_2)_3^+$ is the predominant peak observed. In contrast to CH_4, only trace amounts of $CH_3CH=CH_2$ are required to react with and completely saturate the gold clusters which then also inhibit the co-adsorption of molecular H_2.

Reactions with CH_3X

Koszinowski et al. showed that the reaction of Au_2^+ with methyl halides could be used to 'synthesise' the gold carbene, $Au_2CH_2^+$, in the gas phase (Eq. (32)) [282]:

$$Au_2^+ + CH_3X \rightarrow Au_2CH_2^+ + HX \qquad (32)$$

Lang and Bernhardt studied bare gold cluster cation Au_3^+ and Au_5^+ reaction with $CHCl_3$ and H_2O [285]. With the ion trap at room temperature (300 K), it was found that the reactions of Au_x^+ $(x = 3, 5)$ proceeded quickly. For the trimer, the number of adsorbate molecules of $CHCl_3$ and H_2O never exceeds the number of gold atoms. The distribution of the adsorbate molecules also suggests that there is an equilibrium constraint to direct the reaction channels. The Au_5^+ cluster exhibits an X-shaped, D_{2h} symmetry, with 4 equiv. gold corner atoms around the central gold atoms. Hence, the number of adsorbate molecules does not exceed 4. The reaction behaviour is comparable to Au_3^+.

In a similar method to that described above, the reactions of the bare gold cluster cations Au_x^+ ($x = 1$–3, 5, 7) with CH_3Br have been investigated [286]. Adsorbates of molecular methyl bromide were observed for each of the gold clusters. When the mass-selected cluster was held in the ion trap for longer reaction times, the products of methyl elimination would successively increase to also produce $Au_x(CH_3Br)_y(Br)_z^+$.

Reactions with Other Neutral Substrates

Höckendorf found that small gold cluster anions Au_x^- ($x = 2$–4) were unreactive to a wide range of substrates (methanol, acetonitrile, acetaldehyde, acetone, dimethyl sulphide, methyl mercaptan, benzene, ethynylbenzene and difluoroacetic acid) in an FT-ICR mass spectrometer [283]. The only reactive substrates were dimethyl-disulphide (discussed above) and trifluoroacetic acid, which reacts via adduct formation for $x = 2$ and 4 (Eq. (33)) and via Au displacement for $x = 2$ (Eq. (34)). Au^- was unreactive towards CF_3CO_2H:

$$Au_x^- + CF_3CO_2H \rightarrow Au_x(CF_3CO_2H)^- \tag{33}$$

$$\rightarrow Au_{x-1}(CF_3CO_2H)^- + Au \tag{34}$$

4.1.2 Reactions of Bare Gold Cluster Ions with More than One Substrate: 'Cooperative Effects'

Reactions with CO and O_2

The bare gold cluster anions Au_2^- and Au_3^-, are able to form co-adsorption complexes, $Au_a^-(CO)_b(O)_c$ ($a = 2$, 3; $b = 1$; $c = 1$, 2), when exposed to small partial pressure of CO and O_2 in a temperature-controlled ion trap at cryogenic temperatures [287]. This type of cluster complex has been proposed theoretically to be a key intermediate in a catalytic CO oxidation cycle [288].

Reactions with H_2 and O_2

Lang et al. have used a variable-temperature ion trap to examine the H_2/O_2 co-adsorption behaviour of cationic gold clusters Au_x^+ ($x = 2$–7) at 100 K [271]. They found a striking odd–even alternation reactivity. For the even x cluster sizes ($x = 2,4,6$), cooperative co-adsorption of only one oxygen molecule was observed, with $Au_2H_4O_2^+$, $Au_4H_6O_2^+$ and $Au_6H_6O_2^+$ being observed. No co-adsorption of multiple O_2 molecules occurs, even at higher O_2 pressures and extended reaction times. For the odd x cluster sizes ($x = 3,5,7$) no measurable co-adsorption of O_2 occurred. The lack of reactivity of Au_x^+ ($x = $ odd) can be explained via a valence electron structure model in which the spin-paired valence electron structure of Au_x^+ hampers the interaction of the gold cluster valence electrons with the unpaired electrons of the $2\pi_g*$ antibonding O_2 orbitals. As the temperature of the ion trap is raised the types of product ions observed for Au_4^+ and Au_6^+ change. At 300 K, $Au_4H_2O^+$ and $Au_6H_4O^+$ are observed, which signifies that the oxygen has been activated. Detailed DFT calculations were carried out on the co-adsorption of H_2 and O_2 and their chemical activation on the Au_4^+ and Au_6^+ clusters. These suggest a mechanisms for dissociation of O_2 involving the adsorption at adjacent sites to give $[(Au)_{x-2}Au(H_2)Au(O_2)]^+$ structures, which can then undergo intermolecular hydrogen atom transfer to form $[(Au)_{x-2}Au(H)Au(O_2H)]^+$ intermediates possessing the mixed hydride, hydroperoxide sites, which in turn can eliminate water to form a gold oxide site.

Reactions with N_2/H_2 and N_2/O_2

The reactions of Au_x^+ ($x = 3$ and 5) with mixtures of N_2 and H_2 and N_2 and O_2 have been studied in a variable-temperature ion trap [281]. As noted above, at low temperatures (100 and 200 K) in the presence of pure N_2, both Au_3^+ and Au_5^+ both give $Au_x(N_2)_y^+$. In pure H_2, Au_3^+ and Au_5^+ both give $Au_x(H_2)_y^+$. In contrast, no reactions are observed between Au_x^+ and O_2. At 100 K, mixtures of N_2 and H_2 react with Au_3^+ to yield $Au_3(H_2)_2(N_2)^+$, $Au_3(H_2)(N_2)_2^+$ and $Au_3(N_2)_3^+$. $Au_5(H_2)_3(N_2)^+$, $Au_5(H_2)_2(N_2)_2^+$ and $Au_5(H_2)(N_2)_3^+$ are observed under the same conditions. The authors suggested that these reactions involve competitive co-adsorption. Au_3^+ and Au_5^+ co-adsorb N_2 and O_2 at 100 K, at longer reactions times and with low concentrations of N_2 to, respectively, give: $Au_3(N_2)_3^+$, $Au_3(N_2)_2(O_2)^+$, $Au_3(N_2)(O_2)_2^+$ and $Au_3(O_2)_3^+$; $Au_5(N_2)_3^+$, $Au_5(N_2)_2(O_2)^+$, $Au_5(N_2)(O_2)_2^+$ and $Au_5(O_2)_3^+$; as well as $Au_5(N_2)_4^+$, $Au_5(N_2)_2(O_2)_2^+$, $Au_5(N_2)(O_2)_3^+$ and $Au_5(O_2)_4^+$. Both cooperative and competitive co-adsorption appears to operate for N_2/O_2 mixtures.

Reactions with CH_4 and O_2

Lang and Bernhardt studied the reactions of Au_x^+ ($x = 2$–4) with CH_4 and O_2 in a variable ion trap [289]. Upon reaction with CH_4 only, these clusters appear to adsorb a CH_4 molecule initially to form $Au_x(CH_4)^+$. Interestingly, at 300 K and upon reaction with another molecule of CH_4, Au_2^+ forms $Au_2(C_2H_4)^+$ indicating dehydrogenation of CH_4, shown in this study to be catalytic. Larger Au_x^+ clusters studied adsorb more CH_4. To study the possibility of methane oxidation, the reaction of CH_4 and O_2 on the gold dimer cation Au_2^+ was also examined at 210 K. Two peaks corresponding to $Au_2(CH_4)O_2^+$ and $Au_2(C_3H_8O_2)^+$ were detected in the mass spectrum. The latter ion corresponds to a dehydrogenated product and was tentatively assigned as $Au_2(CH_2O)_2(CH_4)^+$ (i.e. formation of formaldehyde) or a further oxidation product $Au_2(CO_2)(CH_4)_2^+$.

Reactions with CO and Benzene

Adsorption complexes of the form $Au_x(C_6H_6)_y(CO)_z^+$ ($n = 3, 5$) were generated from reactions with benzene and with a mixture of benzene and carbon monoxide with Au_x^+ ($x = 3, 5$) [290]. Benzene was shown to react with all Au_x^+ clusters exhibiting size-dependent coverage, and co-adsorption of benzene and CO was also achieved for both clusters.

Reactions with CH$_3$X and Water

Adsorption complexes of the form $Au_x(H_2O)_y(CHCl_3)_z^+$ ($x = 3, 5$) were generated from reactions with each of water and chloroform, as well as a mixture of the two reagents [285]. The reactions of chloroform with Au_x^+ ($x = 3, 5$) were found to proceed quickly at room temperature where one molecule of CHCl$_3$ adsorbs at each corner of the cluster hindering the adsorption of H$_2$O. Reducing the partial pressure of chloroform in the ion trap resulted in distinct co-adsorption complexes of both H$_2$O and CHCl$_3$ on the bare gold cluster cations.

4.2   Ligated Gold Cluster Ions

Gas phase chemistry on ligand-protected gold clusters has been recently employed [291] to resolve a controversy regarding the catalysis of the Sonogashira coupling reaction by gold nanoparticles as an alternative to the widely used palladium catalyst. Multistage mass spectrometry experiments using an ion trap mass spectrometer together with DFT calculations were used to show that the mononuclear Au (I) complex ligated with two bis-phosphinoalkane ligands (L), cannot induce the oxidative addition of iodobenzene, a crucial step in the Sonogashira coupling. In contrast, the trinuclear cluster, $[Au_3L]^+$, was found to activate the C–I bond. Another key finding from this work is that the linker size, n, of the bisphosphinoalkane ligand, $Ph_2P(CH_2)_nPPh_2$ (L), tunes the rate of the reaction of the cluster with iodobenzene. As the linker size of the ligand (L) increases, the rate of the reaction increases (Fig. 28).

There is growing interest in the role of dinuclear gold species in catalysis [292]. In many cases ligated gold clusters are directly related to well-known reactive intermediates via the isolobal analogy [118, 119, 293, 294]. For example, Scheme 1 shows two gold analogues 2 [295] and 3 [296] of H$_3^+$, 1 [297, 298], first reported by Sir JJ Thomson and now recognised as playing a key role in interstellar chemistry. In the next sections we discuss the chemistry of ligated gold clusters and, where appropriate, use the isolobal analogy to highlight their relationship to known reactive intermediates in organic chemistry. The gas phase is an ideal environment to study the bimolecular and unimolecular reactivity of such species, and we discuss reports on gold hydrides, gold oxides and gem-diaurated ions.

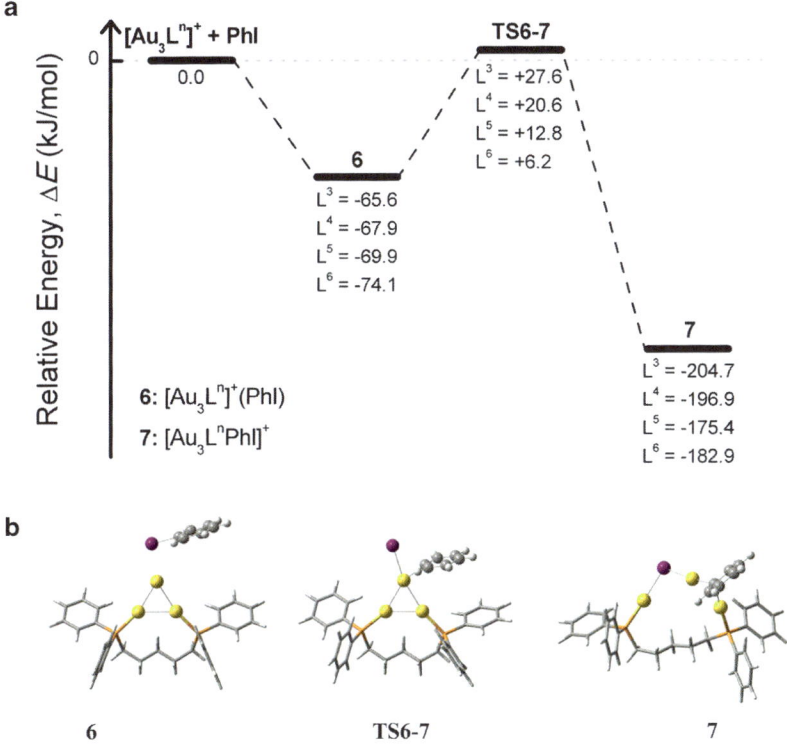

Fig. 28 (**a**) DFT-calculated energy diagrams (M06/SDD6-31G(d,p)//B3LYP/SDD6-31G(d,p) level of theory) for ion–molecule reactions of iodobenzene with $[Au_3L_n]^+$ (L=$Ph_2P(CH_2)_nPPh_2$; $n = 3$–6); (**b**) key species for reaction of $[Au_3L_5]^+$. Figure reproduced from [291]

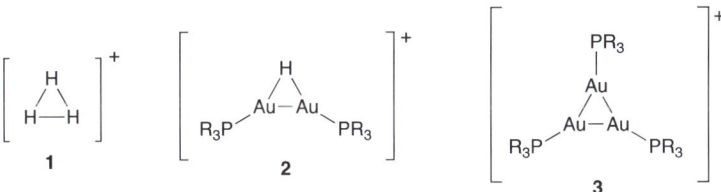

Scheme 1 Isolobal analogy illustrated for H_3^+

4.2.1 Gold Hydride Cluster Ions

Gold hydrides have been implicated as reactive intermediates in the transformation of organic substrates, but few have been isolated and structurally characterised [299]. While the bare coinage metal hydrides M_2H^+ (where M=Cu and Ag) have been formed via CID of suitable precursors such as amino acids [300, 301], Au_2H^+ has not yet be reported. The phosphine-ligated gold hydrides, $(R_3PAu)(R'_3PAu)H^+$ (where R=R'=Ph; R=R'=Me; R=Me and R'=Ph) **2**, can be prepared in the gas

Fig. 29 B3LYP/LanL2D6-
31G(d)-calculated potential
energy diagram for the
competing fragmentation
reactions of the model
system $(H_3PAu)_2H^+$.
Figure reproduced from
[295]

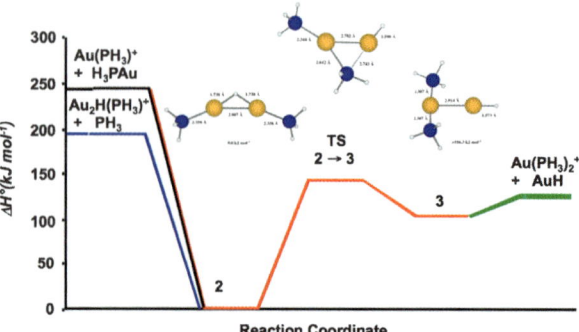

phase via CID of the deprotonated amino acid N,N-dimethylglycine (DMG) cluster
ions $(R_3PAu)(R'_3PAu)(DMG-H)^+$, which are formed via electrospray ionisation of
a mixture of the two gold phosphine chlorides, R_3PAuCl (R=Ph and Me), silver
nitrate and the amino acid N,N-dimethylglycine (DMG). For the homodimeric
precursor ions, four fragment channels are observed and include formation of the
ligated dinuclear gold hydride (Eq. (35)); formation of the bis-ligated gold cation,
$(R_3P)_2Au^+$ (Eq. (36)); loss of the neutral ligated monogold hydride (Eq. (37));
and formation of the monoligated gold cation, $(R_3P)Au^+$ (Eq. (38)). In all cases
formation of **2** is the dominant reaction:

$$[(Me_2NCH_2CO_2) + 2AuPR_3]^+$$

$$\rightarrow (R_3PAu)_2H^+ + [H_7, C_4, N, O_2] \tag{35}$$

$$\rightarrow (R_3P)_2Au^+ + Me_2NCH_2CO_2Au \tag{36}$$

$$\rightarrow [(Me_2NCH_2CO_2) + AuPR_3 - H]^+ + R_3PAuH \tag{37}$$

$$\rightarrow R_3PAu^+ + Me_2NCH_2CO_2AuPR_3 \tag{38}$$

Unlike the bare silver hydride, Ag_2H^+, which reacts with a range of substrates
[300, 302], **2** is unreactive towards a range of neutral reagents including H_2O,
MeOH, 2-propanol, acetonitrile, pyridine, allyl iodide, O_2, N_2O, nitromethane and
DMSO. Low-energy CID of **2** proceeds via loss of AuH (Eq. (39)), rather than loss
of R_3P (Eq. (40)) or direct cluster fragmentation (Eq. (41)). The observed reactivity
is accounted for by the DFT-calculated potential energy diagram of the model
system $(H_3PAu)_2H^+$ shown in Fig. 29. Thus both the barrier for isomerisation of **2** to
3 (140.8 kJ mol^{-1}) and the final reaction endothermicity for loss of AuH
(126.1 kJ mol^{-1}) are lower than the energy required for loss of H_3P (Eq. (40),
194.9 kJ mol^{-1}) or formation of $(R_3P)Au^+$ (Eq. (41), 244.8 kJ mol^{-1}):

$$(R_3PAu)_2H^+ \rightarrow (R_3P)_2Au^+ + AuH \tag{39}$$

$$\rightarrow (R_3P)Au_2H^+ + R_3P \tag{40}$$

$$\rightarrow (R_3P)Au^+ + (R_3P)AuH \tag{41}$$

Sugawara et al. have probed gold hydride clusters formation upon laser ablation of a gold rod in a H_2/He mixture [67]. The hydride cluster cations $Au_nH_m^+$ were produced for $n = 1$–7, while bare Au_n^+ clusters were the main products for $n \geq 8$. The authors suggested that the border between $n = 7$ and 8 hints to the formation of planar gold hydrides as the structure of bare Au_n^+ changes from planar for $n = 7$ to three-dimensional for $n = 8$. Finally, the stoichiometries of the gold hydride cluster cations $Au_nH_m^+$ favour combinations of n and m that yield even electron clusters. Thus the main gold hydride clusters formed were $Au_2H_5^+$, $Au_3H_6^+$, $Au_5H_6^+$, $Au_6H_5^+$ and $Au_7H_4^+$.

4.2.2 Gold Hydroxide Cluster Ions

Wallace et al. [303] have studied the reactions of gold cluster anions with oxygen under humid conditions in a near-atmospheric-pressure flow reactor. In the absence of oxygen and at near room temperature, abundant gold hydroxide cluster anions, Au_xOH^- $(x = 2$–6), are observed. When oxygen is added, adsorption occurs to yield, $Au_xOH(O_2)^-$. An even/odd size effect is observed for Au_xOH^- which is different to the bare gold cluster anion Au_x^- reactions with O_2. Thus the binding of an OH group enhances the reactivity towards molecular oxygen on odd-sized anionic gold clusters but lowers the reactivity on even-sized ones.

4.2.3 Gold Oxide Cluster Ions

Castleman's group has widely studied the reactions of gold oxide cluster ions in the gas phase, and much of that work has been reviewed [304]. This work has been inspired by Haruta's discovery of CO oxidation by gold clusters (Eq. (3)). Indeed the focus of all of the bimolecular work has been on developing an understanding of how charge and cluster stoichiometry influence the reactions of gold oxide cluster ions, $Au_xO_y^{-/+}$, with CO. For the anions, there are three types of reactions that can occur [305–309]: association (Eq. (42)), replacement (Eq. (43)) and oxidation of CO (Eq. (44)). Which of these reactions occur depends on both x and y:

$$Au_xO_y^- + CO \rightarrow Au_xO_y(CO)^- \tag{42}$$

$$\rightarrow Au_xO_{y-2}(CO)^- + O_2 \tag{43}$$

$$\rightarrow Au_xO_{y-1}^- + CO_2 \tag{44}$$

Detailed DFT calculations and molecular dynamics calculations were carried out on several systems to help rationalise the experimental observations. The results of DFT calculations on $Au_2O_y^-$ and $Au_3O_y^-$ $(y = 1$–5), which are shown in Fig. 30, reveal that there are three types of oxygen centres in these cluster anions: peripheral O atoms (e.g. Au_2O^-), bridging O atoms (e.g. $Au_2O_2^-$) and molecular O_2 (e.g. Au_3O^-). Experimentally it was found that association (Eq. (42)) occurs for

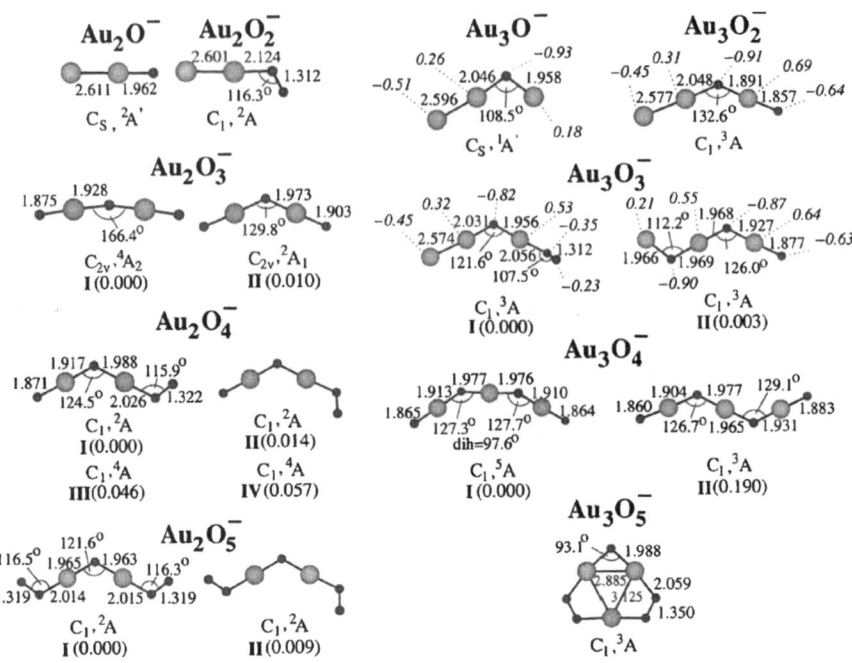

Fig. 30 Lowest-energy structures of $Au_2O_y^-$ and $Au_3O_y^-$ ($y = 1-5$). All structures fall within an energy range of 0.2 eV. The energy differences in eV with respect to the most stable structure are given in round brackets. Labels of the symmetry group and the ground electronic state are also given. Bond distances are in Å. Figure reproduced from reference [307]

Au_3O^-, replacement occurs for $Au_3O_3^-$ (Eq. (43)) and oxidation (Eq. (44)) occurs for Au_2O^-, $Au_2O_3^-$ and $Au_2O_4^-$. Theoretical results suggest that oxidation reactions mainly occur at the peripheral O atoms [309].

Gold oxide cluster cations react with CO via four types of reactions that also depend on both x and y [306, 310, 311]: association (Eq. (45)), replacement (Eq. (46)), oxidation of CO (Eq. (47)) and cluster fragmentation (Eq. (48)). Clusters with one oxygen promoted oxidation (Eq. (47)), with Au_3O^+ reacting more rapidly than Au_2O^+ [311]. Clusters with larger numbers of oxygen favoured adsorption of CO and loss of O_2 (Eqs. (45) and (46)), while $Au_2O_3^+$, $Au_3O_3^+$, $Au_4O_2^+$ and $Au_4O_3^+$ all underwent some cluster fragmentation (Eq. (48)):

$$Au_xO_y^+ + CO \rightarrow Au_xO_y(CO)^+ \tag{45}$$

$$\rightarrow Au_xO_{y-2}(CO)^+ \tag{46}$$

$$\rightarrow Au_xO_{y-1}^+ + CO_2 \tag{47}$$

$$\rightarrow Au_{x-1}O_{y-2}(CO)^+ + AuO_2 \tag{48}$$

A combination of experiments and theory provided detailed insights into how the charge state of gold oxide cluster ions influences the mechanism of oxidation of CO

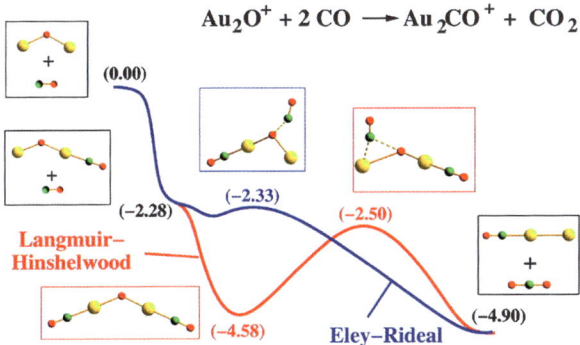

Fig. 31 Energy profiles for the reaction of CO with Au_2O^+. All energies are given in electron volts (eV) relative to the energy of the reactant at 0 K. Shown are ER-type (*blue*) and LH-type (*red*) mechanisms. For the ER-like mechanism, a barrier of 0.2 eV and an excess energy of 2.5 eV have been calculated for the rate-determining step. For the LH-like mechanism, the calculated barrier and excess energy are 2.08 and 4.58 eV, respectively. Figure reproduced from [306]

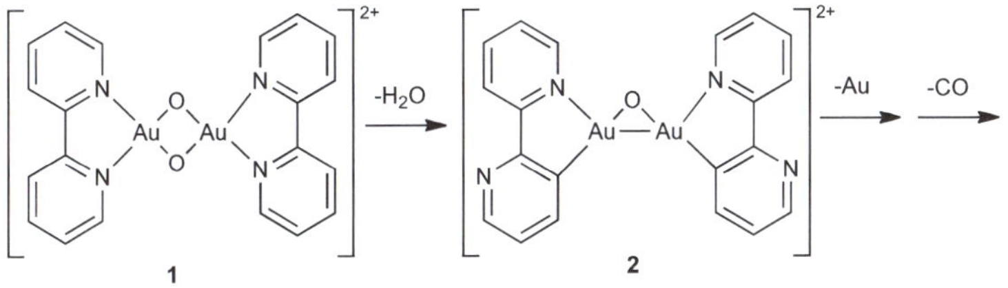

Scheme 2 Loss of H_2O from dinuclear gold(III) oxo cluster coordinated by 2,2′-bipyridyl ligands

[306]. Au_2O^- reacts with one molecule of CO to yield Au_2^-. Theoretical calculations reveal that the CO directly attacks the O atom of the cluster. Thus CO oxidation of CO by Au_2O^- proceeds via an Eley–Rideal (ER)-type mechanism rather than attacks at Au via a Langmuir–Hinshelwood (LH)-type mechanism. In contrast, Au_2O^+ requires adsorption of two molecules of CO for oxidation to proceed. The DFT calculations highlight that both ER- and LH-type mechanisms can operate (Fig. 31).

Tyo et al. have reported on the gas-phase fragmentation reactions of dinuclear gold(III) oxo clusters possessing a rhombic Au_2O_2 core and 2,2′-bipyridyl ligands with substituents in the 6-position (bipyR) [312]. The clusters were synthesised as hexafluorophosphate salts in solution and were transferred to the gas phase via ESI-MS to yield the free dications $[(bipyR)Au(\mu-O)_2Au(bipyR)]^{2+}$. A noteworthy aspect of this report is the use of condensed phase IR spectroscopy, gas-phase IRMPD and DFT calculations to provide evidence that isomerisation of the substituted complexes does not occur in the gas phase. CID of this series of complexes revealed a significant effect of substitution. The parent, un-substituted complex, **1**, fragments via loss of H_2O, presumably forming the organometallic ion **2**, (Scheme 2) via a rollover cyclometalation reaction [313]. This ion subsequently undergoes loss of Au and CO (Scheme 2).

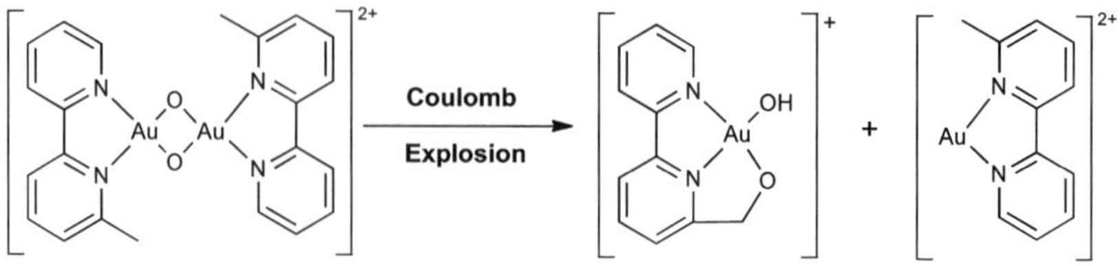

Scheme 3 Coloumb explosion pathway for dinuclear gold(III) oxo cluster coordinated by 6-substituted 2,2'-bipyridyl ligands

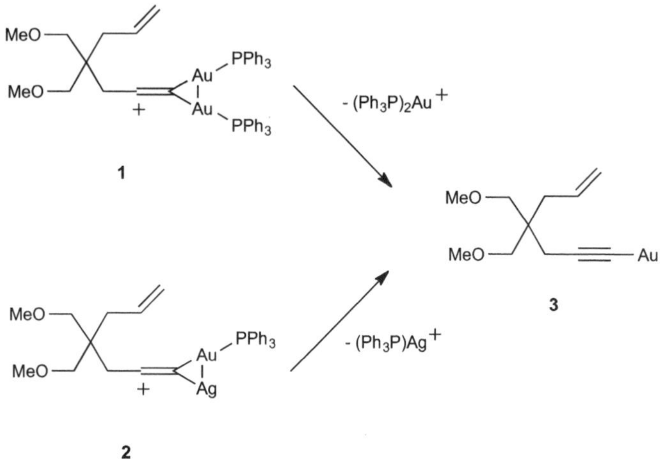

Scheme 4 Gas-phase fragmentation of homo (1) and hetero (2) dinuclear acetylides [315]

In contrast, all complexes with alkyl substituents in the 6-position of the ligands undergo Coulomb explosion to produce two monocationic fragments, as illustrated for the methyl derivative in Scheme 3. The difference in behaviour was rationalised by the combined effects of steric strain introduced to the central Au_2O_2 core by the substituents on the bipyridine ligand and the presence of oxidisable C–H bonds in the substituents.

4.2.4 Gem-Diaurated Cluster Ions

There has been considerable interest in the possibility of cooperation of two gold centres in catalytic reaction pathways, which has been termed 'dual gold catalysis' [314]. A key question is whether the two gold centres are remote from each other or whether they directly interact via aurophilic interactions. Two studies have appeared in which mass spectrometry experiments and DFT calculations have been carried out to locate the site(s) of the Au atoms. In the first, Simonneau et al. demonstrated that gold acetylides formed the homo (1), and hetero (2) dinuclear systems, in Scheme 4 under ESI-MS conditions and when subjected to CID, these fragmented via loss of either $(Ph_3P)_2Au^+$ or $(Ph_3P)Ag^+$ [315]. On the

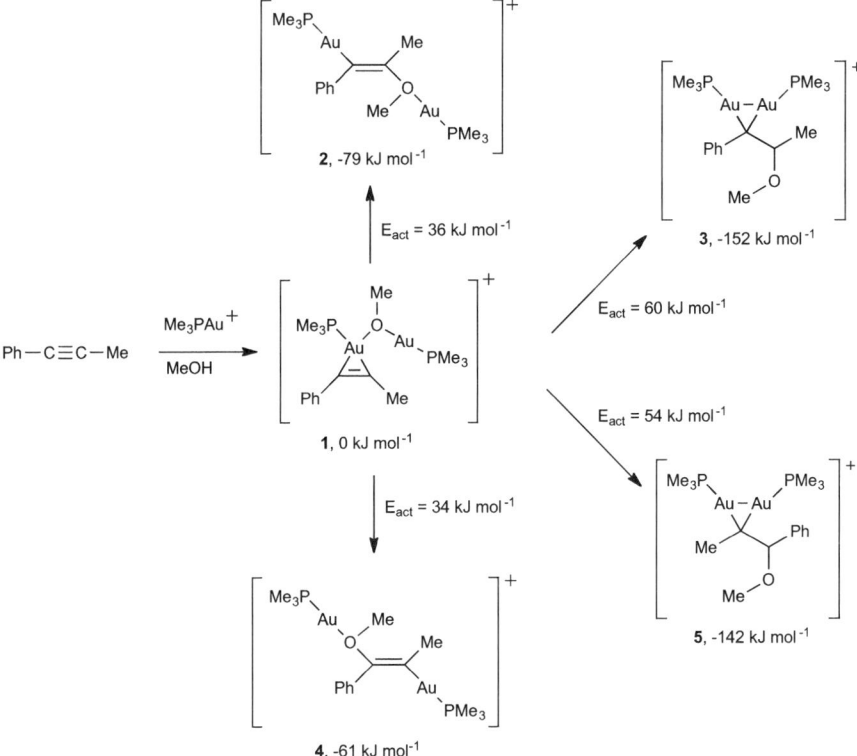

Scheme 5 DFT-calculated potential mechanism for gold–gold cooperation in the addition of methanol to an internal alkyne, 1-phenylpropyne [316]

basis of DFT calculations, the neutral product (**2**) was assigned as terminal acetylide with no intramolecular interaction between the Au and alkene site. Presumably these fragmentation reactions involve phosphine ligand migration prior to fragmentation in an analogous fashion to $(H_3PAu)_2H^+$ (Fig. 29).

Roithová et al. have reported an interesting example of gold–gold cooperation in the addition of methanol to an internal alkyne, 1-phenylpropyne [316]. Using a combination of electrospray ionisation, isotope-labelling experiments and gas-phase IR spectroscopy, they were able to find evidence for the formation of the gem-diaurated intermediates **3** and **5**. Based on DFT calculations, the authors proposed a possible mechanism for the formation of these gem-diaurated intermediates, which is shown in Scheme 5.

4.3 Gold-Containing Bimetallic Cluster Ions

The synergistic or cooperative effects of bimetallic clusters containing gold atom (s) and other metal atoms towards substrates have been demonstrated in several studies. To model the Degussa process of formation of HCN, Schwarz's group have employed

FT-ICR MS to investigate metal cluster ion-mediated bond coupling reaction between CH_4 and NH_3 [317]. While Pt is a known catalyst for this reaction, Pt_n^+ clusters ($n \geq 2$) do not induce the C–N coupling step and the Au_n^+ clusters do not activate the C–H bond ($n \geq 2$). In contrast, the mixed metal cluster $PtAu^+$ was successful since it combines the reactivity of both metals needed for each step (Eqs. (49) and (50)). Deuterium labelling confirmed that N–H bond activation occurs in the second step:

$$PtAu^+ + CH_4 \rightarrow PtAuCH_2^+ + H_2 \tag{49}$$

$$PtAuCH_2^+ + NH_2 \rightarrow [PtAuCH_3N]^+ + H_2 \tag{50}$$

In a follow-up study, the C–N coupling of methane and NH_3 was examined for larger bimetallic platinum–gold cluster cations [282]. Unlike the heterodimer, $PtAuCH_2^+$, which undergoes C–N bond coupling, $Pt_2AuCH_2^+$ reacts with NH_3 via carbide formation (Eq. (51)), as does $PtAu_2CH_2^+$ (Eq. (52)), which also undergoes Au atom displacement (Eq. (53)). The reactivity of the bare bimetallic platinum–gold cluster cations towards ammonia and methylamine was also examined. The reactivity was found to depend on the nature of both the metal cluster and the amine. For example, $PtAu^+$ reacts with NH_3 via displacement of Au (Eq. (54)), while for methylamine three channels are observed: hydride abstraction (Eq. (55)), dehydrogenation (Eq. (56)) and double dehydrogenation (Eq. (57)):

$$Pt_2AuCH_2^+ + NH_3 \rightarrow Pt_2AuC(NH_3)^+ + H_2 \tag{51}$$

$$PtAu_2CH_2^+ + NH_3 \rightarrow PtAu_2C(NH_3)^+ + H_2 \tag{52}$$

$$\rightarrow [PtAuCH_5N]^+ + Au \tag{53}$$

$$PtAu^+ + NH_3 \rightarrow Pt(NH_3)^+ + Au \tag{54}$$

$$PtAu^+ + CH_3NH_2 \rightarrow CH_2 = NH_2^+ + PtAuH \tag{55}$$

$$\rightarrow [PtAuCH_3N]^+ + H_2 \tag{56}$$

$$\rightarrow [PtAuCHN]^+ + 2H_2 \tag{57}$$

The reactivity of bimetallic $Pt_xAu_y^+$ clusters ($x + y \leq 4$) towards O_2 and CH_4 highlights the role of cluster composition in controlling the preferred reaction channel(s) [318]. While platinum-rich clusters behave similarly to Pt_x^+ clusters, gold-rich clusters like their pure gold counterparts are inert. For the oxygen substrate, platinum-only and platinum-rich clusters react via cluster decomposition, as illustrated for the trimer clusters in Eqs. (58)–(62). When methane is used as a substrate, pure and platinum-rich clusters react via dehydrogenation (Eq. (63)). Indeed, this reaction only occurs for $y = 1$ and $x \geq 1$. The carbene $Pt_xAu_yCH_2^+$ mainly loses H_2 upon CID (Eq. (64)) and undergoes dehydrogenation in reactions with a second molecule of methane (Eq. (65)):

$$Pt_3^+ + O_2 \rightarrow Pt_2^+ + PtO_2 \tag{58}$$

$$\rightarrow Pt_2O^+ + PtO \tag{59}$$

$$\rightarrow Pt_2O_2^+ + Pt \tag{60}$$

$$Pt_2Au^+ + O_2 \rightarrow PtAu^+ + PtO_2 \tag{61}$$

$$PtAu_2^+ + O_2 \rightarrow NO\ REACTION \tag{62}$$

$$Pt_xAu_y^+ + CH_4 \rightarrow Pt_xAu_yCH_2^+ + H_2 \tag{63}$$

$$PtAuCH_2^+ \rightarrow PtAuC^+ + H_2 \tag{64}$$

$$PtAuCH_2^+ + CH_4 \rightarrow PtAuC_2H_4^+ + H_2 \tag{65}$$

At the higher pressures of an ion-trap mass spectrometer, Lang et al. have examined the reactions of $PdAu_2^+$, Pd_2Au^+ and $Pd_2Au_2^+$ with CD_4 [319]. All ions react to sequentially add CD_4. The palladium-rich trimer, Pd_2Au^+, also undergoes loss of two D_2 upon addition of the second CD_4 (Eq. (66)). Deuterated ethene is liberated upon addition of a third molecule of CD_4, thereby closing a catalytic cycle, which is related to that observed for Au_2^+ which is discussed in detail in Sect. 4.4:

$$Pd_2Au(CD_4)^+ + CD_4 \rightarrow Pd_2Au(C_2D_4)^+ + 2D_2 \tag{66}$$

To model the requirements of co-adsorption of both CO and water in the water–gas shift reactions (Eq. (67)), Fleischer et al. have used a combination of ion-trap mass spectrometry experiments and DFT calculations to compare the reactions of water, CO and their mixtures with the pure and mixed silver–gold trimer cations, Au_3^+, Ag_3^+ and Ag_2Au^+ [320]. Figure 32 clearly highlights that cluster composition plays a key role in the outcomes of these reactions. Au_3^+ rapidly reacts via sequential addition of three CO (Fig. 32a) or water molecules (Fig. 32d), while Ag_3^+ and Ag_2Au^+ mainly add a single CO (Fig. 32b, c) or water molecule (Fig. 32e, f). The cooperative effect becomes apparent when the reactions of all three trimers with a mixture of water and carbon monoxide are compared. Au_3^+ preferentially adsorbs three CO molecules (Fig. 32g) while Ag_3^+ adsorbs up to two water molecules (Fig. 32i). This is consistent with the high affinity of gold cluster cations for CO discussed in Sect. 3.7. In contrast, the replacement of two gold atoms by silver in Ag_2Au^+ frees up sites for the adsorption of H_2O since the co-adsorption complexes $Ag_2Au(CO)(H_2O)^+$ and $Ag_2Au(CO)(H_2O)_2^+$ are observed (Fig. 32h). Detailed kinetic modelling of the experimental data reveals that the process of co-adsorption involves the initial adsorption of CO (Eq. (68)) followed by sequential adsorption of water (Eqs. (69) and (70)). DFT calculations on possible isomers of the co-adsorption complexes $Ag_2Au(CO)(H_2O)^+$ and $Ag_2Au(CO)(H_2O)_2^+$ uncovered 3 and 3 isomers, respectively, all lying within around 0.2 eV (≈ 2 kJ mol^{-1}):

$$H_2O + CO \rightarrow CO_2 + H_2 \tag{67}$$

$$Ag_2Au^+ + CO \rightarrow Ag_2Au(CO)^+ \tag{68}$$

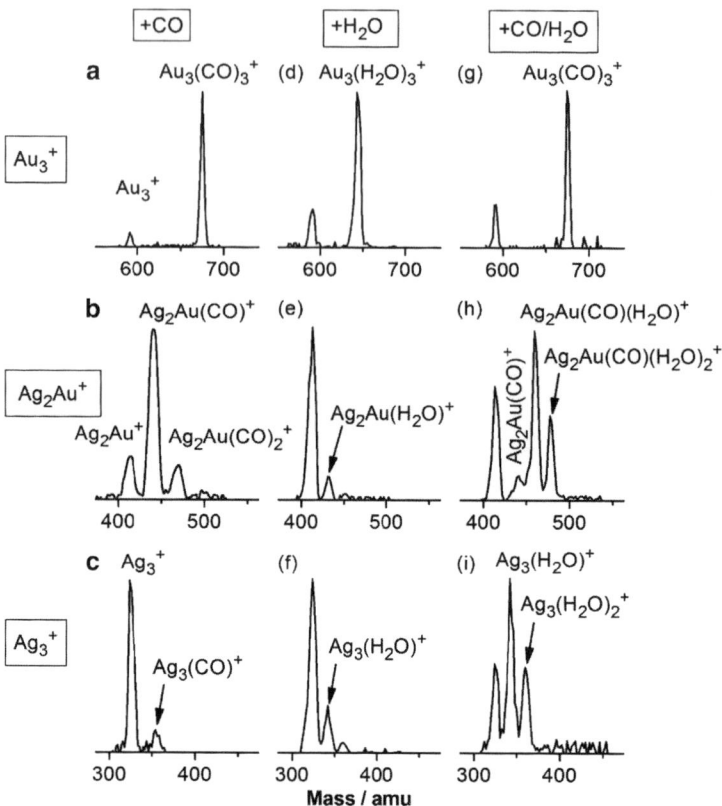

Fig. 32 Ionic products formed after the reaction of Au_3^+ (*a, d, g*), Ag_2Au^+ (*b, e, h*) and Ag_3^+ (*c, f, i*) with CO (*a–c*) and H_2O (*d–f*) at 300 K (p(He) = 1 Pa; p(CO) = 0.02 Pa, except in (*b*) p(CO) = 0.22 Pa; $p(H_2O)$ = 0.003 Pa). The spectra in the right column display product ion mass distributions obtained when both reactive molecules CO and H_2O were present in the ion trap (p(CO) = 0.04 Pa, $p(H_2O)$ = 0.004, 300 K) for Au_3^+ (*g*), Ag_2Au^+ (*h*) and Ag_3^+ *i*). All the spectra were obtained after a reaction time of 500 ms. Figure reproduced from [320]

$$Ag_2Au(CO)^+ + H_2O \rightarrow Ag_2Au(CO)(H_2O)^+ \qquad (69)$$

$$Ag_2Au(CO)(H_2O)^+ + H_2O \rightarrow Ag_2Au(CO)(H_2O)_2^+ \qquad (70)$$

4.4 Catalysis by Gold Cluster Ions

Trapping mass spectrometers are uniquely suited to study complete catalytic cycles [44]. In 1981, Kappes and Staley reported groundbreaking research on the first examples of transition-metal-catalysed reactions in the gas phase using an ICR mass spectrometer [321]. A key reaction they studied was the oxidation of CO (Eq. (71)), which is exothermic ($\Delta H° = -107$ kcal mol^{-1}) but does not occur at room temperature in the absence of a catalyst. They described a simple two-step catalytic cycle for the oxidation of CO catalysed by the atomic iron cation

Scheme 6 Fe^+-catalysed oxidation of CO [321]. Step 1 is oxygen atom abstraction; step 2 is oxygen atom transfer

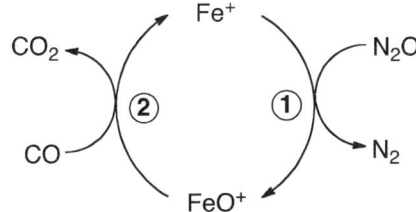

(Scheme 6). In the first step, Fe^+ reacted with N_2O by oxygen atom transfer to yield FeO^+, which transferred an oxygen atom to CO in the second step to yield neutral CO_2 and regenerate the atomic Fe^+ catalyst. The exploration of gas-phase transition-metal catalytic cycles has continued in the intervening 4 decades and been reviewed on several occasions [28, 29, 287, 322, 323]. Here we focus on complete gas-phase catalytic cycles that are catalysed by gold nanoclusters. Readers are also directed to Sect. 4 for data relevant for potential catalytic cycles whose final step (s) has not yet been realised.

$$CO + N_2O \rightarrow CO_2 + N_2 \tag{71}$$

Bernhardt's group has used a variable-temperature ion trap to examine related oxidation of CO by N_2O occurring on the coinage metal cluster cations Au_3^+ (at 300 K) and Ag_3^+ (at 230 and 250 K) [287]. They studied the reactions of Au_3^+ and Ag_3^+ with N_2O only, CO only and with both N_2O and CO in the ion trap. Both Au_3^+ and Ag_3^+ react with N_2O via a series of oxygen atom abstraction reactions (Eq. (72), with $n = 1$–3). Additional products formulated as $M_3O_n(N_2O_2)_m^+$ were observed (where $n = 1$–3 for $m = 1$ for M=Ag and Au; $n = 2$ and 3 for $m = 2$ for M=Ag; $n = 2$ for $m = 2$ for M=Au). When only CO was in the ion trap, Au_3^+ reacted to exclusively give $Au_3(CO)_3^+$ (Eq. (73), with $n = 3$) while Ag_3^+ gave a combination of $Ag_3(CO)^+$ and $Ag_3(CO)_2^+$ (Eq. (73), with $n = 1$ and 2). When both N_2O and CO were in the ion trap two main products were observed for Au_3^+: $Au_3(CO)_3^+$ and $Au_3(CO_2)_3^+$. The observation of the latter product suggests that oxidation of all each oxygen atom in the oxide $Au_3O_3^+$ reacted with one CO molecule to produce CO_2 that remained bound to the cluster (Eq. (74)). While the authors did not examine the CID reactions of the product $Au_3(CO_2)_3^+$, if this were to undergo losses of three molecules of CO_2 (Eq. (75)), then this would formally close a catalytic cycle for the oxidation of CO (Eq. (71)). Finally, oxidation of CO when N_2O was present in the trap proceeded via a different process for Ag_3^+, which is shown in Eq. (76):

$$M_3^+ + nN_2O \rightarrow M_3O_3^+ + nN_2 \tag{72}$$

$$M_3^+ + nCO \rightarrow M_3(CO)_n^+ \tag{73}$$

$$Au_3O_3^+ + 3CO \rightarrow Au_3(CO_2)_3^+ \tag{74}$$

Scheme 7 Au_2^--catalysed oxidation of CO [287, 321]. Step 1 is oxygen addition; step 2 is CO addition; step 3 is oxygen atom transfer

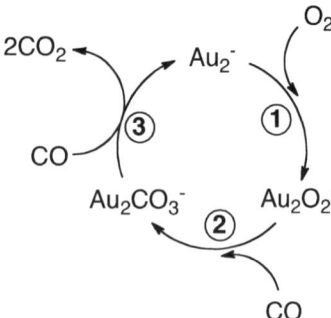

$$Au_3(CO_2)_3^+ \rightarrow Au_3^+ + 3CO_2 \tag{75}$$

$$Ag_3O_n^+ + nN_2O + nCO \rightarrow Ag_3O_n(CO_2)_n^+ + nN_2 \tag{76}$$

Bernhardt's group has used the same experimental set-up to study the related oxidation of CO (Eq. (77)) by O_2 catalysed by the gold dimer anion Au_2^- over a temperature range of 100–300 K (Scheme 7) [324, 325]. DFT calculations were used to examine key intermediates associated with this cycle [288, 325]. The first step of the cycle involves adsorption of O_2 onto Au_2^-, with the resultant complex stabilised by further collisions with the helium bath gas. DFT calculations predict that $Au_2O_2^-$ consists of dioxygen bound molecularly to Au_2^-, with partial electron transfer to the adsorbed O_2 molecule resulting in a 'superoxo-like' species (cf. Sect. 3.4). In step 2, $Au_2O_2^-$ reacts with CO to yield the metastable complex $Au_2CO_3^-$, which was not observed at room temperature but could be detected at lower temperatures (100 K). DFT calculations of potential isomers of '$Au_2CO_3^-$' revealed the digold-carbonate species of connectivity $[Au-Au-OCO_2]^-$ to be the most stable species, although a second structural isomer was found and suggested to be relevant to the catalytic process. The final step of the cycle in which $Au_2CO_3^-$ reacts with CO resulted in the liberation of two molecules of CO_2 and the regeneration of Au_2^-, thus closing a three-step catalytic cycle for the oxidation of CO to CO_2 with O_2 as the terminal oxidant (Scheme 7). Related work by other groups on the reactions of gold clusters with O_2 and CO was highlighted in previous sections.

$$2CO + O_2 \rightarrow 2CO_2 \tag{77}$$

Two competing, temperature-dependent catalytic cycles for the oxidation of methane by O_2 catalysed by Au_2^+ have been described (Scheme 8) [326, 327]. Variable-temperature ion-trap mass spectrometry experiments were carried out in which the partial pressures of CH_4 and O_2 were varied and product distributions at different temperature were examined. In combination with these detailed experiments, kinetic modelling of the product ion abundances was used to establish the order of the individual steps of each of the catalytic cycles, and DFT calculations were used to shed light on possible mechanisms. Reversible absorption of methane onto Au_2^+ (step 1 of Scheme 8) represents the key first step and provides the shared entry point into both catalytic cycles, which also share the same next step involving

Scheme 8 Competing catalytic cycles for oxidation of methane catalysed by Au_2^+: Cycle 1 involves dehydrocoupling of methane to yield ethane (Eq. (78)) [326, 327]; Cycle 2 involves oxidation of methane to formaldehyde (Eq. (79)) [327]

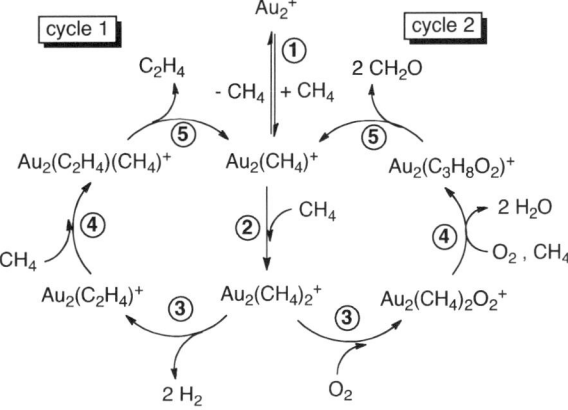

reaction with a second molecule of methane (step 2 of Scheme 8). The subsequent reactions of $Au_2(CH_4)_2^+$ represent the branch points for the competing cycles.

Either in the absence of O_2 or at higher temperatures that disfavour the adsorption of O_2, cycle 1 proceeds via elimination of H_2 (step 3) [326, 327]. Desorption of ethene from $Au_2(C_2H_4)^+$ does not occur spontaneously, but rather is triggered by reaction with a third molecule of methane (step 5), which closes catalytic cycle 1 via the regeneration of $Au_2(CH_4)^+$. This cycle corresponds to the dehydrocoupling of two molecules of methane to form ethene (Eq. (78)). By studying the relative abundances of the various product ions as a function of temperature, information was gleaned on the relative barriers associated with key steps. Thus the barrier for dehydrogenation (step 3) can only be surmounted at 250 K and above, while loss of ethene from $Au_2(C_2H_4)(CH_4)^+$ only occurs at temperatures above 270 K. At the lowest temperature studied (200 K), a minor product formulated as $Au_2(C_4H_{12})^+$ was observed, while the ion $Au(C_2H_4)(CH_4)^+$, which corresponds to loss of an Au atom from $Au_2(C_2H_4)(CH_4)^+$, appears at 250 K. The DFT calculations reveal that the mechanistic sequence for dehydrogenation of $Au_2(CH_4)_2^+$ involves stepwise losses of H_2 and involves 11 intermediates and 9 transition states. The first loss of H_2 gives rise to the organometallic cluster ion, $(CH_3Au)_2^+$ as a key intermediate. A related catalytic cycle was recently reported for the reaction of mixed metal cluster Pd_2Au^+ with CD_4 [319]:

$$2CH_4 \rightarrow CH_2{=}CH_2 + 2H_2 \tag{78}$$

Cycle 2, which corresponds to the oxidation of methane to formaldehyde (Eq. (79)), dominates at 210 K, while cycle 2 becomes competitive at 240 K and then dominates at temperatures of 250 K and above [327]. The cooperative action of multiple substrate and oxidant molecules is vital for the successful progress of cycle 2. Only $Au_2(CH_4)_2^+$ can react with a molecule of O_2 to form $Au_2(CH_4)_2O_2^+$ (step 3), which reacts with a third molecule methane and a second molecule of O_2 to form $Au_2(C_3H_8O_2)^+$ and 2 equiv. of water (step 4) [327]. The final step involves loss of two molecules of formaldehyde from $Au_2(C_3H_8O_2)^+$ to regenerate the catalyst $Au_2(CH_4)^+$. A multistep process for oxidation of both adsorbed methane molecules to formaldehyde was predicted via DFT calculations. Each formaldehyde is formed at a different end of the gold dimer, and they are formed in the following related steps: oxygen absorbs at a gold site to form a peroxo intermediate, $(CH_4)Au$-Au $(O$-$O)(CH_4)^+$, which then abstracts a hydrogen atom to yield an organogold-hydroperoxide intermediate, $(CH_4)Au$-$Au(O$-$OH)(CH_3)^+$, which then loses water to form a oxo-gold-carbene, $(CH_4)Au$-$Au(O)(CH_2)^+$, that can then rearrange to the O bound formaldehyde complex, $(CH_4)Au$-$Au(OCH_2)^+$. The final step involves absorption of the third methane molecule onto the bisformaldehyde complex, $(CH_2O)Au$-$Au(OCH_2)^+$, which acts as a trigger to release both formaldehydes:

$$2CH_4 + 2O_2 \rightarrow 2CH_2O + 2H_2O \tag{79}$$

5   From the Gas Phase to Materials

Mass spectrometry-based studies offer exciting opportunities to direct the synthesis of new materials, as dramatically highlighted by the discovery and subsequent isolation of bulk fullerenes. Initial studies using a cluster beam source coupled to a mass spectrometer led to the discovery of the magic number of C_{60} [328]. Subsequently Kroto et al. proposed the Buckminsterfullerene structure [329]. These exciting results inspired other researchers to devise synthetic strategies aimed at isolating samples of C_{60} in order to structurally characterise them, but it was not until 5 years later that bulk samples became available [330].

Within the context of AuNCs, mass spectrometry has had an impact in the generation of materials in two main ways: (1) the use of MS to 'direct' the synthesis of gold NCs and (2) the use of MS as a preparative tool to mass select gold NCs and soft land them onto surfaces. Both of these approaches are briefly described below.

5.1   MS-Directed Synthesis of Gold Clusters

As noted in Sect. 2, mass spectrometry has been used to monitor the growth and processing of gold clusters. In several cases, MS had been used to direct the synthesis and isolation of bulk material of the gold cluster for subsequent structure and property studies. Recent cases include work from the group of Konishi et al. [58, 59]. For example, the isolation of $[Au_{11}(Ph_2P(CH_2)_2PPh_2)_6]^{3+}$ [59], as the SbF_6^- salt, was found to be composed of an Au_9 core with two gold atoms located at the *exo* position. In brief, the gold complex $[Au_2(Ph_2P(CH_2)_2PPh_2)]Cl_2$ [331, 332] (200 μmol) in 75 mL ethanol was treated with a 5 mL ethanolic solution of sodium borohydride (400 μmol). After 1 h of stirring crude material was precipitated using excess $NaSbF_6$. ESI-MS analysis of the crude precipitate revealed monodispersed $[Au_{11}(Ph_2P(CH_2)_2PPh_2)_6]^{3+}$, $m/z = 1519$. The isolation of crystalline material suitable for structural studies by X-ray crystallography was prompted by ESI-MS analysis of the crude material and crystals successfully grown from ethanol/dichloromethane.

5.2   MS-Selected Deposition of Gold Cluster Ions

Preparative mass spectrometry, whereby a species is ionised, mass selected and collected as a bulk sample for subsequent use, has a history dating back to the use of calutrons during World War II for separation and isolation of ^{235}U [333]. The low-energy collisions of ions with surfaces have been comprehensively reviewed and, depending on the conditions used, can be exploited for the deposition of mass-selected clusters onto surfaces [334]. This is an active area of research since it holds great promise for the modification of surfaces and the synthesis of novel structures. Soft landing (SL) experiments allow intact, mass-selected clusters to be applied to surfaces. The ability to control the size, density and in some cases the morphology of the deposited particles allows unprecedented flexibility in the creation of new types of nanostructures and as such holds considerable promise in the development of new catalysts. This dynamic field has been reviewed [335], including a consideration of the types of instruments that have been developed, which are essentially all 'homebuilt' [334]. The challenges of characterising the resultant surface covered nanoclusters have also been discussed [336].

As noted by Johnson et al., mass spectrometry offers several unique capabilities for deposition experiments, including (1) multiple ways of forming gas-phase ions such as laser ablation and ESI of intact clusters; (2) the ability to mass select ions from a complex mixture – this allows deposition of a monodisperse cluster from a polydisperse mixture; and (3) since ions are used, it is possible to focus and pattern the ion beam, thereby controlling the landing process. The surface onto which the clusters are deposited can also be varied to examine how the support influences the reactivity of the landed cluster.

The next two sections consider the role of the surface on the structure and properties of surface landed gold nanoclusters and the reactions of surface landed gold nanoclusters with substrates.

5.2.1   Determining the Role of Surface on the Structure and Properties of Surface Landed Gold Nanoclusters

Table 11 highlights the range of different surfaces that have been used in soft landing of mass-selected gold NC ions, which include: graphite, highly ordered pyrolytic graphite (HOPG); silica and alumina; titania; $TiO_2(110)$; and MgO. Some key questions that have been addressed when assessing the outcomes of these experiments include: Have the AuNCs remained intact, or have they fragmented [345, 346] Are the AuNCs mobile [341] Has agglomeration (or sintering) occurred [337, 341] What happens to the charge [345, 346] Has the shape of the cluster changed [341] What role do surface defects play on the structure, reactivity and properties of soft landed AuNCs [348] In order to address the questions, techniques have been developed to examine the structure and properties of surface landed AuNCs. Interested readers in the details of these techniques are directed to key reviews [334–336, 359]. Table 11 also highlights some of the key findings for the growing literature of soft landed AuNPs.

Table 11 Summary of key literature on the soft landing of AuNC ions onto surfaces

AuNC ion	Surface	Key findings	References
Au^+	Amorphous carbon	An example of 'bottom-up synthesis' via ion beam deposition, which requires agglomeration. The ion beam energy plays a key role in the resultant cluster size	[337]
Au_x^+ (x = 5,7, 27,33)	Amorphous carbon	No significant fragmentation or agglomeration of the clusters was observed	[338]
Au_{250}^+	Graphite	Clusters move on the surface but do not always agglomerate when then touch	[339]
Au_x^+	Au(111)	Neither diffusion nor coalescence of the SL AuNCs was detected	[340]
Au_x^+ (x = 1–8)	TiO_2(110)	Au^+ undergoes sintering. For x = 2–8, SL Au_x are not mobile and do not sinter. The SL clusters adopt a different shape to the gas phase due to charge transfer from the surface and ligation by the surface	[341]
Au_x^+ (x = 2–10)	SiO_2/Si(111) and Al_2O_3/ SiO_2/ Si(111)	2D SL AuNCs undergo a structural change on the surface from vertical into horizontal orientations. The horizontal clusters can then undergo diffusion and agglomeration	[342]
$Au_{330 \pm 17}^+$ and $Au_{10000 \pm 500}^+$	Amorphous carbon	The role of temperature and cluster size on agglomeration were studied. At higher temperatures smaller clusters are more mobile and agglomerate more rapidly	[343]
Au_x^+	Graphite	The role of background pressure of agglomeration of NCs was examined	[344]
$Au_{11}L_5^{3+a}$	SAMs[a]	The cluster remains intact, but the SAM used influences the charge state of $Au_{11}L_5^{n+}$: FSAM: n = 3; COOH-SAM n = 2, HSAM: n = 1	[345]
$Au_{11}L_5^{3+a}$	SAMs[a]	The charge state of soft landed $Au_{11}L_5^{n+}$ is influenced by both the SAM and the surface coverage. At higher coverages, pronounced reduction is observed	[346]
Au_x^+	Silica	Photoelectron spectroscopy was used to examine how the valence electronic states changes as a function of cluster size	[347]
Au_4^+, Au_8^+ and Au_3Sr^+	MgO(100)	A review focussing on the role of oxide support defects, cluster size dependence, cluster structural fluxionality and impurity doping on the catalytic properties of size-selected metal clusters on surfaces	[348]
Au_x^+ ($1 < x \leq 20$)	MgO(100)	Low-temperature oxidation of CO was studied as a function of cluster size. Au_8 was found to be the smallest catalytically active cluster	[349]

(continued)

Table 11 (continued)

AuNC ion	Surface	Key findings	References
Au_x^+ $(1 < x \leq 20)$	MgO(100)	Low-temperature oxidation of CO was studied. SL NCs of Au_x, Pt_x, Pd_x and Rh_x were compared. Rh_{20} showed the highest reactivity	[350]
Au_x^+ $(x = 1,2, 3,4,7)$	TiO$_2$(110)	Au_7 NCs are >50 times more reactive in catalysing CO oxidation than samples prepared by Au or Au_2 deposition. $^{18}O_2$ and mass spectrometry were used to analyse desorbed $C^{16}O$, $C^{18}O$ and $^{16}OC^{18}O$. A lack of $C^{18}O$ formation highlights that CO does not decompose on the AuNCs	[351]
Au_8^+	MgO(100)	Oxide support defects play a key role in enhancing cluster reactivity for the oxidation of CO	[352]
Au_{17}^+	Graphite	Sub-monolayer films of size-selected Au clusters present binding sites to stabilise individual protein molecules and complexes for single-molecule measurements	[353]
Au_x^+ $(x = 2–10)$	Sputter-damaged HOPG	Oxidation of the SL AuNCs by atomic oxygen was studied as a function of cluster size. The resultant oxides were allowed to react with CO. Only Au_8 shows significant reactivity	[354]
Au_x^+ $(x = 2–10)$	Silica	An additional Au atom can significantly change the oxidation of SL AuNCs by atomic oxygen. Au_5 and Au_7 are resistant to oxidation	[355]
Au_x^+ $(x = 5–8)$	Silica	SL AuNCs were first treated with either water or NaOH, and their oxidation by atomic oxygen was then studied. Water has no effect, while oxidation resistant Au_5 and Au_7 become more reactive towards oxidation upon treatment with NaOH	[356]
Au_x^+ $(x = 6–10)$	Amorphous alumina	SL AuNCs catalyse the epoxidation of propene. The highest selectivity for epoxide formation over acrolein formation was found for gas mixtures involving oxygen and water, thereby circumventing the use of hydrogen	[357]
Au_{20}^+	MgO(100)	The role of the thickness and stoichiometry of the MgO films on the catalytic activity of SL Au_{20} towards CO oxidation was examined. Theoretical calculations were used to probe the mechanism of CO oxidation	[358]

[a]L = 1,3-bis(diphenylphosphino)propane, different self-assembled monolayers on gold (SAMs) were used: 1H,1H,2H,2H-perfluorodecanethiol (FSAM); 16-mercaptohexadecanoic acid (COOH-SAM); 1-dodecanethiol (HSAM)
This table is adapted from [334]

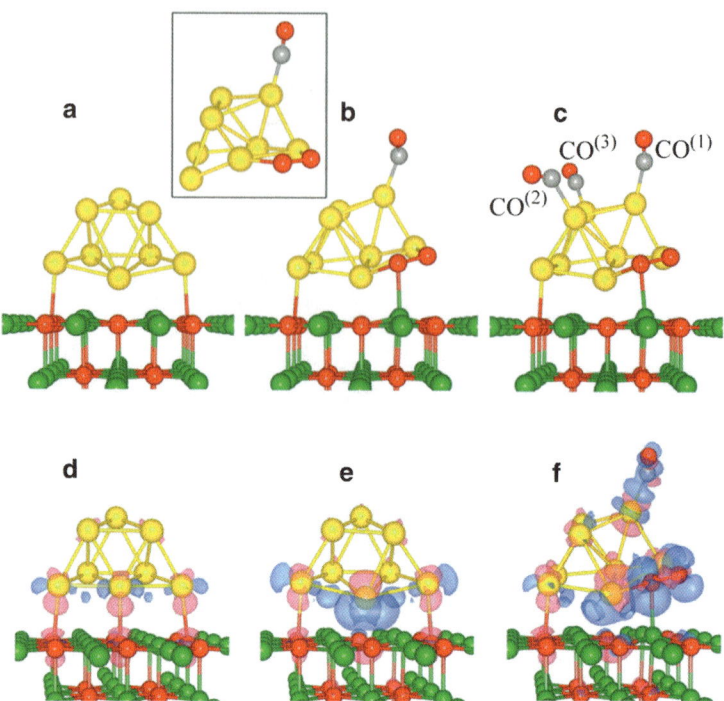

Fig. 33 Optimised structures of (*A*) bare Au$_8$ cluster (*yellow spheres*) adsorbed on an *F* centre of a MgO(001) surface (O atoms are in red and Mg atoms in green); (*B*) a surface-supported gold octamer with O$_2$ adsorbed at the interface between the Au$_8$ cluster and the magnesia surface and a CO molecule adsorbed on the top triangular facet (the C atom is depicted in grey). The inset between (*A*) and (*B*) shows a local-energy-minimum structure of the free Au$_8$ cluster in the three-dimensional (3D) isomeric form with co-adsorbed O$_2$ and CO molecules. (*C*) Au$_8$ on the magnesia surface [MgO(FC)] with three CO molecules adsorbed on the top facet of the cluster and an O$_2$ molecule preadsorbed at the interface between the cluster and the magnesia surface. Isosurfaces of charge differences ($\delta\rho$) are as follows: (*D*) Au$_8$ cluster adsorbed on defect-free MgO; (*E*) Au$_8$ cluster anchored to a surface *F* centre of MgO; (*F*) same as (*E*) but with O$_2$ and CO molecules adsorbed on the gold cluster. Pink isosurfaces represent $\delta\rho < 0$ (depletion) and blue ones correspond to $\delta\rho > 0$ (excess). Figure reproduced from reference [352]

5.2.2   Reactions of Surface Landed Gold Nanoclusters

Given Haruta's discovery that gold clusters can catalyse the oxidation of CO (Eq. (3)), it is not surprising that several studies have examined how surface landed gold nanoclusters react with mixtures of CO and O$_2$ as well as other substrates. Table 11 highlighted key findings of these studies, while earlier work has been previously reviewed [348]. Here we briefly discuss two important studies that use a combination of experiments and theoretical calculations to shed light on how surface defects and impurities can influence reactivity.

Yoon et al. have shown that F-centre surface defects in the MgO support play a key role in the enhanced catalytic activity of SL Au_8 clusters towards the oxidation of CO [352]. Temperature-programmed desorption using mass spectrometry detection of CO_2 showed a higher yield of CO_2 at low temperature (140 K) for Au_8 clusters supported on defect-rich MgO substrates compared to defect-poor surfaces. In situ IR spectroscopy of CO molecules bound to the Au_8 NC on the defect-rich MgO surface showed a redshift of the CO band, suggesting electron back-donation from the cluster to the CO antibonding orbitals. Figure 33 shows the results of detailed DFT calculations on the structures of the AuNCs and their reaction intermediates for CO oxidation. Figure 33a–c shows the changes to the Au_8 NC upon co-adsorbed O_2 and CO molecules. The change in electron density on the Au_8 NC, shown in Fig. 33d–e, highlights that the surface defect plays a key role in transferring electron density to the cluster.

Lee et al. have explored how the catalytic activity and selectivity of Au_{6-10} clusters towards the epoxidation of propene is influenced by adding either hydrogen or water [357]. In the absence of these additives, propene is oxidised, but the selectivity is poor as the major product is acrolein (formed in a 2:1 ratio). This selectivity improves, with a ratio for formation of acrolein/propene oxide of up to 1:2 for H_2 as an additive and 1:14 for water. DFT calculations were used to identify key reaction intermediates and reaction pathways. These calculations highlight that (1) the high catalyst activity arose from the formation of propene oxide metallacycles and (2) the key role of the H_2 and H_2O additives is to maintain a hydroxy-terminated alumina surface, which is critical to the enhanced selectivity.

6 Conclusions

Over the past 2 decades mass spectrometry-based methods have played important roles in the analysis of AuNCs and in defining the structures, properties and reactivities of AuNCs in the gas phase. An emerging role for mass spectrometry is its use in the synthesis of bulk materials. A common theme from fundamental gas-phase studies is that the spectroscopy, stability and reactivity of charged AuNCs often exhibit properties which depend on whether there are even or odd number metal atoms. The study of homogeneous and heterogeneous catalysts based on gold has highlighted the need for continued research to better define and understand the role of AuNCs. The mass spectrometric-based methods described in this chapter are likely to play an important role in developing that understanding.

Research on gold cluster ions continues unabated. Since this chapter was submitted several new studies have appeared that are relevant to the following sections:

Section 2: ESI-MS has been used to characterize histidine stabilized AuNCs [360] and the novel $Au_{22}(L^8)_6$ nanocluster (where L^8=1,8-bis-(diphenyl-phosphino) octane) [361]. Ligand exchange reactions of the $Au_{25}(2\text{-PET})_{18}$ nanocluster (where 2-PET=2-phenylethylthiolate) have been examined using MALDI-TOF-MS [362]. The role of the ligand (1,3-bis(dicyclohexylphosphino)propane versus 1,3-bis(diphenyl-phosphino)propane) in the growth of AuNCs and their fragmentation reactions have been examined via ESI-MS and CID [363]. ESI-MS was used to assign the novel AuNCs $Au_{10}(HSPh\text{-}pNH_2)_{10}$ (where HSPh-pNH_2=4-aminothiophenol) by the formation of charged adducts such as $[Au_{10}(HSPh\text{-}pNH_2)_{10}+H^+]^+$ [364] and $Au_{30}(tert\text{-thiol})_{18}$ (tert-thiol=tert-butanethiol and 1-adamantanethiol) [365] which also used MALDI-MS. Laser ablation has been used to synthesize gold carbides [366] and gold arsenides [367]. MALDI of lysozyme-Au adducts produces the following bare gas phase AuNC cations: Au_{18}^+, Au_{25}^+, Au_{38}^+, and Au_{102}^+. Bare alloy clusters of the type $Au_{24}Pd^+$ are formed for mixtures of gold and palladium lysozyme adducts [368].

Section 3: A recent photoelectron spectroscopy and theoretical study on Au_n^- (n = 36, 37, 38) found that the most stable structures of Au_{36}^- to Au_{38}^- exhibit core-shell type structures all with a highly robust tetrahedral four-atom core. [369]

Section 4: Other gem aurated acetylides have been studied (cf. Scheme 5) [370]. Au_2^+ has been shown to be a superior catalyst compared to Pd_2^+ in the dehydro-coupling of methane (Scheme 8) [371].

Section 5: ESI-MS has been used to characterize Ph_3P protected Au nanoclusters on a range of surfaces [372].

Acknowledgements We thank the ARC for generously funding our work on coinage metals.

References

1. Castleman AW, Jena P (2006) Clusters: a bridge between disciplines. Proc Natl Acad Sci 103(28):10552–10553. doi:10.1073/pnas.0601783103
2. Castleman AW, Jena P (2006) Clusters: a bridge across the disciplines of environment, materials science, and biology. Proc Natl Acad Sci 103(28):10554–10559. doi:10.1073/pnas.0601780103
3. Jena P, Castleman AW (2006) Clusters: a bridge across the disciplines of physics and chemistry. Proc Natl Acad Sci 103(28):10560–10569. doi:10.1073/pnas.0601782103
4. Evans DG, Mingos DMP (1985) Molecular orbital analysis of the bonding in penta- and hepta-nuclear gold tertiary phosphine clusters. J Organomet Chem 295(3):389–400
5. Mingos DMP (1983) Polyhedral skeletal electron pair approach. A generalised principle for condensed polyhedra. J Chem Soc Chem Commun 12:706–708
6. Mingos DMP (1984) Polyhedral skeletal electron pair approach. Acc Chem Res 17(9):311–319
7. Mingos DMP (1984) Gold cluster compounds – are they metals in miniature? Gold Bull 17(1):5–12
8. Mingos DMP (1996) Gold – a flexible friend in cluster chemistry. J Chem Soc Dalton Trans 5:561–566
9. Mingos DMP, Slee T, Zhenyang L (1990) Bonding models for ligated and bare clusters. Chem Rev 90(2):383–402
10. Mingos DMP (2014) Structural and bonding issues in clusters and nano-clusters. Struct Bond. doi:10.1007/430_2014_141

11. Bond GC, Sermon PA, Webb G, Buchanan DA, Wells PB (1973) Hydrogenation over supported gold catalysts. J Chem Soc Chem Commun 13:444b–445b
12. Hutchings GJ (1985) Vapor phase hydrochlorination of acetylene: correlation of catalytic activity of supported metal chloride catalysts. J Catal 96(1):292–295
13. Haruta M, Yamada N, Kobayashi T, Iijima S (1989) Gold catalysts prepared by coprecipitation for low-temperature oxidation of hydrogen and of carbon monoxide. J Catal 115(2):301–309. doi:10.1016/0021-9517(89)90034-1
14. Haruta M (2005) Catalysis: gold rush. Nature 437(7062):1098–1099
15. Hutchings GJ (2005) Catalysis by gold. Catal Today 100(1–2):55–61
16. Bond GC, Louis CD, Thompson DTD (2006) Catalysis by gold. Catalytic science series: v. 6. Imperial College Press, London
17. Crabtree RH (2011) Resolving heterogeneity problems and impurity artifacts in operationally homogeneous transition metal catalysts. Chem Rev 112(3):1536–1554. doi:10.1021/cr2002905
18. Oliver-Meseguer J, Cabrero-Antonino JR, Domínguez I, Leyva-Pérez A, Corma A (2012) Small gold clusters formed in solution give reaction turnover numbers of 107 at room temperature. Science 338(6113):1452–1455
19. Hashmi ASK (2012) Sub-nanosized gold catalysts. Science 338(6113):1434
20. Johnston RL (2002) Atomic and molecular clusters. Master's series in physics and astronomy. Taylor and Francis, London
21. Boyle R (1661) The sceptical chymist: or Chymico-physical doubts & paradoxes, touching the spagyrist's principles commonly call'd hypostatical, as they are wont to be propos'd and defended by the generality of alchymists. Whereunto is praemis'd part of another discourse relating to the same subject. London, Printed by J. Cadwell for J. Crooke, 1661
22. Todd JFJ (1991) Recommendations for nomenclature and symbolism for mass spectroscopy (including an appendix of terms used in vacuum technology). Pure App Chem 63:1541. doi:10.1351/pac199163101541
23. Alonso JA (2005) Structure and properties of atomic nanoclusters. Imperial College Press, London
24. Murray KK, Boyd RK, Eberlin MN, Langley GJ, Li L, Naito Y (2013) Definitions of terms relating to mass spectrometry (IUPAC Recommendations 2013). Pure Appl Chem ASAP 85(7):1515–1609
25. Eleanor C (2012) Atom cluster. McGraw-Hill. Available via EBSCOhost. https://ezp.lib.unimelb.edu.au/login?url=https://search.ebscohost.com/login.aspx?direct=true&db=edsasc&AN=edsasc.059850&scope=site
26. Kreibig U, Vollmer M (1995) Optical properties of metal clusters. Springer series in materials science: 25. Springer, Berlin
27. Ott LS, Finke RG (2007) Transition-metal nanocluster stabilization for catalysis: a critical review of ranking methods and putative stabilizers. Coord Chem Rev 251(9–10):1075–1100
28. O'Hair RAJ, Khairallah GN (2004) Gas phase ion chemistry of transition metal clusters: production, reactivity, and catalysis. J Clust Sci 15(3):331–363
29. Böhme DK, Schwarz H (2005) Gas-phase catalysis by atomic and cluster metal ions: the ultimate single-site catalysts. Angew Chem Int Ed 44(16):2336–2354
30. Ervin KM (2001) Metal-ligand interactions: gas-phase transition metal cluster carbonyls. Int Rev Phys Chem 20(2):127–164
31. Armentrout PB (2001) Reactions and thermochemistry of small transition metal cluster ions. Ann Rev Phys Chem 52:423–461
32. Parent DC, Anderson SL (1992) Chemistry of metal and semimetal cluster ions. Chem Rev 92(7):1541–1565
33. Harkness KM, Cliffel DE, McLean JA (2010) Characterization of thiolate-protected gold nanoparticles by mass spectrometry. Analyst 135(5):868–874. doi:10.1039/b922291j
34. Bernhardt TM (2005) Gas-phase kinetics and catalytic reactions of small silver and gold clusters. Int J Mass Spectrom 243(1):1–29

35. Spectrometry ASfM (1995) What is mass spectrometry? Am Soc Mass Spectrom. http://www.asms.org

36. Henderson W, McIndoe JS (2005) Mass spectrometry of inorganic, coordination and organometallic compounds: tools – techniques – tips. Inorganic chemistry. Chichester, Wiley, Hoboken

37. McLuckey SA, Wells JM (2001) Mass analysis at the advent of the 21st century. Chem Rev 101(2):571–606

38. Farrar JM, Saunders WH (1988) Techniques for the study of ion-molecule reactions, vol 20. Wiley, New York, NY

39. Freiser BS (1996) Gas-phase metal ion chemistry. J Mass Spectrom 31(7):703–715

40. Asamoto B (ed) (1991) FT-ICR/MS: analytical applications of fourier transform ion cyclotron resonance mass spectrometry. vol Copyright (C) 2013 American Chemical Society (ACS). All Rights Reserved. VCH

41. Hammad LA, Gerdes G, Chen P (2005) Electrospray ionization tandem mass spectrometric determination of ligand binding energies in platinum(II) complexes. Organometallics 24(8): 1907–1913. doi:10.1021/om0491793

42. Schoen E, Zhang X, Zhou Z, Chisholm MH, Chen P (2004) Gas-phase and solution-phase polymerization of epoxides by cr(salen) complexes: evidence for a dinuclear cationic mechanism. Inorg Chem 43(23):7278–7280. doi:10.1021/ic049120o

43. Damrauer R (2004) Organometallic chemistry in the flowing afterglow: a review. Organometallics 23(7):1462–1479. doi:10.1021/om030591c

44. O'Hair RAJ (2006) The 3D quadrupole ion trap mass spectrometer as a complete chemical laboratory for fundamental gas-phase studies of metal mediated chemistry. Chem Commun (Cambridge) 14:1469–1481. doi:10.1039/b516348j

45. De Haeck J, Veldeman N, Claes P, Janssens E, Andersson M, Lievens P (2011) Carbon monoxide adsorption on silver doped gold clusters. J Phys Chem A 115(11):2103–2109

46. Xie Y, Dong F, Bernstein ER (2011) Experimental and theory studies of the oxidation reaction of neutral gold carbonyl clusters in the gas phase. Catal Today 177(1):64–71

47. Knickelbein MB (1999) Reactions of transition metal clusters with small molecules. Ann Rev Phys Chem 50:79–155

48. Yin S, Bernstein ER (2012) Gas phase chemistry of neutral metal clusters: distribution, reactivity and catalysis. Int J Mass Spectrom 321–322:49–65

49. Colaianni L, Kung SC, Taggart D, De Giorgio V, Greaves J, Cioffil N, Penner RM (2009) Gold nanowires: deposition, characterization and application to the mass spectrometry detection of low-molecular weight analytes. In: Trani 2009. 3rd International workshop on advances in sensors and interfaces, IWASI, train, Italy, pp 20–24

50. McLean JA, Stumpo KA, Russel DH (2005) Size-selected (2–10 nm) gold nanoparticles for matrix assisted laser desorption ionization of peptides. J Am Chem Soc 127(15):5304–5305

51. Huang Y-F, Chang H-T (2007) Analysis of adenosine triphosphate and glutathione through gold nanoparticles assisted laser desorption/ionization mass spectrometry. Anal Chem 79 (13):4852–4859. doi:10.1021/ac070023x

52. Wu HP, Su CL, Chang HC, Tseng WL (2007) Sample-first preparation: a method for surface-assisted laser desorption/ionization time-of-flight mass spectrometry analysis of cyclic oligosaccharides. Anal Chem 79(16):6215–6221

53. Su C-L, Tseng W-L (2007) Gold nanoparticles as assisted matrix for determining neutral small carbohydrates through laser desorption/ionization time-of-flight mass spectrometry. Anal Chem 79(4):1626–1633. doi:10.1021/ac061747w

54. Bruce MI, Liddell MJ (1987) Applications of fast-atom-bombardment mass spectrometry (FAB MS) to organometallic and coordination chemistry. Appl Organomet Chem 1(3): 191–226. doi:10.1002/aoc.590010302

55. Boyle PD, Johnson BJ, Alexander BD, Casalnuovo JA, Gannon PR, Johnson SM, Larka EA, Mueting AM, Pignolet LH (1987) Characterization of large cationic transition-metal-gold clusters by fast atom bombardment mass spectroscopy (FABMS). New rhenium–gold and

platinum–gold clusters: [Au$_4$Re(H)$_4$[P(p-tol)$_3$]$_2$(PPh$_3$)$_4$]$^+$, [Au$_2$Re$_2$(H)$_6$(PPh$_3$)$_6$]$^+$, and [Au$_6$Pt (PPh$_3$)$_7$]$^{2+}$. Inorg Chem 26(9):1346–1350. doi:10.1021/ic00256a002

56. Chisholm DM, Scott McIndoe J (2008) Charged ligands for catalyst immobilisation and analysis. Dalton Trans 30:3933–3945. doi:10.1039/B800371H

57. Van der Velden JWA, Bour JJ, Vollenbroek FA, Beurskens PT, Smits JMM (1979) Synthesis of a new pentanuclear gold cluster by metal evaporation. Preparation and X-ray structure determination of [tris{bis(diphenylphosphino)methane}][bis(diphenylphosphino)methanido] pentagold dinitrate. J Chem Soc Chem Commun 24:1162–1163. doi:10.1039/c39790001162

58. Shichibu Y, Konishi K (2010) HCL-induced nuclearity convergence in diphosphine-protected ultrasmall gold clusters: a novel synthetic route to "Magic-Number" Au$_{13}$ clusters. Small 6(11):1216–1220

59. Shichibu Y, Kamei Y, Konishi K (2012) Unique [core+two] structure and optical property of a dodeca-ligated undecagold cluster: critical contribution of the exo gold atoms to the electronic structure. Chem Commun 48(61):7559–7561

60. Heaven MW, Dass A, White PS, Holt KM, Murray RW (2008) Crystal structure of the gold nanoparticle [N(C$_8$H$_{17}$)$_4$][Au$_{25}$(SCH$_2$CH$_2$Ph)$_{18}$]. J Am Chem Soc 130(12):3754–3755. doi:10.1021/ja800561b

61. Zeng C, Qian H, Li T, Li G, Rosi NL, Yoon B, Barnett RN, Whetten RL, Landman U, Jin R (2012) Total structure and electronic properties of the gold nanocrystal Au$_{36}$(SR)$_{24}$. Angew Chem Int Ed 51(52):13114–13118. doi:10.1002/anie.201207098

62. Liu J, Lee T, Janes DB, Walsh BL, Melloch MR, Woodall JM, Reifenberger R, Andres RP (2000) Guided self-assembly of Au nanocluster arrays electronically coupled to semiconductor device layers. Appl Phys Lett 77(3):373–375

63. Kadossov E, Cabrini S, Burghaus U (2010) Adsorption kinetics and dynamics of CO on silica supported Au nanoclusters – utilizing physical vapor deposition and electron beam lithography. J Mol Catal A Chem 321(1–2):101–109. doi:10.1016/j.molcata.2010.02.009

64. Ishida T, Kinoshita N, Okatsu H, Akita T, Takei T, Haruta M (2008) Influence of the support and the size of gold clusters on catalytic activity for glucose oxidation. Angew Chem Int Ed 47(48):9265–9268

65. Gibson JK (1998) Laser ablation and gas-phase reactions of small gold cluster ions, Au$^+$n (1≤n≤7). J Vac Sci Technol A Vac Surfaces Films 16(2):653–659

66. Hu CW, Kasuya A, Wawro A, Horiguchi N, Czajka R, Nishina Y, Saito Y, Fujita H (1996) Gold clusters deposited on highly oriented pyrolytic graphite by pulse laser ablation and liquid metal ion source. Mater Sci Eng A 217:28:103–107

67. Sugawara KI, Sobott F, Vakhtin AB (2003) Reactions of gold cluster cations Au$_n$$^+$(n = 1–12) with H$_2$S and H$_2$. J Chem Phys 118(17):7808–7816

68. Guo W, Yuan J, Wang E (2012) Organic-soluble fluorescent Au$_8$ clusters generated from heterophase ligand-exchange induced etching of gold nanoparticles and their electrochemiluminescence. Chem Commun 48(25):3076–3078

69. Zhou R, Shi M, Chen X, Wang M, Chen H (2009) Atomically monodispersed and fluorescent sub-nanometer gold clusters created by biomolecule-assisted etching of nanometer-sized gold particles and rods. Chem Eur J 15(19):4944–4951. doi:10.1002/chem.200802743

70. Duan H, Nie S (2007) Etching colloidal gold nanocrystals with hyperbranched and multivalent polymers: a new route to fluorescent and water-soluble atomic clusters. J Am Chem Soc 129(9):2412–2413. doi:10.1021/ja067727t

71. De Heer WA (1993) The physics of simple metal clusters: experimental aspects and simple models. Rev Mod Phys 65(3):611–676

72. Duncan MA (2012) Invited Review Article: Laser vaporization cluster sources. Rev Sci Instrum 83(4):041101

73. Turkevich J, Stevenson PC, Hillier J (1951) A study of the nucleation and growth processes in the synthesis of colloidal gold. Discuss Faraday Soc 11:55–75

74. Rao CNR, Kulkarni GU, Thomas PJ, Edwards PP (2000) Metal nanoparticles and their assemblies. Chem Soc Rev 29(1):27–35

75. Wilcoxon JP, Abrams BL (2006) Synthesis, structure and properties of metal nanoclusters. Chem Soc Rev 35(11):1162–1194. doi:10.1039/B517312B

76. Guo S, Wang E (2011) Noble metal nanomaterials: controllable synthesis and application in fuel cells and analytical sensors. Nano Today 6(3):240–264

77. Lu Y, Chen W (2012) Sub-nanometre sized metal clusters: from synthetic challenges to the unique property discoveries. Chem Soc Rev 41(9):3594–3623. doi:10.1039/C2CS15325D

78. Ayela C, Lalo H, Kuhn A (2013) Introducing a well-ordered volume porosity in 3-dimensional gold microcantilevers. Appl Phys Lett 102(5)

79. Daniel M-C, Astruc D (2003) Gold nanoparticles: assembly, supramolecular chemistry, quantum-size-related properties, and applications toward biology, catalysis, and nanotechnology. Chem Rev 104(1):293–346. doi:10.1021/cr030698+

80. Zhao P, Li N, Astruc D (2013) State of the art in gold nanoparticle synthesis. Coord Chem Rev 257(3–4):638–665. doi:10.1016/j.ccr.2012.09.002

81. Pei Y, Zeng XC (2012) Investigating the structural evolution of thiolate protected gold clusters from first-principles. Nanoscale 4(14):4054–4072. doi:10.1039/c2nr30685a

82. Jiang D-E (2010) Understanding and predicting thiolated gold nanoclusters from first principles. Wuli Huaxue Xuebao 26(4):999–1016

83. Jin R (2010) Quantum sized, thiolate-protected gold nanoclusters. Nanoscale 2(3):343–362. doi:10.1039/b9nr00160c

84. Brust M, Walker M, Bethell D, Schiffrin DJ, Whyman R (1994) Synthesis of thiol-derivatised gold nanoparticles in a two-phase Liquid-Liquid system. J Chem Soc Chem Commun 7: 801–802. doi:10.1039/C39940000801

85. Briñas RP, Hu M, Qian L, Lymar ES, Hainfeld JF (2008) Gold nanoparticle size controlled by polymeric Au(I) thiolate precursor size. J Am Chem Soc 130(3):975–982

86. Simpson CA, Farrow CL, Tian P, Billinge SJL, Huffman BJ, Harkness KM, Cliffel DE (2010) Tiopronin gold nanoparticle precursor forms aurophilic ring tetramer. Inorg Chem 49 (23):10858–10866. doi:10.1021/ic101146e

87. Dharmaratne AC, Krick T, Dass A (2009) Nanocluster size evolution studied by mass spectrometry in room temperature $Au_{25}(SR)_{18}$ synthesis. J Am Chem Soc 131(38): 13604–13605. doi:10.1021/ja906087a

88. Gaur S, Miller JT, Stellwagen D, Sanampudi A, Kumar CSSR, Spivey JJ (2012) Synthesis, characterization, and testing of supported Au catalysts prepared from atomically-tailored $Au_{38}(SC_{12}H_{25})_{24}$ clusters. Phys Chem Chem Phys 14(5):1627–1634. doi:10.1039/ C1CP22438G

89. Tlahuice-Flores A, Black DM, Bach SBH, Jose-Yacaman M, Whetten RL (2013) Structure & bonding of the gold-subhalide cluster $I\text{-}Au_{144}C_{160}[z]$. Phys Chem Chem Phys 15(44): 19191–19195. doi:10.1039/c3cp53902d

90. Nimmala PR, Yoon B, Whetten RL, Landman U, Dass A (2013) $Au_{67}(SR)_{35}$ nanomolecules: characteristic size-specific optical, electrochemical, structural properties and first-principles theoretical analysis. J Phys Chem A 117(2):504–517. doi:10.1021/jp311491v

91. Negishi Y, Chaki NK, Shichibu Y, Whetten RL, Tsukuda T (2007) Origin of magic stability of thiolated gold clusters: a case study on $Au_{25}(SC_6H_{13})_{18}$. J Am Chem Soc 129(37): 11322–11323. doi:10.1021/ja073580+

92. Zeng C, Liu C, Pei Y, Jin R (2013) Thiol ligand-induced transformation of Au38(SC2H4Ph) 24 to Au36(SPh-t-Bu)24. ACS Nano 7(7):6138–6145. doi:10.1021/nn401971g

93. Malatesta L, Naldini L, Simonetta G, Cariati F (1966) Triphenylphosphine-gold(0)/gold(I) compounds. Coord Chem Rev 1(1–2):255–262. doi:10.1016/S0010-8545(00)80179-4

94. Naldini L, Cariati F, Simonetta G, Malatesta L (1966) Gold-tertiary phosphine derivatives with intermetallic bonds. Chem Commun 18:647–648. doi:10.1039/c19660000647

95. Cariati F, Naldini L, Simonetta G, Malatesta L (1967) Ethyldiphenylphosphine-gold derivatives with intermetallic bonds. Inorg Chim Acta 1(1):24–26. doi:10.1016/S0020-1693(00) 93133-5

96. Cariati F, Naldini L, Simonetta G, Malatesta L (1967) Clusters of gold compounds with 1,2-bis(diphenylphosphino)ethane. Inorg Chim Acta 1(2):315–318. doi:10.1016/S0020-1693 (00)93194-3

97. McPartlin M, Malatesta L, Mason R (1969) Cluster complexes of gold(0)-gold(I). J Chem Soc D 7:334. doi:10.1039/c29690000334

98. Abu-Salah OM, Al-Ohaly ARA, Knobler CB (1985) Preparation, identification, and X-ray structure of a novel pentanuclear gold-copper cluster complex. J Chem Soc Chem Commun (21):1502–1503. doi:10.1039/c39850001502

99. Briant CE, Hall KP, Mingos DMP (1982) Unusual degradation reaction of icosahedral cluster compounds of gold with chelating diphosphanes and the X-ray structure of di[bis(diphenyl-phosphino)methanido]digold(I), [Au(Ph$_2$P)$_2$CH]$_2$. J Organomet Chem 229(1):C5–C8. doi:10.1016/S0022-328X(00)89123-8

100. Briant CE, Theobald BRC, White JW, Bell LK, Mingos DMP, Welch AJ (1981) Synthesis and x-ray structural characterization of the centered icosahedral gold cluster compound [Au$_{13}$(PMe$_2$Ph)$_{10}$Cl$_2$](PF$_6$)$_3$; the realization of a theoretical prediction. J Chem Soc Chem Commun 5:201–202. doi:10.1039/c39810000201

101. van der Velden JWA, Bour JJ, Bosman WP, Noordik JH (1983) Reactions of cationic gold clusters with Lewis bases. Preparation and X-ray structure investigation of [Au$_8$(PPh$_3$)$_7$] (NO$_3$)2.2CH$_2$Cl$_2$ and Au$_6$(PPh$_3$)$_4$[Co(CO)$_4$]$_2$. Inorg Chem 22(13):1913–1918. doi:10.1021/ic00155a018

102. van der Velden JWA, Bour JJ, Steggerda JJ, Beurskens PT, Roseboom M, Noordik JH (1982) Gold clusters. Tetrakis[1,3-bis(diphenylphosphino)propane]hexagold dinitrate: preparation, X-ray analysis, and gold-197 Moessbauer and phosphorus-31{proton} NMR spectra. Inorg Chem 21(12):4321–4324. doi:10.1021/ic00142a041

103. van der Velden JWA, Vollenbroek FA, Bour JJ, Beurskens PT, Smits JMM, Bosman WP (1981) Gold clusters containing bidentate phosphine ligands. Preparation and X-ray structure investigation of [Au$_5$(dppmH)$_3$(dppm)](NO$_3$)$_2$ and [Au$_{13}$(dppmH)$_6$](NO$_3$)$_n$. Recl Trav Chim Pays Bas 100(4):148–152

104. Demartin F, Manassero M, Naldini L, Ruggeri R, Sansoni M (1981) Synthesis and X-ray characterization of an iodine-bridged tetranuclear gold cluster, di-μ-iodo-tetrakis(triphenyl-phosphine)-tetrahedro-tetragold. J Chem Soc Chem Commun 5:222–223. doi:10.1039/c39810000222

105. Cooper MK, Dennis GR, Henrick K, McPartlin M (1980) A new type of gold cluster compound. The syntheses and x-ray structure analysis of pentakis(tricyclohexylphosphine) tris(thiocyanato)enneagold, [Au$_9${P(C$_6$H$_{11}$)$_3$}$_5$(SCN)$_3$], and bis{tri(cyclohexyl)phosphineto} gold(I) hexafluorophosphate, [Au{P(C$_6$H$_{11}$)$_3$}$_2$][PF$_6$]. Inorg Chim Acta 45(4):L151–L152. doi:10.1016/S0020-1693(00)80129-2

106. Vollenbroek FA, Bosman WP, Bour JJ, Noordik JH, Beurskens PT (1979) Reactions of gold-phosphine cluster compounds. Preparation and X-ray structure determination of octakis (triphenylphosphine)octagold bis(hexafluorophosphate). J Chem Soc Chem Commun 9: 387–388. doi:10.1039/c39790000387

107. Manassero M, Naldini L, Sansoni M (1979) A new class of gold cluster compounds. Synthesis and X-ray structure of the octakis(triphenylphosphinegold) dializarinsulfonate, [Au$_8$(PPh$_3$)$_8$](aliz)$_2$. J Chem Soc Chem Commun 9:385–386. doi:10.1039/c39790000385

108. Bellon PL, Cariati F, Manassero M, Naldini L, Sansoni M (1971) Novel gold clusters. Preparation, properties, and X-ray structure determination of salts of octakis (triarylphosphine)enneagold, [Au$_9$L$_8$]X$_3$. J Chem Soc D 22:1423–1424. doi:10.1039/c29710001423

109. Feld H, Leute A, Rading D, Benninghoven A, Schmid G (1990) Formation of very large gold superclusters (clusters of clusters) as secondary ions up to (Au$_{13}$)$_{55}$ by SIMS. J Am Chem Soc 112(22):8166–8167. doi:10.1021/ja00178a051

110. Van SMPJ, Brom HB, De JLJ, Schmid G (1986) Physical properties of metal cluster compounds II: d.c.-conductivity of the high-nuclearity gold cluster compound $Au_{55}(PPh_3)_{12}Cl_6$. Solid State Commun 60(4):319–322. doi:10.1016/0038-1098(86)90741-6
111. Schmid G (1982) Complex transition metal compounds. EP66287A2
112. Schmid G, Pfeil R, Boese R, Brandermann F, Meyer S, Calis GHM, Van der Velden JWA (1981) $Au_{55}[P(C_6H_5)_3]_{12}Cl_6$ – a gold cluster of unusual size. Chem Ber 114(11):3634–3642
113. Zavras A, Khairallah GN, O'Hair RAJ (2013) Bis(diphenylphosphino)methane ligated gold cluster cations: synthesis and gas-phase unimolecular reactivity. Int J Mass Spectrom 354–355:242–248. doi:10.1016/j.ijms.2013.05.034
114. Robinson PSD, Nguyen TL, Lioe H, O'Hair RAJ, Khairallah GN (2012) Synthesis and gas-phase uni- and bi-molecular reactivity of bisphosphine ligated gold clusters, $[Au_xLy]_n^+$. Int J Mass Spectrom 330–332:109–117
115. Pettibone JM, Hudgens JW (2011) Gold cluster formation with phosphine ligands: etching as a size-selective synthetic pathway for small clusters? ACS Nano 5(4):2989–3002
116. Bergeron DE, Coskuner O, Hudgens JW, Gonzalez CA (2008) Ligand exchange reactions in the formation of diphosphine-protected gold clusters. J Phys Chem C 112(33):12808–12814
117. Hudgens JW, Pettibone JM, Senftle TP, Bratton RN (2011) Reaction mechanism governing formation of 1,3-bis(diphenylphosphino)propane-protected gold nanoclusters. Inorg Chem 50(20):10178–10189. doi:10.1021/ic2018506
118. Evans DG, Mingos DMP (1982) Molecular orbital analysis of the bonding in low nuclearity gold and platinum tertiary phosphine complexes and the development of isolobal analogies for the M(PR3) fragment. J Organomet Chem 232(2):171–191. doi:10.1016/S0022-328X(00)87645-7
119. Hoffmann R (1982) Building bridges between inorganic and organic chemistry (nobel lecture). Angew Chem Int Ed Engl 21(10):711–724. doi:10.1002/anie.198207113
120. Pan QJ, Zhou X, Guo YR, Fu HG, Zhang HX (2009) Inorg Chem 48:2844–2854
121. King RB (1986) Inorg Chim Acta 116:109–117
122. Schwerdtfeger P, Hermann HL, Schmidbaur H (2003) Inorg Chem 42:1334–1342
123. Mingos DMP (1982) Philos Trans R Soc A 308:75–83
124. Lo CTF, Karan K, Davis BR (2007) Ind Eng Chem Res 46:5478–5484
125. Brown HC, Mead EJ, Shoaf CJ (1956) J Am Chem Soc 78:3616–3620
126. Majimel J, Bacinello D, Durand E, Vallee F, Treguer-Delapierre M (2008) Langmuir 24:4289–4294
127. Elian M, Chen MML, Mingos DMP, Hoffmann R (1976) Inorg Chem 15:1148–1155
128. Mingos DMP, Slee T, Lin ZY (1990) Chem Rev 90:383–402
129. Mingos DMP (1976) J Chem Soc Dalton Trans 1163–1169
130. Mingos DMP (1984) Polyhedron 3:1289–1297
131. Mingos DMP (1996) J Chem Soc Dalton Trans 561–566
132. Pyykkö P (2004) Angew Chem Int Ed 43:4412–4456
133. Shafai G, Hong S, Bertino MF, Rahman TS (2009) J Phys Chem C 113:12072–12078
134. Vollenbroek FA, Bour JJ, Vandervelden JWA (1980) Recl Trav Chim Pays Bas 99:137–141
135. Van Der Linden JGM, Paulissen MLH, Schmitz JEJ (1983) J Am Chem Soc 105:1903–1907
136. Cheetham GMT, Harding MM, Haggitt JL, Mingos DMP, Powell HR (1993) J. Chem Soc D Chem Commun 1000–1001
137. Laguna A, Laguna M, Gimeno MC, Jones PG (1992) Organometallics 11:2759–2760
138. Wang BS, Hou H, Yoder LM, Muckerman JT, Fockenberg C (2003) J Phys Chem A 107:11414–11426
139. Hong S, Shafai G, Bertino M, Rahman TS (2011) J Phys Chem C 115:14478–14487
140. Bergeron DE, Hudgens JW (2007) Ligand dissociation and core fission from diphosphine-protected gold clusters. J Phys Chem C 111(23):8195–8201
141. Pettibone JM, Hudgens JW (2012) Reaction network governing diphosphine-protected gold nanocluster formation from nascent cationic platforms. Phys Chem Chem Phys 14(12):4142–4154

142. Pettibone JM, Hudgens JW (2010) Synthetic approach for tunable, size-selective formation of monodisperse, diphosphine-protected gold nanoclusters. J Phys Chem Lett 1(17):2536–2540
143. Colton R, Harrison KL, Mah YA, Traeger JC (1995) Cationic phosphine complexes of gold (I): an electrospray mass spectrometric study. Inorg Chim Acta 231(1–2):65–71
144. Muetterties EL, Alegranti CW (1970) Solution structure of coinage metal–phosphine complexes. J Am Chem Soc 92(13):4114–4115. doi:10.1021/ja00716a052
145. Mays MJ, Vergnano PA (1979) Structure and bonding in gold(I) compounds. Part 4. A phosphorus-31 nuclear magnetic study of the structure of some gold(I) phosphine complexes in solution. J Chem Soc Dalton Trans 6:1112–1115. doi:10.1039/DT9790001112
146. Pettibone JM, Hudgens JW (2012) Predictive gold nanocluster formation controlled by metal-ligand complexes. Small 8(5):715–725
147. Pyykkö P (1997) Strong closed-shell interactions in inorganic chemistry. Chem Rev 97(3): 597–636. doi:10.1021/cr940396v
148. Pyykkö P, Mendizabal F (1998) Theory of d 10 – d 10 closed-shell attraction. III. Rings. Inorg Chem 37(12):3018–3025
149. Bertino MF, Sun ZM, Zhang R, Wang LS (2006) Facile syntheses of monodisperse ultrasmall Au clusters. J Phys Chem B 110(43):21416–21418
150. Golightly JS, Gao L, Castleman AW Jr, Bergeron DE, Hudgens JW, Magyar RJ, Gonzalez CA (2007) Impact of swapping ethyl for phenyl groups on diphosphine-protected undecagold. J Phys Chem C 111(40):14625–14627
151. Yanagimoto Y, Negishi Y, Fujihara H, Tsukuda T (2006) Chiroptical activity of BINAP-stabilized undecagold clusters. J Phys Chem B 110(24):11611–11614
152. Brown LO, Hutchison JE (1999) Controlled growth of gold nanoparticles during ligand exchange [14]. J Am Chem Soc 121(4):882–883
153. Schaaff TG, Whetten RL (1999) Controlled etching of Au:SR cluster compounds. J Phys Chem B 103(44):9394–9396
154. Maye MM, Zheng W, Leibowitz FL, Ly NK, Zhong CJ (2000) Heating-induced evolution of thiolate-encapsulated gold nanoparticles: a strategy for size and shape manipulations. Langmuir 16(2):490–497
155. Shichibu Y, Negishi Y, Tsunoyama H, Kanehara M, Teranishi T, Tsukuda T (2007) Extremely high stability of glutathionate-protected Au$_{25}$ clusters against core etching. Small 3(5):835–839
156. Toikkanen O, Ruiz V, Rönnholm G, Kalkkinen N, Liljeroth P, Quinn BM (2008) Synthesis and stability of monolayer-protected Au$_{38}$ clusters. J Am Chem Soc 130(33):11049–11055
157. Kanehara M, Sakurai JI, Sugimura H, Teranishi T (2009) Room-temperature size evolution of thiol-protected gold nanoparticles assisted by proton acids and halogen anions. J Am Chem Soc 131(5):1630–1631
158. Davis RE, Gottbrath JA (1962) Boron hydrides. V. Methanolysis of sodium borohydride. J Am Chem Soc 84(6):895–898
159. Kamei Y, Shichibu Y, Konishi K (2011) Generation of small gold clusters with unique geometries through cluster-to-cluster transformations: octanuclear clusters with edge-sharing gold tetrahedron motifs. Angew Chem Int Ed 50(32):7442–7445
160. Wen F, Englert U, Gutrath B, Simon U (2008) Crystal structure, electrochemical and optical properties of [Au$_9$(PPh$_3$)$_8$](NO$_3$)$_3$. Eur J Inorg Chem 1:106–111
161. Ticknor BW, Bandyopadhyay B, Duncan MA (2008) Photodissociation of noble metal-doped carbon clusters. J Phys Chem A 112(48):12355–12366
162. Cohen Y, Bernshtein V, Armon E, Bekkerman A, Kolodney E (2011) Formation and emission of gold and silver carbide cluster ions in a single C$_{60}$-surface impact at keV energies: Experiment and calculations. J Chem Phys 134(12):124701
163. Pyykko P, Patzschke M, Suurpere J (2003) Calculated structures of [Au=C=Au]$^{2+}$ and related systems. Chem Phys Lett 381(1,2):45–52. doi:10.1016/j.cplett.2003.09.045

164. Li D-Z, Li S-D (2011) A density functional investigation on $C_2Au_n^+$ (n = 1, 3, 5) and C_2Aun (n = 2, 4, 6): from gold terminals, gold bridges, to gold triangles. J Cluster Sci 22(3): 331–341. doi:10.1007/s10876-011-0383-5

165. Bolbach G, Main DE, Standing KG, Westmore JB (1995) Structures of gas phase oligomeric gold–oxygen–hydrogen negative ions formed by cesium ion bombardment of vapor-deposited gold surfaces. Inorg Chem 34(1):247–253. doi:10.1021/ic00105a040

166. Panyala NR, Pena-Mendez EM, Havel J (2012) Laser ablation synthesis of new gold phosphides using red phosphorus and nanogold as precursors. Laser desorption ionisation time-of-flight mass spectrometry. Rapid Commun Mass Spectrom 26(9):1100–1108. doi:10. 1002/rcm.6207

167. Svihlova K, Prokes L, Skacelova D, Pena-Mendez EM, Havel J (2013) Laser ablation synthesis of new gold tellurides using tellurium and nanogold as precursors. Laser desorption ionisation time-of-flight mass spectrometry. Rapid Commun Mass Spectrom 27(14): 1600–1606. doi:10.1002/rcm.6613

168. Maity P, Tsunoyama H, Yamauchi M, Xie S, Tsukuda T (2011) Organogold clusters protected by phenylacetylene. J Am Chem Soc 133(50):20123–20125

169. Schooss D, Weis P, Hampe O, Kappes MM (2010) Determining the size-dependent structure of ligand-free gold-cluster ions. Philos Trans Royal Soc A Math Phys Eng Sci 368(1915): 1211–1243

170. Becker S, Dietrich G, Hasse HU, Klisch N, Kluge HJ, Kreisle D, Krückeberg S, Lindinger M, Lützenkirchen K, Schweikhard L, Weidele H, Ziegler J (1994) Fragmentation pattern of gold clusters collided with xenon atoms. Comput Mater Sci 2(3–4):633–637. doi:10.1016/0927-0256(94)90099-X

171. Becker S, Dietrich G, Hasse HU, Klisch N, Kluge HJ, Kreisle D, Krueckeberg S, Lindinger M, Luetzenkirchen K et al (1994) Fragmentation of gold clusters stored in a Penning trap. Rapid Commun Mass Spectrom 8(5):401–402. doi:10.1002/rcm.1290080512

172. Schweikhard L, Beiersdorfer P, Bell W, Dietrich G, Krueckeberg S, Luetzenkirchen K, Obst B, Ziegler J (1996) Production and investigation of multiply charged metal clusters in a Penning trap. Hyperfine Interact 99(1–3):97–104. doi:10.1007/BF02274913

173. Schweikhard L, Dietrich G, Kruckeberg S, Lutzenkirchen K, Walther C, Ziegler J (1997) Collision induced dissociation of doubly charged stored metal cluster ions. Rapid Commun Mass Spectrom 11(14):1592–1595. doi:10.1002/(SICI)1097-0231(199709)11:14<1592:: AID-RCM996>3.0.CO;2-1

174. Ziegler J, Dietrich G, Kruckeberg S, Lutzenkirchen K, Schweikhard L, Walther C (1998) Dissociation pathways of doubly and triply charged gold clusters. Hyperfine Interact 115 (1–4):171–179. doi:10.1023/A:1012661008519

175. Weidele H, Vogel M, Herlert A, Kruckeberg S, Lievens P, Silverans RE, Walther C, Schweikhard L (1999) Decay pathways of stored metal-cluster anions after collisional activation. Eur Phys J D 9(1–4):173–177. doi:10.1007/s100530050421

176. Spasov VA, Shi Y, Ervin KM (2000) Time-resolved photodissociation and threshold collision-induced dissociation of anionic gold clusters. Chem Phys 262(1):75–91. doi:10. 1016/S0301-0104(00)00165-8

177. Ziegler J, Dietrich G, Kruckeberg S, Lutzenkirchen K, Schweikhard L, Walther C (2000) Multicollision-induced dissociation of multiply charged gold clusters, Au_n^{2+}, n = 7–35, and Au_n^{3+}, n = 19–35. Int J Mass Spectrom 202(1–3):47–54

178. Herlert A, Schweikhard L (2012) Electron binding energies from collisional activation of metal-cluster dianions. Appl Phys B Lasers Opt 107(4):1131–1143. doi:10.1007/s00340-011-4792-9

179. Walther C, Becker S, Dietrich G, Kluge HJ, Lindlinger M, Luetzenkirchen K, Schweikhard L, Ziegler J (1996) Photofragmentation of metal clusters stored in a penning trap. Z Phys D At Mol Clusters 38(1):51–58. doi:10.1007/s004600050063

180. Lindinger M, Dasgupta K, Dietrich G, Kruckeberg S, Kuznetsov S, Lutzenkirchen K, Schweikhard L, Walther C, Ziegler J (1997) Time resolved photofragmentation of Au_n^+

and Ag_n^+ clusters (n = 9, 21). Z Phys D At Mol Clusters 40(1–4):347–350. doi:10.1007/s004600050225

181. Vogel M, Hansen K, Herlert A, Schweikhard L (2001) Energy dependence of the decay pathways of optically excited small gold clusters. Appl Phys B Lasers Opt 73(4):411–416. doi:10.1007/s003400100719

182. Vogel M, Hansen K, Herlert A, Schweikhard L (2001) Decay pathways of small gold clusters. The competition between monomer and dimer evaporation. Eur Phys J D 16(1–3):73–76. doi:10.1007/s100530170063

183. Vogel M, Hansen K, Herlert A, Schweikhard L (2001) Determination of dissociation energies by use of energy-dependent decay pathway branching ratios. Chem Phys Lett 346(1,2): 117–122. doi:10.1016/S0009-2614(01)00935-6

184. Vogel M, Hansen K, Herlert A, Schweikhard L (2001) Model-free determination of dissociation energies of polyatomic systems. Phys Rev Lett 87(1):013401/013401–013401/013404. doi:10.1103/PhysRevLett.87.013401

185. Herlert A, Schweikhard L, Vogel M (2002) Photoinduced dissociation of anionic and electron detachment of dianionic gold clusters by use of a laser pointer. Int J Mass Spectrom 213(2/3): 157–161. doi:10.1016/S1387-3806(01)00529-2

186. Vogel M, Hansen K, Herlert A, Schweikhard L (2002) Dimer dissociation energies of small odd-size clusters Au_n^+. Eur Phys J D 21(2):163–166. doi:10.1140/epjd/e2002-00190-3

187. Vogel M, Hansen K, Herlert A, Schweikhard L (2002) Multisequential photofragmentation of size-selected gold cluster ions. Phys Rev A At Mol Opt Phys 66(3):033201/033201–033201/033209. doi:10.1103/PhysRevA.66.033201

188. Herlert A, Schweikhard L (2003) Production of dianionic and trianionic noble metal clusters in a Penning trap. Int J Mass Spectrom 229(1–2):19–25. doi:10.1016/S1387-3806(03)00251-3

189. Vogel M, Herlert A, Schweikhard L (2003) Photodissociation of small group-11 metal cluster ions: fragmentation pathways and photoabsorption cross sections. J Am Soc Mass Spectrom 14(6):614–621. doi:10.1016/S1044-0305(03)00203-4

190. Vogel M, Hansen K, Schweikhard L (2004) Signature of cluster isomers in time-resolved photodissociation experiments. Int J Mass Spectrom 233(1–3):117–123. doi:10.1016/j.ijms. 2003.12.027

191. Schweikhard L, Hansen K, Herlert A, Herraiz LMD, Vogel M (2005) Photodissociation of stored metal clusters. Eur Phys J D 36(2):179–185. doi:10.1140/epjd/e2005-00264-8

192. Hansen K, Herlert A, Schweikhard L, Vogel M (2006) Dissociation energies of gold clusters AuN^+, N=7–7. Phys Rev A At Mol Opt Phys 73(6):063202/063201–063202/063214. doi:10.1103/PhysRevA.73.063202

193. Herlert A, Schweikhard L (2006) First observation of delayed electron emission from dianionic metal clusters. Int J Mass Spectrom 252(2):151–156. doi:10.1016/j.ijms.2006.01.051

194. Herlert A, Schweikhard L (2006) Delayed neutral-atom evaporation of photoexcited anionic gold clusters. Int J Mass Spectrom 249(250):215–221. doi:10.1016/j.ijms.2005.12.027

195. Herlert A, Schweikhard L (2012) Two-electron emission after photoexcitation of metal-cluster dianions. New J Phys 14(May):055015/055011–055015/055024. doi:10.1088/1367-2630/14/5/055015

196. Herlert A, Kruckeberg S, Schweikhard L, Vogel M, Walther C (1999) First observation of doubly charged negative gold cluster ions. Phys Scr T T80B (IX International Conference on the Physics of Highly Charged Ions, 1998):200–202. doi:10.1238/Physica.Topical.080a00200

197. Schweikhard L, Herlert A, Kruckeberg S, Vogel M, Walther C (1999) Electronic effects in the production of small dianionic gold clusters by electron attachment on to stored Au_n^-, n = 12–28. Philos Mag B 79(9):1343–1352

198. Herlert A, Kruckeberg S, Schweikhard L, Vogel M, Walther C (2000) Electron impact ionization/dissociation of size selected gold cluster cations. J Electron Spectrosc Relat Phenom 106(2–3):179–186. doi:10.1016/S0368-2048(99)00075-4

199. Yannouleas C, Landman U, Herlert A, Schweikhard L (2001) Trianionic gold clusters. Eur Phys J D 16(1–3):81–85. doi:10.1007/s100530170065
200. Yannouleas C, Landman U, Herlert A, Schweikhard L (2001) Multiply charged metal cluster anions. Phys Rev Lett 86(14):2996–2999. doi:10.1103/PhysRevLett.86.2996
201. Herlert A, Jertz R, Alonso OJ, Gonzalez MAJ, Schweikhard L (2002) The influence of the trapping potential on the attachment of a second electron to stored metal cluster and fullerene anions. Int J Mass Spectrom 218(3):217–225. doi:10.1016/S1387-3806(02)00723-6
202. Zhang HF, Stender M, Zhang R, Wang C, Li J, Wang LS (2004) Toward the solution synthesis of the tetrahedral Au_{20} cluster. J Phys Chem B 108(33):12259–12263
203. Li J, Li X, Zhai HJ, Wang LS (2003) Au_{20}: a tetrahedral cluster. Science 299(5608):864–867
204. Pease LF III, Elliott JT, Tsai D-H, Zachariah MR, Tarlov MJ (2008) Determination of protein aggregation with differential mobility analysis: application to IgG antibody. Biotechnol Bioeng 101(6):1214–1222. doi:10.1002/bit.22017
205. Eiceman GA, Karpas Z (2010) Ion mobility spectrometry. 2nd ed. Taylor and Francis, Boca, Florida
206. Lapthorn C, Pullen F, Chowdhry BZ (2013) Ion mobility spectrometry-mass spectrometry (IMS-MS) of small molecules: separating and assigning structures to ions. Mass Spectrom Rev 32(1):43–71. doi:10.1002/mas.21349
207. Weis P (2005) Structure determination of gaseous metal and semi-metal cluster ions by ion mobility spectrometry. Int J Mass Spectrom 245(1–3):1–13. doi:10.1016/j.ijms.2005.06.005
208. Gilb S, Weis P, Furche F, Alhrichs R, Kappes MM (2002) Structures of small gold cluster cations (Au_n^+, $n<14$): ion mobility measurements versus density functional calculations. J Chem Phys 116(10):4094–4101
209. Furche F, Ahlrichs R, Weis P, Jacob C, Gilb S, Bierweiler T, Kappes MM (2002) The structures of small gold cluster anions as determined by a combination of ion mobility measurements and density functional calculations. J Chem Phys 117(15):6982–6990
210. Lenggoro IW, Xia B, Okuyama K, De la Mora JF (2002) Sizing of colloidal nanoparticles by electrospray and differential mobility analyzer methods. Langmuir 18(12):4584–4591
211. Tsai DH, Pease LF 3rd, Zangmeister RA, Tarlov MJ, Zachariah MR (2009) Aggregation kinetics of colloidal particles measured by gas-phase differential mobility analysis. Langmuir 25(1):140–146. doi:10.1021/la703164j
212. Li M, You R, Mulholland GW, Zachariah MR (2013) Evaluating the mobility of nanorods in electric fields. Aerosol Sci Technol 47(10):1101–1107. doi:10.1080/02786826.2013.819565
213. Tsai D-H, Del RFW, Keene AM, Tyner KM, MacCuspie RI, Cho TJ, Zachariah MR, Hackley VA (2011) Adsorption and conformation of serum albumin protein on gold nanoparticles investigated using dimensional measurements and in situ spectroscopic methods. Langmuir 27(6):2464–2477. doi:10.1021/la104124d
214. Elzey S, Tsai DH, Yu LL, Winchester MR, Kelley ME, Hackley VA (2013) Real-time size discrimination and elemental analysis of gold nanoparticles using ES-DMA coupled to ICP-MS. Anal Bioanal Chem 405(7):2279–2288. doi:10.1007/s00216-012-6617-z
215. Tsai D-H, Cho TJ, Elzey SR, Gigault JC, Hackley VA (2013) Quantitative analysis of dendron-conjugated cisplatin-complexed gold nanoparticles using scanning particle mobility mass spectrometry. Nanoscale 5(12):5390–5395. doi:10.1039/c3nr00543g
216. Angel LA, Majors LT, Dharmaratne AC, Dass A (2010) Ion mobility mass spectrometry of $Au_{25}(SCH_2CH_2Ph)_{18}$ nanoclusters. ACS Nano 4(8):4691–4700
217. Harkness KM, Fenn LS, Cliffel DE, McLean JA (2010) Surface fragmentation of complexes from thiolate protected gold nanoparticles by ion mobility-mass spectrometry. Anal Chem (Washington, DC, U S) 82(7):3061–3066. doi:10.1021/ac100251d
218. Jadzinsky PD, Calero G, Ackerson CJ, Bushnell DA, Kornberg RD (2007) Structure of a thiol monolayer-protected gold nanoparticle at 1.1 Å resolution. Science (Washington, DC, U S) 318(5849):430–433. doi:10.1126/science.1148624
219. Harkness KM, Balinski A, McLean JA, Cliffel DE (2011) Nanoscale phase segregation of mixed thiolates on gold nanoparticles. Angew Chem Int Ed 50(45):10554–10559

220. Asmis KR, Fielicke A, von Helden G, Meijer G (2007) Chapter 8: Vibrational spectroscopy of gas-phase clusters and complexes. Chem Phys of Solid Surfaces, pp 327–375
221. Lapoutre VJF, Redlich B, van der Meer AFG, Oomens J, Bakker JM, Sweeney A, Mookherjee A, Armentrout PB (2013) Structures of the dehydrogenation products of methane activation by 5d transition metal cations. J Phys Chem A 117(20):4115–4126. doi:10.1021/jp400305k
222. Asmis KR, Wende T, Bruemmer M, Gause O, Santambrogio G, Stanca-Kaposta EC, Doebler J, Niedziela A, Sauer J (2012) Structural variability in transition metal oxide clusters: gas phase vibrational spectroscopy of V3O6-8+. Phys Chem Chem Phys 14(26):9377–9388. doi:10.1039/c2cp40245a
223. Asmis KR (2012) Structure characterization of metal oxide clusters by vibrational spectroscopy: possibilities and prospects. Phys Chem Chem Phys 14(26):9270–9281. doi:10.1039/c2cp40762k
224. Ghiringhelli LM, Gruene P, Lyon JT, Rayner DM, Meijer G, Fielicke A, Scheffler M (2013) Not so loosely bound rare gas atoms: finite-temperature vibrational fingerprints of neutral gold-cluster complexes. New J Phys 15(8):083003. doi:10.1088/1367-2630/15/8/083003
225. Gruene P, Rayner DM, Redlich B, van der Meer AFG, Lyon JT, Meijer G, Fielicke A (2008) Structures of neutral Au7, Au19, and Au20 clusters in the gas phase. Science (Washington, DC, U S) 321(5889):674–676. doi:10.1126/science.1161166
226. Woodham AP, Meijer G, Fielicke A (2013) Charge separation promoted activation of molecular oxygen by neutral gold clusters. J Am Chem Soc 135(5):1727–1730. doi:10.1021/ja312223t
227. Fielicke A, Von HG, Meijer G, Pedersen DB, Simard B, Rayner DM (2005) Gold cluster carbonyls: saturated adsorption of CO on gold cluster cations, vibrational spectroscopy, and implications for their structures. J Am Chem Soc 127(23):8416–8423. doi:10.1021/ja0509230
228. Donald WA, O'Hair RAJ (2012) Shapeshifting: ligation by 1,4-cyclohexadiene induces a structural change in Ag5+. Dalton Trans 41(11):3185–3193. doi:10.1039/c2dt11876a
229. Manard MJ, Kemper PR, Bowers MT (2005) Probing the structure of gas-phase metallic clusters via ligation energetics: sequential addition of C2H4 to Agm+ (m = 3–7). J Am Chem Soc 127(28):9994–9995. doi:10.1021/ja052251j
230. Rousseau R, Dietrich G, Kruckeberg S, Lutzenkirchen K, Marx D, Schweikhard L, Walther C (1998) Probing cluster structures with sensor molecules: methanol adsorbed onto gold clusters. Chem Phys Lett 295(1,2):41–46. doi:10.1016/S0009-2614(98)00926-9
231. Fielicke A, von Helden G, Meijer G, Simard B, Rayner DM (2005) Direct observation of size dependent activation of NO on gold clusters. Phys Chem Chem Phys 7(23):3906–3909. doi:10.1039/b511710k
232. Woodham AP, Meijer G, Fielicke A (2012) Activation of molecular oxygen by anionic gold clusters. Angew Chem Int Ed 51(18):4444–4447
233. Collings BA, Athanassenas K, Lacombe D, Rayner DM, Hackett PA (1994) Optical absorption spectra of Au7, Au9, Au11, and Au13, and their cations: gold clusters with 6, 7, 8, 9, 10, 11, 12, and 13 s-electrons. J Chem Phys 101(5):3506–3513. doi:10.1063.1.4675.5
234. Hamouda R, Bellina B, Bertorelle F, Compagnon I, Antoine R, Broyer M, Rayane D, Dugourd P (2010) Electron emission of gas-phase [Au25(SG)18-6H]7 – gold cluster and its action spectroscopy. J Phys Chem Lett 1(21):3189–3194. doi:10.1021/jz101287m
235. Schooss D, Weis P, Hampe O, Kappes MM (2010) Determining the size-dependent structure of ligand-free gold-cluster ions. Philos Trans R Soc A 368(1915):1211–1243. doi:10.1098/rsta.2009.0269
236. Xing X, Yoon B, Landman U, Parks JH (2006) Structural evolution of Au nanoclusters: from planar to cage to tubular motifs. Phys Rev B Condens Matter Mater Phys 74(16):165423/165421–165423/165426. doi:10.1103/PhysRevB.74.165423
237. Johansson MP, Lechtken A, Schooss D, Kappes MM, Furche F (2008) 2D-3D transition of gold cluster anions resolved. Phys Rev A At Mol Opt Phys 77(5):058202

238. Taylor KJ, Pettiette-Hall CL, Cheshnovsky O, Smalley RE (1992) Ultraviolet photoelectron spectra of coinage metal clusters. J Chem Phys 96(4):3319–3329

239. Wang LM, Wang LS (2012) Probing the electronic properties and structural evolution of anionic gold clusters in the gas phase. Nanoscale 4(14):4038–4053

240. Häkkinen H, Yoon B, Landman U, Li X, Zhai HJ, Wang LS (2003) On the electronic and atomic structures of small AuN – (N = 4–14) clusters: a photoelectron spectroscopy and density-functional study. J Phys Chem A 107(32):6168–6175

241. Huang W, Wang LS (2009) Au10: isomerism and structure-dependent O_2 reactivity. Phys Chem Chem Phys 11(15):2663–2667. doi:10.1039/b823159a

242. Stolcic D, Fischer M, Gantefoer G, Kim YD, Sun Q, Jena P (2003) Direct observation of key reaction intermediates on gold clusters. J Am Chem Soc 125(10):2848–2849. doi:10.1021/ja0293406

243. Huang W, Zhai HJ, Wang LS (2010) Probing the interactions of O_2 with small gold cluster anions (Au_n^-, n = 1–7): chemisorption vs physisorption. J Am Chem Soc 132(12):4344–4351

244. Pal R, Wang L-M, Huang W, Wang L-S, Zeng XC (2011) Structure evolution of gold cluster anions between the planar and cage structures by isoelectronic substitution: Au_n- (n = 13–15) and MAu_n- (n = 12–14; M=Ag, Cu). J Chem Phys 134(5):054306. 10.1063.1.35334.3

245. Bulusu S, Li X, Wang LS, Zeng XC (2006) Evidence of hollow golden cages. Proc Natl Acad Sci U S A 103(22):8326–8330

246. Zhu M, Qian H, Jin R (2009) Thiolate-protected Au_{20} clusters with a large energy gap of 2.1 eV. J Am Chem Soc 131(21):7220–7221. doi:10.1021/ja902208h

247. Bulusu S, Li X, Wang LS, Zeng XC (2007) Structural transitions from pyramidal to fused planar to tubular to core/shell compact in gold clusters: Au_n^- (n = 21–25). J Phys Chem C 111(11):4190–4198

248. Ji M, Gu X, Li X, Gong X, Li J, Wang L-S (2005) Experimental and theoretical investigation of the electronic and geometrical structures of the Au_{32} cluster. Angew Chem Int Ed 44(43):7119–7123. doi:10.1002/anie.200502795

249. Gu X, Bulusu S, Li X, Zeng XC, Li J, Gong XG, Wang LS (2007) Au_{34}: a fluxional core-shell cluster. J Phys Chem C 111(23):8228–8232

250. Kebarle P (1992) Ion molecule equilibria, how and why. J Am Soc Mass Spectrom 3(1):1–9. doi:10.1016/1044-0305(92)85012-9

251. Kebarle P (2000) Gas phase ion thermochemistry based on ion-equilibria from the ionosphere to the reactive centers of enzymes. Int J Mass Spectrom 200(1–3):313–330. doi:10.1016/S1387-3806(00)00326-2

252. Peschke M, Blades AT, Kebarle P (2001) Determination of sequential metal ion-ligand binding energies by gas phase equilibria and theoretical calculations: application of results to biochemical processes. Adv Met Semicond Clusters 5 (Metal Ion Solvation and Metal-Ligand Interactions):77–119

253. Armentrout PB (2003) Guided ion beam studies of transition metal-ligand thermochemistry. Int J Mass Spectrom 227(3):289–302

254. Armentrout PB (2003) Threshold collision-induced dissociations for the determination of accurate gas-phase binding energies and reaction barriers. Top Curr Chem (Modern Mass Spectrometry) 225:233–262. doi:10.1007/b10468

255. Rodgers MT (2004) Armentrout PB gas phase coordination chemistry. Elsevier, Oxford, UK, pp 141–158, 141 plate. doi:10.1016/B0-08-043748-6/01119-1

256. Armentrout PB (2003) The thermochemistry of adsorbates on transition metal cluster ions: relationship to bulk-phase properties. Eur J Mass Spectrom 9(6):531–538. doi:10.1255/ejms.585

257. Armentrout PB (2010) Reactivity and thermochemistry of transition metal cluster cations. Elsevier, Amsterdam, Netherland, pp 269–297. doi:10.1016/B978-0-444-53440-8.00006-9

258. Li F, Hinton CS, Citir M, Liu F, Armentrout PB (2011) Guided ion beam and theoretical study of the reactions of Au^+ with H_2, D_2, and HD. J Chem Phys 134(2):024310

259. Li FX, Gorham K, Armentrout PB (2010) Oxidation of atomic gold ions: thermochemistry for the activation of O_2 and N_2O by Au^+(1S 0and 3D). J Phys Chem A 114(42):11043–11052
260. Li FX, Armentrout PB (2006) Activation of methane by gold cations: guided ion beam and theoretical studies. J Chem Phys 125(13):188114
261. Schwarz H (2003) Relativistic effects in gas-phase ion chemistry: an experimentalist's view. Angew Chem Int Ed 42(37):4442–4454
262. Schroeder D, Schwarz H, Hrusak J, Pyykkoe P (1998) Cationic gold(I) complexes of xenon and of ligands containing the donor atoms oxygen, nitrogen, phosphorus, and sulfur. Inorg Chem 37(4):624–632. doi:10.1021/IC970986M
263. Neumaier M, Weigend F, Hampe O, Kappes MM (2005) Binding energies of CO on gold cluster cations Au_n^+ (n = 1–65): a radiative association kinetics study. J Chem Phys 122(10): 104702
264. Neumaier M, Weigend F, Hampe O, Kappes MM (2006) Reactions of mixed silver-gold cluster cations AgmAu n + (m + n = 4,5,6) with CO: Radiative association kinetics and density functional theory computations. J Chem Phys 125(10):104308
265. Lang SM, Bernhardt TM, Barnett RN, Landman U (2010) Size-dependent binding energies of methane to small gold clusters. ChemPhysChem 11(7):1570–1577
266. Popolan DM, Nössler M, Mitrić R, Bernhardt TM, Bonačić-Koutecký V (2011) Tuning cluster reactivity by charge state and composition: experimental and theoretical investigation of CO binding energies to Ag nAu m +/- (n + m = 3). J Phys Chem A 115(6):951–959
267. Popolan DM, Nößler M, Mitrić R, Bernhardt TM, Bonaić-Koutecký V (2010) Composition dependent adsorption of multiple CO molecules on binary silver-gold clusters $Ag_nAu_m^+$ (n + m = 5): theory and experiment. Phys Chem Chem Phys 12(28):7865–7873
268. Bernhardt TM, Hagen J, Lang SM, Popolan DM, Socaciu-Siebert LD, Wöste L (2009) Binding energies of O_2 and CO to small gold, silver, and binary silver–gold cluster anions from temperature dependent reaction kinetics measurements. J Phys Chem A 113(12):2724–2733
269. Cox DM, Brickman R, Creegan K, Kaldor A (1991) Gold clusters: reactions and deuterium uptake. Z Phys D At Mol Clusters 19(1–4):353–355. doi:10.1007/BF01448327
270. Lian L, Hackett PA, Rayner DM (1993) Relativistic effects in reactions of the coinage metal dimers in the gas phase. J Chem Phys 99(4):2583–2590. doi:10.1063.1.4652.1
271. Lang SM, Bernhardt TM, Barnett RN, Yoon B, Landman U (2009) Hydrogen-promoted oxygen activation by free gold cluster cations. J Am Chem Soc 131(25):8939–8951. doi:10.1021/ja9022368
272. Dietrich G, Luetzenkirchen K, Becker S, Hasse HU, Kluge HJ, Lindinger M, Schweikhard L, Ziegler J, Kuznetsov S (1994) Au_n^+-induced decomposition of N_2O. Ber Bunsen Ges 98(12): 1608–1612
273. Lee TH, Ervin KM (1994) Reactions of copper group cluster anions with oxygen and carbon monoxide. J Phys Chem 98(40):10023–10031
274. Cox DM, Brickman RO, Creegan K, Kaldor A (1991) Studies of the chemical properties of size selected metal clusters: kinetics and saturation. Mater Res Soc Symp Proc (Clusters Cluster-Assem Mater) 206:43–48
275. Salisbury BE, Wallace WT, Whetten RL (2000) Low-temperature activation of molecular oxygen by gold clusters: a stoichiometric process correlated to electron affinity. Chem Phys 262(1):131–141. doi:10.1016/S0301-0104(00)00272-X
276. Wallace WT, Whetten RL (2002) Coadsorption of CO and O_2 on selected gold clusters: evidence for efficient room-temperature CO_2 generation. J Am Chem Soc 124(25): 7499–7505. doi:10.1021/ja0175439
277. Su T, Chesnavich WJ (1982) Parametrization of the ion-polar molecule collision rate constant by trajectory calculations. J Chem Phys 76(10):5183–5185. doi:10.1063.1.4428.8
278. Wallace WT, Wyrwas RB, Leavitt AJ, Whetten RL (2005) Adsorption of carbon monoxide on smaller gold-cluster anions in an atmospheric-pressure flow-reactor: temperature and humidity dependence. Phys Chem Chem Phys 7(5):930–937. doi:10.1039/b500398a

279. Wallace WT, Whetten RL (2001) Metastability of gold-carbonyl cluster complexes, AuN(CO)M. Eur Phys J D 16(1–3):123–126. doi:10.1007/s100530170075
280. Wallace WT, Whetten RL (2000) Carbon monoxide adsorption on selected gold clusters. Highly size-dependent activity and saturation compositions. J Phys Chem B 104(47): 10964–10968. doi:10.1021/jp002889b
281. Lang SM, Bernhardt TM (2009) Cooperative and competitive coadsorption of H_2, O_2, and N_2 on Au_x^+(x = 3,5). J Chem Phys 131(2):024310/024311–024310/024318. doi:10.1063.1.31683.6
282. Koszinowski K, Schroeder D, Schwarz H (2004) C–N coupling of methane and ammonia by bimetallic platinum–gold cluster cations. Organometallics 23(5):1132–1139. doi:10.1021/om0306675
283. Höckendorf RF, Cao Y, Beyer MK (2010) Gas-phase ion chemistry of small gold cluster anions. Organometallics 29(13):3001–3006. doi:10.1021/om100228y
284. Lang SM, Bernhardt TM (2009) Reactions of small gold cluster cations with propylene, methane, and hydrogen: permissive and competitive coadsorption effects. Eur Phys J D 52 (1–3):139–142. doi:10.1140/epjd/e2009-00070-4
285. Lang SM, Bernhardt TM (2009) Reactions of free gold cluster cations with H_2O, CH_3Cl, and mixtures thereof. Int J Mass Spectrom 286(1):39–41. doi:10.1016/j.ijms.2009.06.005
286. Popolan DM, Bernhardt TM (2011) Interaction of gold and silver cluster cations with CH3Br: thermal and photoinduced reaction pathways. Eur Phys J D 63(2):251–254. doi:10.1140/epjd/e2010-10588-9
287. Popolan DM, Bernhardt TM (2011) Communication: CO oxidation by silver and gold cluster cations: identification of different active oxygen species. J Chem Phys 134(9):091102/091101–091102/091103
288. Haekkinen H, Landman U (2001) Gas-phase catalytic oxidation of CO by Au_2. J Am Chem Soc 123(39):9704–9705. doi:10.1021/ja0165180
289. Lang SM, Bernhardt TM (2011) Methane activation and partial oxidation on free gold and palladium clusters: mechanistic insights into cooperative and highly selective cluster catalysis. Faraday Discuss 152(Gold):337–351. doi:10.1039/c1fd00025j
290. Popolan DM, Bernhardt TM (2009) Formation and femtosecond photodissociation of Ag_n^+ and Au_n^+ complexes with benzene and carbon monoxide. Chem Phys Lett 470(1–3):44–48
291. Robinson PSD, Khairallah GN, da Silva G, Lioe H, O'Hair RAJ (2012) Gold Mediated C–I Bond Activation of Iodobenzene Angew Chem Int Ed 51:3812–3817
292. Gōmez-Suárez A, Nolan SP (2012) Dinuclear gold catalysis: are two gold centers better than one? Angew Chem Int Ed 51(33):8156–8159
293. Raubenheimer HG, Schmidbaur H (2012) Gold chemistry guided by the isolobality concept. Organometallics 31(7):2507–2522
294. Hoffmann R (1982) Building bridges between inorganic and organic chemistry. Prix Nobel 173–205
295. Khairallah GN, O'Hair RAJ, Bruce MI (2006) Gas-phase synthesis and reactivity of binuclear gold hydride cations, $(R_3PAu)2H^+$ (R=Me and Ph). Dalton Trans 30:3699–3707
296. Robilotto TJ, Bacsa J, Gray TG, Sadighi JP (2012) Synthesis of a trigold monocation: an isolobal analogue of $[H_3]^+$. Angew Chem Int Ed 51(48):12077–12080
297. Thomson JJ (1913) Bakerian lecture: rays of positive electricity. Proc Royal Soc Lond Ser A 89(607):1–20. doi:10.1098/rspa.1913.0057
298. Herbst E, Moller S, Oka T, Watson JKG (eds) (2000) Astronomy, physics and chemistry of H_3^+. In: Papers of a Discussion Meeting held 9–10 February 2000. Philos Trans R Soc London Ser A 358(1774)]. vol Copyright (C) 2013 American Chemical Society (ACS). All Rights Reserved Royal Society
299. Hashmi ASK (2010) Homogeneous gold catalysis beyond assumptions and proposals-characterized intermediates. Angew Chem Int Ed 49(31):5232–5241
300. Khairallah GN, O'Hair RAJ (2005) Gas phase synthesis and reactivity of Ag_n^+ and $Ag_{n-1}H^+$ cluster cations. Dalton Trans 16:2702–2712

301. Blagojevic V, Samad SN, Banu L, Thomas MC, Blanksby SJ, Bohme DK (2011) Mass spectrometric study of the dissociation of Group XI metal complexes with fatty acids and glycerolipids: Ag_2H^+ and Cu_2H^+ ion formation in the presence of a double bond. Int J Mass Spectrom 299(2–3):125–130

302. Khairallah GN, O'Hair RAJ (2005) Gas-phase synthesis of $[Ag_4H]^+$ and its mediation of the C–C coupling of allyl bromide. Angew Chem Int Ed 44(5):728–731. doi:10.1002/anie. 200461328

303. Wallace WT, Wyrwas RB, Whetten RL, Mitric R, Bonacic-Koutecky V (2003) Oxygen adsorption on hydrated gold cluster anions: experiment and theory. J Am Chem Soc 125(27): 8408–8414. doi:10.1021/ja034905z

304. Johnson GE, Mitric R, Bonacic-Koutecky V, Castleman AW (2009) Clusters as model systems for investigating nanoscale oxidation catalysis. Chem Phys Lett 475(1–3):1–9. doi:10.1016/j.cplett.2009.04.003

305. Kimble ML, Castleman AW Jr, Mitric R, Buergel C, Bonacic-Koutecky V (2004) Reactivity of atomic gold anions toward oxygen and the oxidation of CO: experiment and theory. J Am Chem Soc 126(8):2526–2535. doi:10.1021/ja030544b

306. Buergel C, Reilly NM, Johnson GE, Mitric R, Kimble ML, Castleman AW Jr, Bonacic-Koutecky V (2008) Influence of charge state on the mechanism of CO oxidation on gold clusters. J Am Chem Soc 130(5):1694–1698. doi:10.1021/ja0768542

307. Kimble ML, Moore NA, Johnson GE, Castleman AW Jr, Burgel C, Mitric R, Bonacic-Koutecky V (2006) Joint experimental and theoretical investigations of the reactivity of $Au_2O_n^-$ and $Au_3O_n^-$ (n = 1–5) with carbon monoxide. J Chem Phys 125(20):204311/ 204311–204311/204314. doi:10.1063.1.23710.2

308. Kimble ML, Castleman AW, Buergel C, Bonacic-Koutecky V (2006) Interactions of CO with $Au_nO_m^-$ (n ≥ 4). Int J Mass Spectrom 254(3):163–167. doi:10.1016/j.ijms.2006.05.015

309. Kimble ML, Moore NA, Castleman AW Jr, Buergel C, Mitric R, Bonacic-Koutecky V (2007) Reactivity of anionic gold oxide clusters towards CO: experiment and theory. Eur Phys J D 43 (1–3):205–208. doi:10.1140/epjd/e2007-00119-4

310. Kimble ML, Castleman AW (2004) Gas-phase studies of $Au_nO_m^+$ interacting with carbon monoxide. Int J Mass Spectrom 233(1–3):99–101. doi:10.1016/j.ijms.2003.11.018

311. Johnson GE, Reilly NM, Tyo EC, Castleman AW (2008) Gas-phase reactivity of gold oxide cluster cations with CO. J Phys Chem C 112(26):9730–9736. doi:10.1021/jp801514d

312. Tyo EC, Castleman AW Jr, Schröder D, Milko P, Roithova J, Ortega JM, Cinellu MA, Cocco F, Minghetti G (2009) Large effect of a small substitution: competition of dehydration with charge retention and coulomb explosion in gaseous $[(bipyR)Au(\mu-O)\,2Au(bipyR)]^{2+}$ dications. J Am Chem Soc 131(36):13009–13019

313. Butschke B, Schwarz H (2012) "Rollover" cyclometalation – early history, recent developments, mechanistic insights and application aspects. Chem Sci 3(2):308–326. doi:10.1039/ c1sc00651g

314. Hansmann MM, Rudolph M, Rominger F, Hashmi ASK (2013) Mechanistic switch in dual gold catalysis of diynes: $C(sp^3)$-H activation through bifurcation-vinylidene versus carbene pathways. Angew Chem Int Ed 52(9):2593–2598. doi:10.1002/anie.201208777

315. Simonneau A, Jaroschik F, Lesage D, Karanik M, Guillot R, Malacria M, Tabet J-C, Goddard J-P, Fensterbank L, Gandon V, Gimbert Y (2011) Tracking gold acetylides in gold(i)-catalyzed cycloisomerization reactions of enynes. Chem Sci 2(12):2417–2422. doi:10.1039/C1SC00478F

316. Roithová J, Janková Š, Jašíková L, Váňa J, Hybelbauerová S (2012) Gold–gold cooperation in the addition of methanol to alkynes. Angew Chem Int Ed 51(33):8378–8382. doi:10.1002/ anie.201204003

317. Koszinowski K, Schroeder D, Schwarz H (2003) Probing cooperative effects in bimetallic clusters: indications of C–N coupling of CH_4 and NH_3 mediated by the cluster ion $PtAu^+$ in the gas phase. J Am Chem Soc 125(13):3676–3677. doi:10.1021/ja029791q

318. Koszinowski K, Schroder D, Schwarz H (2003) Additivity effects in the reactivities of bimetallic cluster ions $Pt_mAu_n^+$. ChemPhysChem 4(11):1233–1237

319. Lang SM, Frank A, Bernhardt TM (2013) Composition and size dependent methane dehydrogenation on binary gold-palladium clusters. Int J Mass Spectrom 354–355:365–371. doi:10.1016/j.ijms.2013.07.014

320. Fleischer I, Popolan DM, Krstić M, Bonačic-Koutecky V, Bernhardt TM (2013) Composition dependent selectivity in the coadsorption of H_2O and CO on pure and binary silver-gold clusters. Chem Phys Lett 565:74–79

321. Kappes MM, Staley RH (1981) Gas-phase oxidation catalysis by transition-metal cations. J Am Chem Soc 103(5):1286–1287. doi:10.1021/ja00395a080

322. Schlangen M, Schwarz H (2012) Effects of ligands, cluster size, and charge state in gas-phase catalysis: a happy marriage of experimental and computational studies. Catal Lett 142(11): 1265–1278

323. Waters T, O'Hair RAJ (2005) Contribution to "The Encyclopedia of Mass Spectrometry". In: Gross ML, Caprioli R (eds in Chief) Fundamentals of and applications to organic (and organometallic) compounds, vol 4. Metal ion complexes: formation and reactivity, topic: "Organometallic Catalysis in the Gas Phase", Chap 6. (Nibbering, N.M.M. Ed. ISBN: 0-08-043846-6), Elsevier, Amsterdam

324. Hagen J, Socaciu LD, Elijazyfer M, Heiz U, Bernhardt TM, Woeste L (2002) Coadsorption of CO and O_2 on small free gold cluster anions at cryogenic temperatures: model complexes for catalytic CO oxidation. Phys Chem Chem Phys 4(10):1707–1709. doi:10.1039/b201236g

325. Socaciu LD, Hagen J, Bernhardt TM, Woste L, Heiz U, Hakkinen H, Landman U (2003) Catalytic CO oxidation by free Au_2: experiment and theory. J Am Chem Soc 125(34): 10437–10445

326. Lang SM, Bernhardt TM, Barnett RN, Landman U (2010) Methane activation and catalytic ethylene formation on free Au_2^+. Angew Chem Int Ed 49(5):980–983. doi:10.1002/anie.200905643

327. Lang SM, Bernhardt TM, Barnett RN, Landman U (2011) Temperature-tunable selective methane catalysis on Au_2^+: from cryogenic partial oxidation yielding formaldehyde to cold ethylene production. J Phys Chem C 115(14):6788–6795. doi:10.1021/jp200160r

328. Rohlfing EA, Cox DM, Kaldor A (1984) Production and characterization of supersonic carbon cluster beams. J Chem Phys 81(7):3322–3330. doi:10.1063.1.4479.4

329. Kroto HW, Heath JR, O'Brien SC, Curl RF, Smalley RE (1985) C_{60}: buckminsterfullerene. Nature (London) 318(6042):162–163. doi:10.1038.3181.2.0

330. Kraetschmer W, Lamb LD, Fostiropoulos K, Huffman DR (1990) Solid C_{60}: a new form of carbon. Nature (London) 347(6291):354–358. doi:10.1038.3473.4.0

331. Bates PA, Waters JM (1985) The crystal and molecular structure of dichloro-1,2-bis (diphenylphosphino)ethanedigold(I). Inorg Chim Acta 98(2):125–129

332. Brandys MC, Jennings MC, Puddephatt RJ (2000) Luminescent gold(i) macrocycles with diphosphine and 4,4'-bipyridyl ligands. J Chem Soc Dalton Trans 24:4601–4606

333. Parkins WE (2005) The uranium bomb, the calutron, and the space-charge problem. Phys Today 58(5):45–51. doi:10.1063.1.1993.7.7

334. Cyriac J, Pradeep T, Kang H, Souda R, Cooks RG (2012) Low-energy ionic collisions at molecular solids. Chem Rev (Washington, DC, US) 112(10):5356–5411. doi:10.1021/cr200384k

335. Johnson GE, Hu Q, Laskin J (2011) Soft landing of complex molecules on surfaces. Ann Rev Anal Chem 83–104

336. Baer DR, Engelhard MH, Johnson GE, Laskin J, Lai J, Mueller K, Munusamy P, Thevuthasan S, Wang H, Washton N, Elder A, Baisch BL, Karakoti A, Kuchibhatla SVNT, Moon D (2013) Surface characterization of nanomaterials and nanoparticles: important needs and challenging opportunities. J Vac Sci Technol A 31(5):050820/050821–050820/050834. doi:10.1116.1.48184.3

337. Thune E, Carpene E, Sauthoff K, Seibt M, Reinke P (2005) Gold nanoclusters on amorphous carbon synthesized by ion-beam deposition. J Appl Phys 98(3):034304/034301–034304/034309. doi:10.1063.1.19859.7
338. DiCenzo SB, Berry SD, Hartford EH Jr (1988) Photoelectron spectroscopy of single-size Au clusters collected on a substrate. Phys Rev B 38(12):8465–8468
339. Perez A, Melinon P, Dupuis V, Jensen P, Prevel B, Tuaillon J, Bardotti L, Martet C, Treilleux M, Broyer M, Pellarin M, Vaille JL, Palpant B, Lerme J (1997) Cluster assembled materials: a novel class of nanostructured solids with original structures and properties. J Phys D Appl Phys 30(5):709–721. doi:10.1088/0022-3727/30/5/003
340. Bardotti L, Prevel B, Melinon P, Perez A, Hou Q, Hou M (2000) Deposition of AuN clusters on Au(111) surfaces. II. Experimental results and comparison with simulations. Phys Rev B Condens Matter Mater Phys 62(4):2835–2842
341. Tong X, Benz L, Kemper P, Metiu H, Bowers MT, Buratto SK (2005) Intact size-selected Aun clusters on a TiO$_2$(110)-(1 μe 1) surface at room temperature. J Am Chem Soc 127(39):13516–13518. doi:10.1021/ja052778w
342. Vajda S, Winans RE, Elam JW, Lee B, Pellin MJ, Seifert S, Tikhonov GY, Tomczyk NA (2006) Supported gold clusters and cluster-based nanomaterials: characterization, stability and growth studies by in situ GISAXS under vacuum conditions and in the presence of hydrogen. Top Catal 39(3–4):161–166. doi:10.1007/s11244-006-0052-3
343. DiVece M, Young NP, Li Z, Chen Y, Palmer RE (2006) Co-deposition of atomic clusters of different size and composition. Small 2(11):1270–1272. doi:10.1002/smll.200600065
344. Bardotti L, Tournus F, Melinon P, Pellarin M, Broyer M (2011) Mass-selected clusters deposited on graphite. Spontaneous organization controlled by cluster surface reaction. Phys Rev B Condens Matter Mater Phys 83(3):035425/035421–035425/035428. doi:10.1103/PhysRevB.83.035425
345. Johnson GE, Priest T, Laskin J (2012) Charge retention by gold clusters on surfaces prepared using soft landing of mass selected ions. ACS Nano 6(1):573–582. doi:10.1021/nn2039565
346. Johnson GE, Priest T, Laskin J (2012) Coverage-dependent charge reduction of cationic gold clusters on surfaces prepared using soft landing of mass-selected ions. J Phys Chem C 116(47):24977–24986
347. Cox DM, Eberhardt W, Fayet P, Fu Z, Kessler B, Sherwood RD, Sondericker D, Kaldor A (1991) Electronic structure of deposited monosized metal-clusters. Z Phys D At Mol Clusters 20(1–4):385–386. doi:10.1007/BF01544017
348. Heiz U, Bullock EL (2004) Fundamental aspects of catalysis on supported metal clusters. J Mater Chem 14(4):564–577. doi:10.1039/b313560h
349. Sanchez A, Abbet S, Heiz U, Schneider WD, Häkkinen H, Barnett RN, Landman U (1999) When gold is not noble: nanoscale gold catalysts. J Phys Chem A 103(48):9573–9578
350. Heiz U, Sanchez A, Abbet S, Schneider WD (2000) Tuning the oxidation of carbon monoxide using nanoassembled model catalysts. Chem Phys 262(1):189–200. doi:10.1016/S0301-0104(00)00268-8
351. Lee S, Fan C, Wu T, Anderson SL (2004) CO oxidation on Aun/TiO$_2$ catalysts produced by size-selected cluster deposition. J Am Chem Soc 126(18):5682–5683. doi:10.1021/ja049436v
352. Yoon B, Haekkinen H, Landman U, Woerz AS, Antonietti J-M, Abbet S, Judai K, Heiz U (2005) Charging effects on bonding and catalyzed oxidation of CO on Au$_8$ clusters on MgO. Science (Washington, DC, U S) 307(5708):403–407. doi:10.1126/science.1104168
353. Leung C, Xirouchaki C, Berovic N, Palmer RE (2004) Immobilization of protein molecules by size-selected metal clusters on surfaces. Adv Mater (Weinheim, Ger) 16(3):223–226. doi:10.1002/adma.200305756
354. Lim DC, Dietsche R, Bubek M, Ketterer T, Ganteför G, Kim YD (2007) Chemistry of mass-selected Au clusters deposited on sputter-damaged HOPG surfaces: the unique properties of Au$_8$ clusters. Chem Phys Lett 439(4–6):364–368
355. Lim DC, Dietsche R, Bubek M, Gantefor G, Kim YD (2006) Oxidation and reduction of mass-selected Au clusters on SiO$_2$/Si. ChemPhysChem 7(9):1909–1911

356. Lim DC, Dietsche R, Ganteför G, Kim YD (2008) Chemical properties of size-selected Au clusters treated under ambient conditions. Chem Phys Lett 457(4–6):391–395

357. Lee S, Molina LM, Lopez M, Alonso JA, Hammer B, Lee B, Seifert S, Winans RE, Elam JW, Pellin MJ, Vajda S (2009) Selective propene epoxidation on immobilized Au_{6-10} clusters: the effect of hydrogen and water on activity and selectivity. Angew Chem Int Ed 48(8): 1467–1471. doi:10.1002/anie.200804154, S1467/1461-S1467/1462

358. Harding C, Habibpour V, Kunz S, Farnbacher AN-S, Heiz U, Yoon B, Landman U (2009) Control and manipulation of gold nanocatalysis: effects of metal oxide support thickness and composition. J Am Chem Soc 131(2):538–548. doi:10.1021/ja804893b

359. Binns C (2001) Nanoclusters deposited on surfaces. Surf Sci Rep 44(1–2):1–49. doi:10.1016/ S0167-5729(01)00015-2

360. Zhang Y, Hu Q, Paau MC, Xie S, Gao P, Wan C, Choi MMF (2013) Probing histidine-stabilized gold nanoclusters product by high-performance liquid chromatography and mass spectrometry. J Phys Chem C 117(36):18697–18708

361. Chen J, Zhang Q-F, Bonaccorso TA, Williard PG, Wang L-S (2014) Controlling gold nanoclusters by diphosphine ligands. J Am Chem Soc 136(1):92–95

362. Knoppe S, Buergi T (2013) The fate of Au25(SR)18 clusters upon ligand exchange with binaphthyl-dithiol: interstaple binding vs. decomposition. Phys Chem Chem Phys 15(38): 15816–15820

363. Johnson GE, Priest T, Laskin J (2013) Synthesis and characterization of gold clusters ligated with 1,3-bis(dicyclohexylphosphino)propane. ChemPlusChem 78(9):1033–1039

364. Lavenn C, Albrieux F, Tuel A, Demessence A (2014) Synthesis, characterization and optical properties of an amino-functionalized gold thiolate cluster: $Au_{10}(SPh\text{-}pNH_2)_{10}$. J Colloid Interface Sci 418:234

365. Crasto D, Dass A (2014) Green gold: $Au_{30}(S\text{-}t\text{-}C_4H_9)_{18}$. J Phys Chem C 117(42): 22094–22097

366. Havel J, Pena-Mendez EM, Amato F, Panyala NR, Bursikova V (2014) Laser ablation synthesis of new gold carbides. From gold-diamond nano-composite as a precursor to gold-doped diamonds. Time-of-flight mass spectrometric study. Rapid Commun Mass Spectrom 28(3):297–304

367. Prokes L, Pena-Mendez EM, Conde JE, Panyala NR, Alberti M, Havel J (2014) Laser ablation synthesis of new gold arsenides using nano-gold and arsenic as precursors. Laser desorption ionisation time-of-flight mass spectrometry and spectrophotometry. Rapid Commun Mass Spectrom 28(6):577–586

368. Baksi A, Pradeep T (2013) Noble metal alloy clusters in the gas phase derived from protein templates: unusual recognition of palladium by gold. Nanoscale 5(24):12245–12254

369. Shao N, Huang W, Mei W, Wang L-S, Wu Q, Zeng XC (2014) Structural evolution of medium-sized gold clusters Au_n^- (n = 36, 37, 38): appearance of bulk-like FCC fragment. J Phys Chem C. doi:10.1021/jp500582t

370. Jasikova L, Roithova J (2013) Interaction of gold acetylides with Gold(I) or Silver(I) cations. Organometallics 32(23):7025–7033

371. Lang SM, Frank A, Bernhardt TM (2013) Comparison of methane activation and catalytic ethylene formation on free gold and palladium dimer cations: product binding determines the catalytic turnover. Cat Sci Tech 3(11):2926–2933

372. Pettibone JM, Osborn WA, Rykaczewski K, Talin AA, Bonevich JE, Hudgens JW, Allendorf MD (2013) Surface mediated assembly of small, metastable gold nanoclusters. Nanoscale 5(14):6558–6566

Index

A

Aberration-corrected scanning transmission
electron microscopy (ac-STEM), 72
Acetaldehyde, 190
Acetone, 190
Acetonitrile, 190
Acetylene, 11, 142
Acetylides, 198
Adsorption, 91
Ammonia, 70, 187, 200
AuM nanorods, 76
AuPd, 67, 77
AuPt, 67
AuRh, 67, 77

B

Benzene, 92, 190
Bimetallic cluster ions, gold-containing, 199
Bimetallic core–shell structures, 67
BINAP, 157
Bipyramids, 69
Bis(diphenylphosphino)propane, 157
Bisformaldehyde, 206
Bis(phosphino)alkane ligands, 152
Borane tert-butylamine complex (BTBC), 149
Bright-field (BF) imaging, 73

C

CaO, 111–116
Capping principle, 1, 17
Carbides, 164
Carbon monoxide, 70, 84, 91, 192, 201
adsorption, 116

oxidation, 180, 203
Catalysis, 139, 202
Centred icosahedron, 41
Cesium ion bombardment, 164
Cetyltrimethylammonium bromide (CTAB), 69
Charge-mediated growth, 99
Chemical modification, 91
Close-packed arrangement, 53
Clusters, 1, 91
condensed, 1
Collision-induced dissociation (CID), 165
Condensed phase, 145
Cooperative effects, 190
Core fission, 163
Core–shell nanoparticles, 71
Core–shell-segregated nanoalloys, 68
Crown geometry, 18
Cubane, 92
Cubes, 69
Cuboctahedral cluster, 24, 31, 44

D

Dehydrogenation, 200
Density functional theory (DFT), 70, 75
Deposition-precipitation (DP), 129
DFT. *See* Density functional theory (DFT)
Diborane, 3
Difluoroacetic acid, 190
Dimethyl sulfide, 190
Direct light scattering (DLS), 150
Dopants, 91
Doped oxide materials, 110
Doping, 111
Dual gold catalysis, 198

E

EDX. *See* Energy dispersive X-ray
      spectroscopy (EDX)
EELS. *See* Electron energy loss spectroscopy
      (EELS)
Electron capture dissociation (ECD), 169
Electron energy loss spectroscopy (EELS),
      72, 74
Electronic properties, 91
Electron-induced dissociation (EID), 166
Electron microscopy, 67
Electrospray ionisation (ESI), 139, 144
Energy dispersive X-ray spectroscopy (EDX),
      72, 74
Ethynylbenzene, 190

F

Formaldehyde, 205
Full-shell clusters, 3

G

Gas-phase gold nanoclusters, 147, 161
Gas-phase reactions, 139
Gem-diaurated ions, 192
Gold, 1, 76
      acetylides, 198
      cluster ions, bare, 183
            ligated, 192
      clusters, 91
      hydrides, 193
      hydroxide cluster ions, 195
      nanoclusters (AuNCs), 139
      nanorods, 67
      oxides, 192
            cluster ions, 195
Gold–silver–platinum clusters, 9
Gold (I) triphenylphosphine complexes, 4

H

High angle annular dark field (HAADF), 72, 73
Highly ordered pyrolytic graphite (HOPG),
      207
HOMO/LUMO, 100
Hopping integrals, 76
Hydrogen sulfide, 188
Hydroxides, 164
Hydroxylation, 91, 125
Hydroxyl groups, 125

I

Infrared multiphoton dissociation (IRMPD),
      165
Inherent structure rule, 49
Interstitial atoms, 18, 34
Ion cyclotron resonance (ICR), 144
Ion mobility, 170
IR spectroscopy, 172
Isoperimetric inequality, 53

J

Jellium model, 1, 34

L

Laser ablation vaporisation, 148
Ligands, activation, 163
      loss, 163

M

Mass spectrometry, 139
Matrix-assisted laser desorption ionisation
      (MALDI), 139, 144
N-(2-Mercaptopropionyl)-glycine (tiopronin),
      150
Metal islands, two-dimensional, 105
Metal-on-metal growth, 70
Metallo-organothiolato-ligands, 1
Methanol, 190, 199
Methyl halides, 189
Methyl mercaptan, 190
Methylamine, 187
MgO, 91, 207–211
      oxygen vacancies, 94
Molecular dynamic simulations, 67
Monolayer-protected clusters (MPCs), 152
MS-directed synthesis, 206
MS-selected deposition, 207

N

N_2, 186
N_2O, 187
Nanoalloys (NAs), 68–75
Nanoclusters, 1, 142, 168, 210
Nanoparticles, 67
Nanorods (NRs), 67–85, 171
Nosé–Hoover thermostat, 79
Nucleation, 124

O

Oblate, 1
Organothiolato-gold cluster compounds, 1
Oxide films, 91
Oxide surfaces, single metal atoms/small
    clusters, 93
Oxide thin films, single adatoms, charge
    transfer, 95

P

Palladium, 69, 76, 80, 201
PES. *See* Photoelectron spectroscopy (PES)
Phenylacetylene, 165
1-Phenylpropyne, 199
Phosphides, 164
Phosphine, 1
    ligands, 147, 152
Photodissociation, 165
Photoelectron spectroscopy (PES), 165, 175
Platinum, 3, 69, 76, 82, 200
    carbonyl cluster, 30
Polyhedra, 9, 45, 69
    linking, 23
Polyhedral skeletal electron pair theory
    (PSEPT), 1, 12, 92
Prisms, 69
Prolate, 1
Propylene, 189
PSEPT. *See* Polyhedral skeletal electron pair
    theory (PSEPT)

Q

Quantum-well states (QWS) 100

R

Raft-shaped Au islands, 111
Reactivity, 180
Reduction phase, 151, 156
Rhodium, 20, 69, 76, 80
Rods, 69

S

Scanning transmission electron microscopy
    (STEM), 67, 72
Silver, 3, 9, 12, 70, 144, 164, 170, 180, 201
Silver hydride, 194
Sodium borohydride, 149–158, 206
Soft landing (SL) experiments, 207

Sonogashira coupling, 192
Spherical arrangement, 53
Sphericity, 1, 52
STEM. *See* Scanning transmission electron
    microscopy (STEM)
Stranski–Krastanov layer-plus island growth,
    70
Superatom model, 38, 49
Surface-induced dissociation (SID), 165
Synthesis, 35, 67, 161, 206

T

Tellurides, 164
Tensor surface harmonic model (TSHM),
    1, 11
Tersoff–Hamann theory, 97
Thermochemistry, 179
Thiolate ligands, 147, 150
TIED. *See* Trapped ion electron diffraction
    (TIED)
Tolman cone angle, 4
p-Tolylacetylene, 165
Toroidal topology, 1
Trapped ion electron diffraction (TIED),
    165, 175
Triangles, 69
Trifluoroacetic acid, 190
TSHM. *See* Tensor surface harmonic model
    (TSHM)

U

Ultrahigh vacuum (UHV)-based
    techniques, 93
UV–Vis spectroscopy, 173

V

Valence shell electron pair repulsion theory
    (VSEPR), 92

W

Wet impregnation, 91

X

X-ray absorption spectroscopy (XAS), 70
X-ray crystallography, 145, 152
X-ray powder diffraction radial distribution
    functions, 56

Printed by Printforce, the Netherlands